Wave Propagation Analysis of Smart Nanostructures

Wave Propagation Analysis of Smart Nanostructures

Farzad Ebrahimi and Ali Dabbagh

CRC Press is an imprint of the
Taylor & Francis Group, an **informa** business

First published in paperback 2024

CRC Press
2385 NW Executive Center Drive, Suite 320, Boca Raton FL 33431

and by CRC Press
4 Park Square, Milton Park, Abingdon, Oxon, OX14 4RN

CRC Press is an imprint of Taylor & Francis Group, LLC

ISBN: 978-0-367-22695-4 (hbk)
ISBN: 978-1-03-283954-7 (pbk)
ISBN: 978-0-429-27922-5 (ebk)

DOI: 10.1201/9780429279225

Visit the Taylor & Francis Web site at
http://www.taylorandfrancis.com

and the CRC Press Web site at
http://www.crcpress.com

To my parents
Farzad Ebrahimi

To my parents and my sisters
Ali Dabbagh

Contents

Preface

The development of carbon-based tiny elements in the 1990s is basically the main reason for the extensive use of nanoscale elements in various engineering applications (practices). When a piece of bulk material is divided into smaller and smaller pieces until we can obtain nanoparticles of the initial material, the nanoparticles do not exhibit the same mechanical behavior as the bulk. The so-called behavioral difference stems from the unbelievable role of small forces, which are generally assumed to be negligible at the macroscale, in the nanoscale elements. On the other hand, there are a large number of nanoelectromechanical systems (NEMSs) which consist of nanosize devices such as nanobeams and nanoplates. Hence, it is of high importance to try and keep an open mind and form a roughly comprehensive idea about the behavior of nanosize elements with the aim of assessing how these small-scale devices can behave in the aforementioned NEMSs. The information concerning the mechanical characteristics of nanostructures can be obtained either by experimental approaches or numerical ones. Due to inherent complexities and challenges of any experimental research dealing with nanoscale structures, one of the best alternatives for confidently predicting the mechanical behaviors of nanostructures is to use the continuum-mechanics-based approach.

Also, the phenomenon of dispersion of elastic waves inside a continuum can reveal lots of data about that specimen which can either be obtained with extreme difficulty using other approaches or not obtained at all. For example, one of the most known applications of wave propagation technique is its efficiency in the prediction of defects in solids. This approach can be used when no nondestructive testing (NDT) can be performed on the specimen under observation. It is also worth mentioning that the dispersion of elastic waves in nanosize structures can generate natural frequencies thousand times greater than those of the vibration phenomenon in the same specimen. Therefore, it is critically important to gain more knowledge about the mechanical behaviors of waves scattered in nanostructures. The present research mainly aims at solving the problem of lack of analytical investigations about the dispersion responses of smart and composite nanostructures in the framework of higher-order shear deformation kinematic hypotheses incorporated with the nonlocal continuum theories developed for the analysis of nanostructures.

The first three chapters deal with the introduction of preliminary assumptions and definitions and also mathematical tools required for a mechanical engineer to modify the constitutive equations of small-scale nanodevices and to solve the wave propagation problem of a continuous system in case the set of governing equations were in hand. Chapters 4–11 are devoted to derive the equations of motion of the wave propagation problem in nanosize beams and plates subjected to various types of mechanical, thermal, electrical and magnetic loadings.

Farzad Ebrahimi
Ali Dabbagh
Imam Khomeini International University (IKIU), Qazvin, Iran, July 2019

Acknowledgments

The authors are eager to extend their gratitude to those who played significant roles for bringing this book to life. The authors would also like to thank the Taylor & Francis Group/CRC Press because of their invaluable support during the preparation of the proposed book. The authors greatly relied upon their insight and kindness. The authors' special gratitude goes to Jonathan Plant and Bhavna Saxena, executive editor and editorial assistant at CRC Press, for their indispensable efforts during the publication procedure.

Authors

Farzad Ebrahimi is an Associate Professor in the Department of Mechanical Engineering at the Imam Khomeini International University (IKIU), Qazvin, Iran. In 2011, he earned a PhD at the School of Mechanical Engineering at the University of Tehran. In 2012, he joined IKIU as an Assistant Professor and in 2017 was elected as Associate Professor. His research interests include mechanics of nanostructures and nanocomposites, smart materials and structures, viscoelasticity, composite materials, functionally graded materials (FGMs) and continuum plate and shell theories. He has published more than 300 international research papers, and he is the author of two books about smart materials. He has also edited three books for international publishers.

Ali Dabbagh is an MSc student at the School of Mechanical Engineering, College of Engineering, University of Tehran, Tehran, Iran. His research interests include solid mechanics, smart materials and structures, composites and nanocomposites, functionally graded materials (FGMs), nanostructures and continuum plate and shell theories. He has published more than 40 international papers in his research area. His MSc thesis is concerned with the mechanical behaviors of hybrid nanocomposite structures subjected to various static and dynamic excitations.

1

An Introduction to Wave Theory and Propagation Analysis

1.1 Introduction

Nearly everybody can assume the propagation of different types of waves by seeing the water waves of the sea or talking about the radio waves and sound waves. In fact, the transition of any external generated disturbance from a desirable point of a substrate to the other contiguous points in the continua is the basic concept of the wave propagation phenomenon in solids. To be honest with you, the wave propagation phenomenon is directly involved with the interatomic interactions of tiny particles like atoms and molecules. However, in ordinary problems in the field of mechanical engineering, either solid or fluid mechanics, the chosen system which is assumed to be the host of the dispersion is regarded to be a continuous system. Due to this assumption, the continuum mechanics relations are applied in order to mathematically formulate the aforementioned phenomenon in common. The aforesaid assumption can be clarified by reviewing the fundamental principles of continuous systems, particularly in solid mechanics, which is the scope of this book. Actually, the mathematical modeling of a continua in the continuum theories corresponds with assuming any element of the continuous system to be consisting of a series of discrete masses coupled together via connective stiff springs. Thus, the produced wave in a solid can be dispersed once the energy transmission among the mass–spring elements is completed. Indeed, whenever the disturbance propagates in the solid, it carries both kinetic and potential energies with itself. The motion of elastic waves is powerful enough to transfer a large amount of energy over remarkable distances in the media. Moreover, it must be considered that wave propagation is a dynamic phenomenon, and it requires a set of necessities to take place. In other words, waves are not able to propagate in a rigid body. Therefore, the first requirement for dispersion of waves in a desirable media is the existence of inertia. From another point of view, wave propagation can happen in a deformable media. The second reason that waves can propagate in elastic bodies is the transition of a forced disturbance from a part of the solid to the adjacent sides. So, it is of importance to use deformable solids when one is purposed to generate a wave in a desirable media. Waves can be classified into different groups including mechanical, electric and magnetic. It is worth mentioning that the mechanical waves can be divided into three major groups in general, namely elastic, plastic and viscoelastic types. Herein in this book, we focus on the elastic waves. In fact, in investigations in the elastic domain, the constitutive equations of a solid can be presented on the basis of the Hook's law for linear elastic solids. However, while the materials seem to be viscoelastic, the effects of time domain as well as those of elastic domain must be covered. In addition, plastic materials are completely different from the elastic ones in their yield behavior and must be studied in other complementary references. Again focusing on the issue of waves' classification, the waves can be classified based on their oscillation direction with respect to the dimension of their dispersion. In a rough sorting style, the waves can be divided in two general groups, namely longitudinal and transverse (flexural).

Besides, one should pay attention to the fact that the waves generally have two types of motion when they are dispersed in a solid. In type-1, the behavior of the wave in solids is similar to that in fluids. Actually, in this type, the solid will be able to carry axial stresses, maybe tension or compression, and the direction of motion of the continua is same as that of the waves' motion. This type of wave motion can be related to longitudinal waves. However, in type-2, the solid can transmit shear stress, and hence the direction of motion of the continua is transverse to the direction of the wave's motion. This type of motion in the solids cannot be observed in the fluids and corresponds with the flexural waves introduced before.

Wave propagation in an elastic solid is a very important phenomenon in a mechanical engineer's point of view. The reason for this reality is the widespread application of wave propagation in practical problems. In many applications, a wave input is sent into a system, and many pieces of information are obtained from the waves' reflection or growth in the media. For instance, propagation of elastic waves is one of the most efficient ways of finding the defects and porosities in a desirable media. This procedure can be completed by sending an elastic wave into a media and computing the differences between the features of the sent and reflected branches. In other words, in many applications, nondestructive tests (NDTs) are used to detect imperfections generated in the fabrication process; however, in some cases, such examinations are not applicable. Due to this fact, wave propagation approach is utilized in order to detect the defects. Precisely, thermal and ultrasonic high-frequency waves are propagated in the examined structure, and the wave amplitude, reflection angle and reflection time of the reflexed waves can show the features of the ruptures of the structure. One of the most important advantages of this method is its high capability in diagnosis of small ruptures due to the high frequency of propagating waves. In this method, different wave modes may be involved that are proportional with the thickness of the structure.

1.2   Practical Applications of Waves

In this section, we aim to emphasize the crucial role of the wave phenomenon in predicting the characteristics of a system in various practical cases. For instance, in many conditions, the system may be subjected to huge impact loadings. As you know, in such loading, common elastic theories are not able to completely satisfy the response of the system, whereas the elastic wave theory is able to estimate the system's response under the aforementioned situation. Actually, in the cases of enormous impact loadings, the elastic wave theory can justify the fracture or perforation of the structure which cannot be clarified by means of previously known elastic theorems.

Another application which is more related to the field of aerospace or military belongs to the materials which are subjected to severe loadings that are powerful enough to generate a permanent damage in the specimen. The investigations in such problems fall into the scope of wave propagation in elastic materials like steel and its derivatives.

In addition, once dispersion of cracks in the structures are studied or the interaction of stress fields in a material possessing cracks and voids is considered to be solved, the wave propagation technique is one of the most applicable and efficient methods. In fact, the mentioned problems can be analyzed by means of the dispersion and diffusion concepts in the field of either acoustic or electromagnetic waves.

The next application is based on ultrasonic wave dispersion in the smart piezoelectric materials. Indeed, the basic concept of the mentioned application relates to the specific feature of the piezoelectric materials that is associated with reciprocal electromechanical energy conversion. Actually, the mechanical signal can be fed into an element, and once the sent signal reaches the crystals and ceramics with piezoelectric effect, the transformation of the initial signal from elastic to electric occurs. Whenever the explained procedure is completed, the final electric signal shows the features of the studied structure.

On the other hand, gaining adequate knowledge about the characteristics of dispersion of waves under the ground can significantly help the physicists to predict the behaviors of the earthquakes. The earthquakes can be considered as waves that are able to move under the ground for about thousands of kilometers. In fact, sometimes a large amount of compressed air generates a very powerful wave which is capable of producing earthquakes. Therefore, another crucial issue is to know about the blast waves that may be one of the most probable candidates for generating an earthquake under the ground. For this reason, it can be so useful to predict the above-discussed phenomena within the frameworks of the wave-propagation-based approaches.

As mentioned before, ultrasonic wave propagation methods can be implemented for the purpose of detecting defects in the structures consisting of various materials. In such approaches, which can be classified in the group of NDTs, the differences between the characteristics of the initial signal and those of the final reflected or propagated signals can reveal the features of the defects and their probable coordination. In this method, it is common to utilize each of the longitudinal, shear and surface waves. Also, it is of importance to point that the elastic and viscoelastic material properties of the materials can be calculated throughout the wave propagation methods. For more information about the mentioned applications, readers are advised to see reference [1].

1.3 Wave Propagation Solution

In this section, a basic introduction about the solution of the wave propagation problem in a continuous system is presented. Herein, the continuous system can be beam or plate. The main idea of the proposed method is based upon the well-known concept of the separation of the variables in advanced mathematics. Clearly, due to the dynamic nature of wave propagation, this phenomenon depends on the time domain and spatial coordination. Thus, the chosen solution function for such a problem must consist of a time-dependent part plus the geometry-dependent part. In the following sections, analytical solution of the wave dispersion problem will be introduced for all the mentioned structural elements.

1.3.1 Beam-Type Solution

As we know, beams are elements in which the width and thickness are very small compared with the length. Thus, in mechanical analyses, spatial analysis will be performed along the beam's length. In other words, when using the general notation of the continuum mechanics for the components of the position vector as $\mathbf{x} = (x_1, x_2, x_3)$, only x_1 component will be involved in solving any problem for a beam. As stated before, solving a wave propagation problem cannot be summarized in the spatial analysis. In fact, the time dependency must be included too. In the following sections, the solution of both classical and refined higher-order shear deformable beams will be discussed in detail.

1.3.1.1   Classical Beams' Solution

In the classical beam theory, known as Euler–Bernoulli beam theory, the effects of shear stress and strain are ignored, and only deflections produced from the bending stresses are included. Thus, only two independent variables are available, namely u, which represents the axial displacement of the neutral axis, and w, which denotes the bending deflection of the beam. The analytical solution of these components can be formulated as [2,3]

$$u(x,t) = U e^{i(\beta x - \omega t)}, \tag{1.1}$$

$$w(x,t) = W e^{i(\beta x - \omega t)}, \tag{1.2}$$

where U and W are the wave amplitudes, β is the wave number in the x direction and ω stands for the circular frequency of the dispersed waves. In order to use the aforementioned equations for solving the wave propagation problem of an Euler–Bernoulli beam, one should try to find the final form of the governing equations in terms of the displacement field. Thereafter, once Eqs. (1.1) and (1.2) are substituted in the governing equations of the beam, the following equation can be obtained:

$$([K]_{2\times 2} - \omega^2 [M]_{2\times 2})[\Delta] = 0 \tag{1.3}$$

in which $[\Delta] = [U, W]^T$ corresponds to the amplitude vector. Also, $[K]$ and $[M]$ are stiffness and mass matrices, respectively. When the eigenvalue problem obtained (Eq. (1.3)) is solved, the circular frequency can be calculated. One should pay attention to the fact that one of the most crucial variants that must be studied in each wave propagation analysis is the velocity of the propagated waves. For this purpose, many definitions can be found in the literature; however, the most important one is that of the phase velocity of the waves that are propagating in a beam. This parameter, which is usually shown by c_p, can be calculated as follows:

$$c_p = \frac{\omega}{\beta}. \tag{1.4}$$

Moreover, some other definitions can be found in wave propagation analyses in the literature. For example, by setting $\beta = 0$, the frequency that is achieved is called cut-off frequency. This value is generally used in order to find the minimum frequency of the waves that are able to propagate in the system. In other words, waves with frequencies lower than this value cannot propagate in the structure. Besides, by tending the wave number to infinity $(\beta \to \infty)$, the escape frequency can be obtained. Beyond the escape frequency, the flexural waves will not be able to disperse anymore.

1.3.1.2   Shear Deformable Beams' Solution

In this section, the effects of the shear deformation will be included when probing a solution for the wave propagation problem of a beam. It can be noted that classical theory of the beams is able to present relatively acceptable responses in the case of using thin beams with high slenderness ratios. However, when the slenderness ratio is not too big, the effects of the shear deformation will be more important compared with the previous situation. So, it is necessary to utilize higher-order beam theories for the purpose of providing reliable results. In the refined higher-order beam models, the deficiency of the Euler–Bernoulli beam model is covered. In such theories, the independent variables will be 3 instead of 2. The axial displacement of the neutral axis is again u; however, the deflection will be divided into two components, bending and shear, denoted by w_b and w_s, respectively. So, the wave propagation solution for such theories can be expressed in the following forms [4,5]:

$$u(x,t) = U e^{i(\beta x - \omega t)} \tag{1.5}$$

$$w_b(x,t) = W_b e^{i(\beta x - \omega t)} \tag{1.6}$$

$$w_s(x,t) = W_s e^{i(\beta x - \omega t)} \tag{1.7}$$

in which U, W_b and W_s are the wave amplitudes. Other parameters are just the same as those introduced in Eqs. (1.1) and(1.2). As in the previous section, the final eigenvalue form of the wave dispersion analysis can be obtained for a refined higher-order beam as follows:

$$([K]_{3\times3} - \omega^2 [M]_{3\times3})[\Delta] = 0, \tag{1.8}$$

where $[\Delta] = [U, W_b, W_s]^T$ is the amplitude vector of the refined higher-order beams. The rest of the analysis is similar to that of the previous section, developed for the Euler–Bernoulli beams. So, the phase velocity, cut-off frequency and escape frequency parameters can be computed based on the definitions provided above.

1.3.2 Plate-Type Solution

In this section, the solution functions of plate elements will be developed on the basis of both classical and refined higher-order plate theories. The plate elements' analysis is more complex than the beams' analysis because in the plates, we are not allowed to neglect the x_2 coordinate. As mentioned in Section 1.3.1, in the beam elements, the width of the structure can be considered to be equal to its thickness. However, in the plates, the dimension x_2- must be treated similar to x_1- direction. Actually, the primary plane stress assumption, which is usually used for analyzing a plate, does not allow us to treat a plate in the same way we treat a beam. Thus, in the case of studying a rectangular plate, one should consider y as well as x, in the Cartesian coordinate system. In the following paragraphs, the mathematical representation of these expressions will be introduced for the sake of clarity.

1.3.2.1 Classical Plates' Solution

As in the classical theory of the beams, the shear deformation effects are assumed to be negligible in the classical plate theory. In the classical plate theory, known as Kirchhoff–Love plate theory, there are three independent parameters. In this hypothesis, u, v and w denote axial displacement in the $x-$ direction, transverse displacement in the $y-$ direction and bending deflection in the $z-$ direction, respectively. According to the introduced notation, the solutions of a classical plate can be written as [6,7]

$$u(x,y,t) = U e^{i(\beta_1 x + \beta_2 y - \omega t)}, \tag{1.9}$$

$$v(x,y,t) = V e^{i(\beta_1 x + \beta_2 y - \omega t)}, \tag{1.10}$$

$$w(x,y,t) = W e^{i(\beta_1 x + \beta_2 y - \omega t)}, \tag{1.11}$$

where β_1 and β_2 are the wave numbers in the x and y directions, respectively. Moreover, U, V and W are the wave amplitudes, and ω stands for the circular frequency of the propagated waves in the plate. In general, the wave numbers β_1 and β_2 are not identical. However, in many references, these two wave numbers are assumed to be identical in order to make the propagation analysis easier. In this condition ($\beta_1 = \beta_2 = \beta$), all the previous parametric studies can be again performed for the classical plates subjected to elastic waves. Henceforward, one can use the definitions of phase velocity, escape frequency and cut-off frequency in this case in order to complete one's analysis. It is worth mentioning that due to the introduced functions in the above equations, the final eigenvalue equation of classical plates is a 3×3 matrix equation.

1.3.2.2 Shear Deformable Plates' Solution

Now that we are familiar with the analysis of a wave propagation problem in the plates on the basis of the Kirchhoff–Love theory, it is time to learn about such an analysis in the framework of refined shear deformable plate theories. In the refined higher-order theorems, an additional component will be added to the previously introduced ones in order to account for the shear deformation effects. Indeed, three variables will be replaced by four independent variables. In these theories, u and v are longitudinal and transverse displacements, respectively. Furthermore, w_b and w_s denote the bending and shear deflections across the thickness, respectively. The displacements can be expressed in the following forms [8,9]:

$$u(x,y,t) = U e^{i(\beta_1 x + \beta_2 y - \omega t)}, \tag{1.12}$$

$$v(x,y,t) = V e^{i(\beta_1 x + \beta_2 y - \omega t)}, \tag{1.13}$$

$$w_b(x,y,t) = W_b e^{i(\beta_1 x + \beta_2 y - \omega t)}, \tag{1.14}$$

$$w_s(x,y,t) = W_s e^{i(\beta_1 x + \beta_2 y - \omega t)}. \tag{1.15}$$

Again, U, V, W_b and W_s are the wave amplitudes. The rest of the wave propagation analysis can be performed on the basis of the previously discussed fundamentals explained in the previous section for the classical plates. Herein, the final eigenvalue problem will be a matrix equation of order 4×4.

2

An Introduction to Nonlocal Elasticity Theories and Scale-Dependent Analysis in Nanostructures

2.1 Size Dependency: Fundamentals and Literature Review

Without any overstatement, the most crucial phenomenon that could generate an evolution in industrial designs is the invention of nanotechnology and nanosize elements. In fact, modern designs are moving as fast as possible toward using the nanosize elements instead of conventional macrosize ones in order to lessen the size of the designed devices as well as gain benefits from the marvelous features of nanosize particles. Application of nanostructures in mechanical engineering is growing just as in other engineering fields. Henceforward, due to this reason, it is of high importance to gain adequate knowledge about the mechanical behaviors of nanosize structures like beams, plates and shells.

In the small scale, the behaviors of the structures cannot be guessed by the means of classical continuum mechanics approaches. Actually, the smaller size of an element, the more important are the interatomic interactions between the particles such as atoms or molecules. In other words, whenever a macrosize structure is assumed to be analyzed, one can neglect the interatomic forces in comparison with the giant loadings that are applied on the structure. However, this point of view is not able to support elements with dimensions in nanoscale. In the range of nanometers, due to the tiny dimensions of the structure, any small force can play an unbelievable role in determining the stress–strain behaviors of the structure. Thus, it is crucial to cover all the previously neglected forces when we try to analyze a nanostructure. In order to solve this issue, a modified version of continuum mechanics is required for the goal of formulizing the constitutive equations of a nanosize element. Reviewing the history of the endeavors performed by many researchers working in this field of interest reveals that Eringen [10,11] was the first person who was able to present a size-dependent formulation which was able to approximate the scale effects. Eringen showed that the stress state inside a nanosize medium must be considered as a multivariable function of the strains of the adjacent points and the strain of the desired point itself. Via this assumption, the nonlocal effects can be considered when probing the constitutive equations of a nanostructure. Eringen suggested a kernel function which is presented in order to bridge the local stress and the nonlocal one. After Eringen's remarkable demodulation, many researchers tried to use this theory in order to analyze the mechanical behaviors of the nanostructures. Now, the most remarkable researches in this field of interest will be reviewed.

For example, buckling analysis of embedded single-layered graphene sheets (SLGSs) was carried out by Pradhan and Murmu [12]. Ansari et al. [13] probed the vibration problem of a multilayered nanoplate via nonlocal elasticity. Mahmoud et al. [14] could analyze the vibration problem of a nanobeam based on the nonlocal theory. The combined effects of moisture and temperature are included in a bending analysis performed by Alzahrani et al. [15] on nanoplates. Furthermore, thermal vibration analysis of a single-walled carbon nanotube (SWCNT) was performed by Ebrahimi and Salari [16]. Zenkour [17] explored the transient dynamic characteristics of graphene sheets via nonlocal elasticity.

Ebrahimi and Shafiei [18] utilized the nonlocal theory of Eringen in order to probe the effect of existence of initial pre-stress on the vibrational characteristics of SLGSs. Ebrahimi and Barati [19] probed the dynamic characteristics of smart flexoelectric nanobeams via the Eringen's nonlocal elasticity. In another endeavor, Ebrahimi and Karimiasl [20] used the nonlocal elasticity for investigating the stability conditions of a flexoelectric nanobeam.

Even though Eringen's theory was a glorious finding in the field of nanomechanics, some experimental examinations have proven that the aforementioned theory is not complete enough to account for scale effects whenever either static or dynamic behaviors of a nanostructure are going to be analyzed [21,22]. In the nonlocal theory [11], a nonlocal parameter is used, which represents a stiffness-softening phenomenon that exists in the elements with dimensions in the range of nanometers. However, scientists have shown that the scale effects are something more than those reported by Eringen [11]. Certainly, the Eringen's hypothesis about the softening effect is valid in the nanostructures; but the scale effect possesses a hardening effect too. Indeed, the nonlocal parameter is a requisite for the nanomechanical analysis, however, it is not a sufficient condition for such an analysis. In other words, a new theory, from a combination of experimental tests and Eringen's theory, seems to be efficient enough to be able to capture small-scale effects completely. For this reason, Lim et al. [23] presented a new theory which presents both softening and hardening effects of nanoscale together. In this novel theory, called nonlocal strain gradient theory, two scale parameters are considered for both of the aforementioned effects.

Until now, some of the researchers have tried to use the nonlocal strain gradient elasticity instead of Eringen's theory to investigate the vibration, buckling, bending and wave propagation responses of nanosize mechanical elements. Ebrahimi and Hosseini [24] performed a nonlinear dynamic analysis on the viscoelastic SLGSs by means of the nonlocal strain gradient theory. Ebrahimi and Dabbagh [7,25] examined the wave dispersion curves of nanoplates made from smart magnetostrictive materials with nonlocal strain gradient elasticity with respect to elastic, thermal, and magnetic effects. The effect of applying a nonuniform in-plane loading on the vibrational behaviors of SLGSs is included in a research performed by Ebrahimi and Barati [26] according to the stress–strain gradient elasticity. Furthermore, the differential quadrature method (DQM) is incorporated with the nonlocal strain gradient theorem by Ebrahimi and Barati [27] in order to analyze the damped dynamic problem of an SLGS. A hygrothermomechanical vibration analysis was undertaken by Ebrahimi and Barati [28] to realize the responses of nonlocal strain gradient SLGSs under severe loading. Again, the nonlocal stress–strain gradient theory of elasticity was used by Ebrahimi and Dabbagh [29] to observe the thermo-elastic wave dispersion behaviors of SLGSs. The same authors could perform another nonlocal-strain-gradient-based analysis of the wave propagation characteristics of SLGSs in hygrothermal environments [30]. In another article, Ebrahimi and Barati [31] investigated the vibration problem of axially loaded double-layered graphene sheets (DLGSs) via nonlocal strain gradient elasticity theory. Effects of temperature gradient and moisture concentration are covered in an analysis by Ebrahimi and Barati [32] based on the nonlocal strain gradient theory. Also, the same side effects were considered by Ebrahimi and Dabbagh [33,34] on the wave propagation problem of DLGSs. Lately, Ebrahimi and Dabbagh [35] studied the wave propagation behaviors of DLGSs with respect to the effects of external magnetic field on the basis of the nonlocal-strain-gradient elasticity theorem.

According to the above literature review, it is of great significance to gain enough knowledge about the size-dependent continuum theories which are used by researchers to perform static and dynamic analyses on the beams, plates and shells in nanoscale. In the following sections, we will consider the mathematical formulation of each of the nonlocal elasticity and stress–strain gradient theories in order to derive the constitutive equations of nanoscale elements.

2.2 Mathematical Formulation of the Nonlocal Elasticity

2.2.1 Constitutive Equation for Linear Elastic Solids

According to the Eringen's nonlocal elasticity for linear elastic solids, the components of the stress tensor in a nanosize continua can be formulated in the following form:

$$\sigma_{ij} = \int_V \alpha(|\mathbf{x} - \mathbf{x}'|, \tau) C_{ijkl} \varepsilon_{kl}(\mathbf{x}') dV(\mathbf{x}') \tag{2.1}$$

in which α is the presented nonlocal kernel function that is a function of an Euclidean distance, $|\mathbf{x} - \mathbf{x}'|$, and a size-dependent coefficient, τ. Indeed, the kernel function is responsible for capturing the impact of the strain at the point x' on the stress at the point x. Furthermore, the coefficient τ can be calculated as

$$\tau = \frac{e_0 a}{l}, \tag{2.2}$$

where $e_0 a$ is a scale coefficient used to estimate the small-scale effects on the constitutive behaviors of nanostructures. It is worth mentioning that the term e_0 is not a general value for various materials. In other words, its value is different for different materials, and it should be determined experimentally. However, due to the difficulties in experimental tests that should be performed to determine this value for each material, it is more common to choose this parameter as a coefficient of the nanostructure's dimension. Eringen [11] reported a 0.39 value for the term e_0 by fitting the dispersion curves obtained from the nonlocal elasticity for plane wave and Born–Karman model of lattice dynamics at the end of the Brillouin zone. In this condition, the relation $\beta a = p$ is valid, in which a is the atomic distance and β stands for wave number in the phonon analysis.

Besides, a and l are internal and external characteristic lengths, respectively. The internal characteristic length can be the grain size of the material, the lattice parameter or the granular distance [11]. Moreover, each of the wavelength, crack length and the sample size can be selected for the external characteristic length in the Eringen's theorem. Also, it is easy to realize that the term $C_{ijkl}\varepsilon_{kl}$ refers to the double contraction product of the fourth-order tensor of elasticity, $\mathbf{C}$, and second-order strain tensor, ε.

When the kernel function is chosen, the integral form of the Eringen's theory can be transferred to its differential form for the linear elastic solids as follows:

$$(1 - (e_0 a)^2 \nabla^2) \sigma_{ij} = C_{ijkl} \varepsilon_{kl}, \tag{2.3}$$

where ∇^2 is the Laplacian operator. By means of Eq. (2.3), the constitutive equations of a nanosize element can be modified. The above equation is commonly written in the following form:

$$(1 - \mu^2 \nabla^2) \sigma_{ij} = C_{ijkl} \varepsilon_{kl}, \tag{2.4}$$

where $\mu = e_0 a$.

2.2.2 Constitutive Equations of Piezoelectric Materials

In many applications, smart nanosize elements are used in order to generate a multiobjective device. In the most common type of smart materials, a coupling appears between the elastic deformation and the electric potential. In these smart materials, named piezoelectric materials, a mechanical excitation results in a reciprocal electrical effect and vice versa.

In other words, in the elements made from piezoelectric materials, the structure is able to generate an electric voltage when it is subjected to an elastic loading, either tension or compression. This coupling effect can be connected reversely, meaning one can apply an electric field to a structure and receive mechanical strain in response. So, it is of high significance to formulate such an important phenomenon in the nanoscale. In this section, Eq. (2.4) will be extended for a piezoelectric material in order to derive the constitutive equations of such a smart material.

Prior to the derivation, it is useful to review the constitutive equations of piezoelectric materials in the macroscale to sense the differences between these materials and the conventional linear elastic solids. The constitutive equations in piezomaterials involve four parameters. Two of them are the components of stress and strain tensors; however, the rest are related to the components of the electric displacement and electric field. The constitutive equations of smart piezoelectric materials can be written as follows:

$$\sigma_{ij} = C_{ijkl}\varepsilon_{kl} - e_{mij}E_m, \tag{2.5}$$

$$D_i = e_{ikl}\varepsilon_{kl} + s_{im}E_m, \tag{2.6}$$

where σ_{ij} and D_i stand for the components of stress tensor and electric displacement vector, respectively, whereas ε_{kl} and E_m are the components of linear strain tensor and electric field vector, respectively. Also, C_{ijkl}, e_{mij} and s_{im} are the components of elasticity tensor, piezoelectric coefficient and dielectric permittivity, respectively. According to Eqs. (2.5) and (2.6), each of the elastic or electric excitations can generate the other one for a piezoelectric material.

Now, we are familiar with the behaviors of the smart piezoelectric materials, so, it is time to develop the constitutive equations for a piezomaterial in nanoscale. Just as in Eq. (2.4), the nonlocal constitutive equations of piezoelectric materials can be expressed in the following forms:

$$(1 - \mu^2\nabla^2)\sigma_{ij} = C_{ijkl}\varepsilon_{kl} - e_{mij}E_m, \tag{2.7}$$

$$(1 - \mu^2\nabla^2)D_i = e_{ikl}\varepsilon_{kl} + s_{im}E_m, \tag{2.8}$$

2.2.3 Constitutive Equations of Magnetoelectroelastic (MEE) Materials

In Sections 2.2.1 and 2.2.2, the nonlocal constitutive equations were developed for both linear elastic and piezoelectric materials, respectively. Herein, it is aimed to introduce another type of smart materials and extend the Eringen's theory to such materials in order to find a set of size-dependent constitutive equations for these materials. In fact, the main functionality of this type of materials is just the same as that of smart piezoelectric ones. However, these materials are the advanced versions of the previously introduced piezoelectric materials.

Remembering the behaviors of piezoelectric materials, they had the capability of transforming a mechanical load into an electric field and vice versa. A more complicated type of smart materials is detected which is able to change this dual energy transformation to a triple one. These materials, which can change each of the elastic, electric or magnetic excitations to the other one, are called magnetoelectroelastic (MEE) materials. These materials are extensively utilized in modern engineering designs. In recent years, a large number of MEE elements are used in nanoelectromechanical systems (NEMSs). Here, the constitutive behaviors of these materials will be reviewed in the nanoscale. As in Section 2.2.2, at first, the behaviors of these materials will be studied in the macroscale; afterward, their size-dependent constitutive behavior will be developed within the framework of Eringen's nonlocal theory.

The constitutive equations of MEE materials can be formulated in the following forms:

$$\sigma_{ij} = C_{ijkl}\varepsilon_{kl} - e_{mij}E_m - q_{nij}H_n, \tag{2.9}$$

$$D_i = e_{ikl}\varepsilon_{kl} + s_{im}E_m + d_{in}H_n, \tag{2.10}$$

$$B_i = q_{ikl}\varepsilon_{kl} + d_{im}E_m + \chi_{in}H_n, \tag{2.11}$$

where B_i and H_n are the components of magnetic induction and magnetic field, respectively. Furthermore, q_{nij}, d_{in} and χ_{in} are piezomagnetic, magnetoelastic and magnetic constants, respectively. Now, the size-dependency effects must be included. By extending Eq. (2.4) for MEE materials, the constitutive equations of MEE solids can be written as

$$(1 - \mu^2\nabla^2)\sigma_{ij} = C_{ijkl}\varepsilon_{kl} - e_{mij}E_m - q_{nij}H_n, \tag{2.12}$$

$$(1 - \mu^2\nabla^2)D_i = e_{ikl}\varepsilon_{kl} + s_{im}E_m + d_{in}H_n, \tag{2.13}$$

$$(1 - \mu^2\nabla^2)B_i = q_{ikl}\varepsilon_{kl} + d_{im}E_m + \chi_{in}H_n. \tag{2.14}$$

2.3 Mathematical Formulation of the Nonlocal Strain Gradient Elasticity

2.3.1 Constitutive Equation for Linear Elastic Solids

In this section, the constitutive equation for linear elastic solids will be derived on the basis of the nonlocal strain gradient elasticity theory. The integral representation of the nonlocal strain gradient elasticity is in the following form:

$$\sigma_{ij} = \int_V \alpha_0(|\mathbf{x} - \mathbf{x}'|, e_0a)C_{ijkl}\varepsilon_{kl}(\mathbf{x}')dV(\mathbf{x}')$$
$$- l^2 \int_V \alpha_1(|\mathbf{x} - \mathbf{x}'|, e_1a)C_{ijkl}\varepsilon_{kl,m}(\mathbf{x}')dV(\mathbf{x}'), \tag{2.15}$$

where $\varepsilon_{kl,m}$ denotes the gradient of strain tensor's components with respect to x_m. It is worth mentioning that in this theory, two kernel functions are presented, namely $\alpha_0(|\mathbf{x} - \mathbf{x}'|, e_0a)$ and $\alpha_1(|\mathbf{x} - \mathbf{x}'|, e_1a)$. The first kernel is indeed the Eringen's kernel which was shown with $\alpha(|\mathbf{x} - \mathbf{x}'|, \tau)$ in Eq. (2.1), and the second one is the kernel function which is responsible for considering the effects of first gradient of the strain tensor on the stress components in the nanoscale.

In Eq. (2.15), a material length scale, l, is introduced to account for the higher-order strain gradient field. As in Section 2.2.1 and by assuming $e_1a = e_0a$, the differential form of the nonlocal strain gradient elasticity can be written in the following form:

$$(1 - (e_0a)^2\nabla^2)\sigma_{ij} = (1 - l^2\nabla^2)C_{ijkl}\varepsilon_{kl}. \tag{2.16}$$

Equation (2.16) can be rewritten in the following form:

$$(1 - \mu^2\nabla^2)\sigma_{ij} = (1 - \lambda^2\nabla^2)C_{ijkl}\varepsilon_{kl} \tag{2.17}$$

in which $\mu = e_0a$ and $\lambda = l$ are nonlocal and length-scale parameters, respectively.

2.3.2   Constitutive Equations of Piezoelectric Materials

In Section 2.3.1, the nonlocal stress–strain gradient theory was extended in order to determine the constitutive equations of linear elastic solids in a nanosize element.

As discussed in Sections 2.2.2 and 2.2.3, the scale effects in smart materials can be applied by inserting the scale parameters in the constitutive equations of the bulk of such materials. Due to this trend, the constitutive equations of a piezoelectric material can be presented in the following form on the basis of the nonlocal strain gradient piezoelectricity:

$$(1 - \mu^2\nabla^2)\sigma_{ij} = (1 - \lambda^2\nabla^2)[C_{ijkl}\varepsilon_{kl} - e_{mij}E_m], \tag{2.18}$$

$$(1 - \mu^2\nabla^2)D_i = (1 - \lambda^2\nabla^2)[e_{ikl}\varepsilon_{kl} + s_{im}E_m]. \tag{2.19}$$

2.3.3   Constitutive Equations of MEE Materials

As in the previous section, the nonlocal-strain-gradient-based constitutive equations of an MEE material can be formulated in the following form:

$$(1 - \mu^2\nabla^2)\sigma_{ij} = (1 - \lambda^2\nabla^2)[C_{ijkl}\varepsilon_{kl} - e_{mij}E_m - q_{nij}H_n], \tag{2.20}$$

$$(1 - \mu^2\nabla^2)D_i = (1 - \lambda^2\nabla^2)[e_{ikl}\varepsilon_{kl} + s_{im}E_m + d_{in}H_n], \tag{2.21}$$

$$(1 - \mu^2\nabla^2)B_i = (1 - \lambda^2\nabla^2)[q_{ikl}\varepsilon_{kl} + d_{im}E_m + \chi_{in}H_n]. \tag{2.22}$$

3

Si e-Dependent Effects on Wave Propagation in Nanostructures

This chapter is dedicated to probe the nanodimensional behaviors of materials. Prior to any discussion about the behaviors of the materials of tiny sizes, it is better to review the effects of the interatomic forces in both macro- and nanoscales. Indeed, when a structure with usual macroscale dimensions is selected for study, the effects of the interatomic forces will be usually dismissed. The main reason for this decision is that the magnitude of these forces is small enough for them to be neglected when analyzing the macroscale structures. In other words, the external loading which is going to be applied on the structure is much bigger than the interatomic forces. So, it is a logical decision to exclude these tiny forces in macroscale mechanical analyses. However, if the dimensions of a structure lie in the nanoscale, this logic does not apply. In this case, due to the small magnitude of all the other forces which are applied on the nanosize element, the interatomic forces must be included in the investigations for a reliable estimation of behaviors of the nanostructure. Generally, nanoscale elements possess a very high surface-to-volume ratio, and due to this fact, each of the nanostructures has a potential to attract the others. This reciprocal attractive force which exists in materials of small sizes is usually called as van der Waals (vdW) force. The aforementioned feature of the nanostructures results in the improved mechanical properties of a nanostructure in comparison with those of a similar element in the macroscale. In other words, vdW force, which is of great importance in nanosize structures and materials, is one of the involved variables which cannot be neglected in nanoscale studies at all. In what follows, more detailed explanations will be presented about the features and importance of the dispersion phenomenon in nanoscale elements.

3.1   Importance of Wave Dispersion in Nanostructures

It is obvious that in modern systems, nanostructures are used instead of the conventional macrosize elements in different devices more than before. Therefore, it is clear that the behaviors of such tiny elements must be completely studied for designing an efficient and suitable nanoscale device. Also, from Chapter 1, the wave dispersion technique is one of the best non-destructive tests (NDTs) which can predict the material properties of any desired structure. This technique can reveal interesting results if the studied structure possesses dimensions in the range of nanometers. Also, it is worth mentioning that because of the improved material properties of the nanostructures, these elements are excellent candidates for generating high domain dynamic responses. In fact, such tiny elements can tolerate frequencies up to terahertz (THz) order. However, this remarkable limit of the natural frequency can be achieved only when a wave is dispersed in a desired tiny medium. In other words, although waves are naturally classified as fluctuating phenomena, they can have improved frequencies even better than those produced by the vibration phenomenon. The nanostructures are generally able to produce frequencies of gigahertz (GHz) order when subjected to a harmonic vibrating excitation, but they will be able to generate frequencies 1,000 times more powerful than those generated by the vibrating excitation. Due to the aforementioned reasons, many researchers allocated their time and effort to probe the mechanical properties of nanosize elements by means of the dispersion technique. Herein, some of the examples will be presented to show the works in this field of interest. Hernandez et al. [36] used a wave-based acoustic technique for the purpose of characterizing bilayer nanostructures made from aluminium and silicon-nitride. Also, Philip et al. [37] could experimentally determine the properties of nanocrystalline diamond nanofilms with the aid of the well-known wave-based Raman spectroscopy method. The transverse acoustic wave equations were solved by Ramprasad and Shi [38] for a multilayer stack to obtain the length scales of the studied nanostructure. Sampathkumar et al. [39] generated an out-of-plane transverse excitation in a clamped–clamped nanobeam to monitor its photothermal operation in the framework of a dispersion-assisted approach. In another research, motivated by the deficiency in employment of waves of short wavelength in determination of the material properties of tiny nanostructures, Huang and Song [40] presented a high-frequency ultrasonic study to characterize the ultra-thin nanostructures. Hepplestone and Srivastava [41] performed a wave-propagation-based analysis of the frequency gaps of composite semiconductors and highlighted the changes that appeared in the answers while the location and size of the gaps were varied. Furthermore, Parsons and Andrews [42] reported the results of their observations from wave scattering in porous silicon superlattices, but their experimental results are limited to wavelengths between 37 and 167 nanometers. Later, Chen and Wang [43] surveyed the band structure problem of a phononic nanosize crystal by means of propagation of waves through the crystal. Besides, Zhen et al. [44] analyzed the effect of surface/interface on the band structure of phononic nanocrystals on the basis of a wave-dispersion-based approach.

3.2   Wave Dispersion in Smart Nanodevices

Herein, some of the conducted researches dealing with the application of dispersion of waves in nanoscale materials and structures will be reviewed to present the significance of wave propagation in nanotechnology. These researches include various applications such as mechanics, biomechanics, medicine and electronics. For example, focusing on the features of the nanotechnology in the solid mechanics, Chong et al. [45] introduced the wave propagation technique as one of the most applicable methods in solid mechanics for the purpose of defect detection in nanoscale structures. In another study, Wang et al. [46] realized that the output power of a nanogenerator (NG) is not independent of the excitation circumstances and the geometrical parameters. Indeed, it was shown in their paper that the amount of power can be changed when the distance from the ultrasonic dispersion resource is varied. Thereafter, Cha et al. [47] utilized a sound-wave-type smart piezoelectric nanowire source for the generation of initial excitation in a novel NG and proved the suitability of such piezoelectric nanostructures to be employed in NGs. The THz frequency range of the nanotechnology was used by Akyildiz and Jornet [48] to describe a smart wireless nanosensor network. Furthermore, He et al. [49] employed the concept of wave propagation incorporated with the use of nanostructures in order to utilize such tiny elements to develop new mechanisms for sensing, imaging, tissue engineering and so on to be employed in medicine. The reflection and refraction coefficients of smart piezoelectric nanocomposites were determined by Al-Hossain et al. [50] on the basis of a wave-propagation-based method using quasi-vertical waves at the interface of the nanocomposite. Abd-Alla et al. [51] probed the characteristics of both reflection and transmission waves in piezoelectric nanocomposites. Just as Al-Hossain et al. [50] conducted their study, Abd-Alla et al. [51] studied the linear problem. Besides, the nanocomposite's interface is considered to be made from two anisotropic hexagonal symmetric smart piezoelectric materials. A novel microwave absorber consisting of a thin and multilayered structure is depicted by Danlée et al. [52] utilizing CNT in the structure as well as the use of polymeric materials to satisfy the electric conduction and dielectric features, respectively, in the frequency range of 8–40 GHz. Attention of the researchers investigating biomechanical problems has been recently attracted by micro-electro-mechanical- and/or nano-electro-mechanical- systems hosting acoustical waves to be employed as biosensors. In such tiny devices, film bulk acoustic waves and surface acoustic waves are used more than other kinds of waves [53]. On the other hand, Shin et al. [54] developed a metamaterial efficient microwave to cover frequencies of order 10–12 GHz for enabling the optical devices to exhibit the object in an invisible manner. The dispersion equations of Love wave scattered inside a smart piezoelectric nanofilm, which is assumed to be seated on a semi-infinite medium, were solved by Zhang et al. [55]. They could correctly capture the effects of the surface of the nanostructure on the wave propagation responses. Another attempt was performed by Lang et al. [56] in designing an acoustic sensor with a notably high sensitivity to capture low-frequency sounds. In another study in the field of biomechanics, Elayan et al. [57] introduced an analytical solution for the wave dispersion analysis of the human tissues with respect to discrepancies and thickness effects.

3.3 Crucial Parameters in Accurate Appro imation of the Wave Propagation Responses in Nanostructures

Another issue which must be discussed is about the effect of the wave characteristics on the wave propagation response in a nanostructure. In other words, the wave dispersion response depends on more than one factor related to the wave's nature directly. Indeed, the wave number can affect the dynamic responses which are determined by means of a wave dispersion analysis. This variable can change the amount of the wave frequency or its dispersion velocity whenever it is varied. This fact can be observed clearly when one studies the wave propagation problem within nanostructures (i.e., either nanobeams, nanoplates or nanoshells). In addition, it must be considered that the microstructure of the surveyed nanosize element can affect the dispersion responses of the system in the nanoscale. In other words, the accuracy of the wave propagation analysis conducted by a researcher deeply depends on the way in which the problem is formulated or simulated. As stated in this chapter and Chapter 2, investigation of wave dispersion in nanostructures requires the application of the size-dependent continuum theories when it is purposed to formulate the problem from the solid mechanics point of view instead of using computers for the nanoscale simulations. In this condition, using proper characteristic lengths plays a crucial role in the determination of the dynamic responses of the waves scattered in nanostructures. Moreover, it is important to point out that in some cases, although the proper characteristic length is not utilized, the correct response can be achieved. This trend again is due to the incredible role of the wave number in the estimation of the wave propagation response of nanoscale structures. In other words, it is proven that in the case of small wave numbers, nearly smaller than $\beta = 0.1$ (1/nm), the wave propagation responses obtained from both classical and nonlocal continuum mechanics hypotheses will be the same. However, the difference between the responses obtained from these aforementioned theorems will be remarkable in the case of wave numbers greater than the aforesaid boundary value. Actually, the reason for this phenomenon is the direct effect of wave number on the stiffness of the nanostructures due to the fact that both of the well-known nonlocal elasticity and nonlocal strain gradient theories have the Laplacian operator in their constitutive equations. As you know, the usual analytical wave solution considers the dependency of the wave propagation on the spatial variables by means of wave numbers, and applying the effect of the Laplacian operator corresponds to appearance of the wave numbers in the equations which results in the high effect of the wave number in the determination of the nanostructures' equivalent stiffness. Hence, depending on the value of the wave number, the stiffness-softening and/or stiffness-hardening impact of the nonlocal and length-scale parameters will vary. In other words, the wave number is thorough-paced enough to determine how powerful the softening and hardening effects of the nonlocal and length-scale parameters on the stiffness of the nanostructure are, respectively. So, it is not strange to observe more clear changes in the wave dispersion curves at high wave numbers whenever the scale parameters are varied.

4

Wave Propagation Characteristics of Inhomogeneous Nanostructures

In this chapter, the wave propagation problem of heterogeneous functionally graded (FG) nanosize beams and plates are surveyed in the framework of a nonlocal strain gradient higher-order beam theory. The effects of shear deformation are completely included free from any additional shear correction coefficient by means of a refined sinusoidal beam theorem. Furthermore, small-scale effects are covered based on the nonlocal strain gradient elasticity theory. It is proven that the aforementioned theory is powerful enough to guesstimate the behavior of the nanostructures better than previously presented ones. As a matter of fact, both stiffness-softening and stiffness-hardening characteristics of nanosize elements are coupled together in this theory. On the other hand, the equivalent material properties are obtained utilizing the rule of the mixture for the nanobeam and Mori–Tanaka method for the nanoplate. Equations of motion are derived extending the dynamic version of the principle of virtual work. Thereafter, the size-dependent partial differential equations of the problem are solved based on an exponential solution function to obtain the circular wave frequency. Next, a set of numerical illustrations is presented to emphasize the effect of various parameters on the mechanical responses of FG nanostructures.

4.1 Introduction

4.1.1 Functionally Graded Materials (FGMs)

In the 1990s, the growing demand for using materials in giant thermal conditions resulted in the invention of a novel type of composites, named functionally graded materials (FGMs). These composites are better candidates for mechanical designs in comparison with laminated composites due to their improved corrosion resistance, toughness and thermal resistance and also their lower stress concentration [58]. Thus, it is crucial to analyze the mechanical behaviors of FG structures. For example, Ebrahimi and Rastgoo [59] surveyed the vibrational responses of FG circular plates, surrounded by piezoactuators, via Kirchhoff plate theory. Shen [60] carried out a comparison study between the buckling and postbuckling responses of an FG plate with smart actuators. In another attempt, a new iterative method was introduced which is powerful enough to solve the vibration problem of axially FG (AFG) beams with nonuniform cross-sections [61]. Also, Alshorbagy et al. [62] performed a finite element analysis (FEA) on the dynamic behaviors of FG beams by means of polynomial shape functions. Şimşek et al. [63] could study the dynamic responses of an AFG beam when a moving harmonic load is imposed. Thai and Choi [64] investigated the vibration problem of an FG plate rested on an elastic substrate based on a refined higher-order plate theory. Ebrahimi [65] analyzed the electromechanically actuated vibrational responses of FG plates. The thermomechanical buckling characteristics of FG circular/annular plates were studied by Ghiasian et al. [66]. Moreover, both free and forced vibration analyses of bidirectional FG beams were accomplished by Şimşek [67]. Ghiasian et al. [68] probed the nonlinear thermoelastic dynamic buckling behaviors of FG beams. Jafarinezhad and Eslami [69] could solve the thermomechanical responses of an FG plate subjected to a flexural thermal shock utilizing the first-order shear deformation theory (FSDT). The stress analysis of FG pressure vessels was carried out by Gharibi et al. [70] in the framework of Frobenius series method. Tang and Yang [71] investigated the postbuckling and nonlinear vibration problems of FG fluid-conveying pipes.

4.1.2   FG Nanostructures

Combination of outstanding features of FGMs with enhanced properties of nanosize beams, plates and shells is one of the most crucial issues which should be well studied. In fact, such compositionally gradient nanostructures are proper choices for engineering designs. Henceforward, some researchers made an effort to probe the static and dynamic responses of FG nanostructures based upon the size-dependent continuum mechanics. Actually, a large number of researches can be found in this field of interest; however, only some of these studies will be reviewed here. For example, Eltaher et al. [72] could investigate the vibrational responses. Rahmani and Pedram [73] presented a nonlocal first-order shear deformable beam theorem for vibration analysis of FG nanobeams. In addition, the nonlinear vibrational responses of FG nanobeams were solved by Nazemnezhad and Hosseini-Hashemi [74]. Ebrahimi and Salari [75] studied thermomechanical vibration and buckling behaviors of FG nanobeams based on the Timoshenko beam theory. Li et al. [76] presented a nonlocal strain gradient Euler–Bernoulli beam model for the purpose of investigating the wave propagation responses of FG nanobeams. Ebrahimi and Barati [77] investigated the hygrothermomechanically affected stability responses of FG nanobeams. Zamani Nejad et al. [78] presented a nonlocal Euler–Bernoulli beam model for vibration analysis of bidirectionally graded nanobeams. Ebrahimi and his coworkers analyzed the wave propagation problem of smart FG nanobeams and nanoplates [79,80]. On the other hand, Ebrahimi and his coworkers [81,82] analyzed the thermally affected dynamic frequency and dispersion characteristics of FG nanobeams in the framework of size-dependent elasticity theorems. Barati and Zenkour [83] analyzed the wave dispersion characteristics of double-layered FG nanobeams based on the Euler–Bernoulli beam model. Also, the issue of investigating the coupled effects of moisture concentration and temperature gradient on the dynamic responses of FG nanobeams was probed by Ebrahimi and Barati [84,85]. They analyzed both elastic and viscoelastic FG nanobeams in their studies. Furthermore, the stability responses of a curved FG nanobeam was surveyed by Ebrahimi and Barati [86] by means of the nonlocal stress–strain gradient elasticity hypothesis. Lately, Srividhya et al. [87] surveyed nonlinear deflection responses of FG nanoplates via a finite element method (FEM).

4.2   Homogenization of FGMs

4.2.1   Power-Law Model

In this section, the equivalent material properties of FGMs are going to be computed based on the power-law relations. Herein, the top and bottom surfaces of the beam are considered to be made from pure ceramic and pure metal, respectively. The volume fraction of each phase is considered to be a function of z-axis in order to satisfy the gradual variation of equivalent material properties in the thickness direction. In this method, the volume fraction of ceramic phase can be presented via the following formula:

$$V_c = \left(\frac{z}{h} + \frac{1}{2}\right)^p, \tag{4.1}$$

where p is the gradient index or power-law exponent which is introduced to control the percentage of each phase with its change. Also, h denotes the structure's thickness. When referring to the initial concept of the rule of mixtures, it can be figured out that volume fraction of metallic phase can be related to that of ceramic phase by

$$V_m = 1 - V_c. \tag{4.2}$$

Now, the equivalent material properties of FGMs can be calculated based on the following relation:

$$P_{eq}(z) = P_c V_c + P_m V_m \tag{4.3}$$

in which P refers to each of the material properties including Young's modulus, Poisson's ratio, and density. Now, the simplified form of equivalent material properties can be written by substituting Eqs. (4.1) and (4.2) in (4.3) as follows:

$$P_{eq}(z) = (P_c - P_m)V_c + P_m. \tag{4.4}$$

4.2.2 Mori Tanaka Model

Based on this well-known homogenization scheme, the local bulk and shear moduli of an FGM medium can be written in the following forms:

$$\frac{K_e - K_m}{K_c - K_m} = \frac{V_c}{1 + \dfrac{V_m(K_c - K_m)}{K_m + 4G_m/3}}, \tag{4.5}$$

$$\frac{G_e - G_m}{G_c - G_m} = \frac{V_c}{1 + \dfrac{V_m(G_c - G_m)}{G_m + G_m\dfrac{9K_m + 8G_m}{6K_m + 12G_m}}}, \tag{4.6}$$

where the subscripts c and m denote the ceramic and metal phases, respectively. It is worth mentioning that the volume fraction of the ceramic phase is same as that introduced in Eq. (4.1). Moreover, the relation between the volume fractions of ceramic and metal phases is similar to Eq. (4.2). Now, on the basis of the implemented homogenization procedure, the Young's modulus and Poisson's ratio of the FGM can be presented as follows:

$$E(z) = \frac{9K_e G_e}{3K_e + G_e}, \tag{4.7}$$

$$\nu(z) = \frac{3K_e - 2G_e}{6K_e + 2G_e}. \tag{4.8}$$

It is worth mentioning that the mass density of the FGM can be calculated by substituting ρ instead of P in Eq. (4.4).

4.3   Analysis of FG Nanobeams

4.3.1   Kinematic Relations of Beams

Now, it is time to derive the equations of motion of a beam-type structure. For this purpose, the strain–displacement relations are required. In this chapter, we aim to combine the displacement field of desirable beam theories with the definition of infinitesimal strains in continuum mechanics in order to obtain the strain–displacement relations.

First, the kinematic relations are derived based on the classical theory of beams, known as Euler–Bernoulli beam theory. In this theory, the effects of shear deformations are not included, and the theorem can be used only in beams with high slenderness ratios. To cover this deficiency, refined higher-order beam models are introduced in another section to take into account shear deformations via a shape function without employing any shear correction factor. In the following sections, the aforementioned theories will be discussed thoroughly.

4.3.1.1   Euler Bernoulli Beam Theory

According to this theory, the distortions of the beam's geometry are neglected. Also, the normal line which is perpendicular to the cross-section of the beam before loading remains perpendicular after loading. In this theory, the displacement fields can be formulated in the following forms:

$$u_x(x, z, t) = u(x, t) - z\frac{\partial w(x, t)}{\partial x}, \tag{4.9}$$

$$u\ (x, z, t) = w(x, t), \tag{4.10}$$

where u and w denote displacement components of the mid-surface in axial and transverse directions, respectively. The infinitesimal strains of a continua can be calculated via the following equation:

$$\varepsilon = \frac{1}{2}\left(\nabla\mathbf{u} + (\nabla\mathbf{u})^T\right). \tag{4.11}$$

Substituting Eq. (4.11) in Eqs. (4.9) and (4.10), the nonzero strains of an Euler–Bernoulli beam can be written as follows:

$$\varepsilon_{xx} = \varepsilon_{xx}^0 - z\kappa_{xx}^0, \tag{4.12}$$

where

$$\varepsilon_{xx}^0 = \frac{\partial u(x, t)}{\partial x}, \kappa_{xx}^0 = \frac{\partial^2 w(x, t)}{\partial x^2}. \tag{4.13}$$

4.3.1.2   Refined Sinusoidal Beam Theory

As stated before, higher-order beam theories are better candidates to be utilized in structural analyses for the purpose of deriving the equations of motion due to their ability to capture shear deformation. However, the conventional form of higher-order theories requires a degree of freedom (DOF) for the rotation of the cross-section as well as those of axial and transverse displacements. When implementing a refined higher-order theory, the aforementioned problem can be solved by dividing the deflection into bending and shearing ones. In other words, the term w in classical theory has been divided into separate terms for bending and shear deformations. According to such theories, the displacement field of a refined higher-order beam can be presented in the following forms:

$$u_x(x, z, t) = u(x, t) - z\frac{\partial w_b(x, t)}{\partial x} - f(z)\frac{\partial w_s(x, t)}{\partial x}, \tag{4.14}$$

$$u\ (x, z, t) = w_b(x, t) + w_s(x, t), \tag{4.15}$$

where w_b and w_s are bending and shear deflections, respectively. Also, $f(z)$ is the shape function which is pursued to govern the profile of shear stress and strain through the thickness direction. In this section, the shape function is supposed to be

$$f(z) = z - \frac{\sin(\xi z)}{\xi}, \tag{4.16}$$

where $\xi = \dfrac{z}{h}$. Next, the nonzero strains of the above theory can be derived by using Eq. (4.11) as follows:

$$\varepsilon_{xx} = \varepsilon_{xx}^0 - z\kappa_{xx}^b - f(z)\kappa_{xx}^s, \varepsilon_x = g(z)\varepsilon_x^0 , \tag{4.17}$$

where

$$\varepsilon_{xx}^0 = \frac{\partial u(x, z, t)}{\partial x}, \kappa_{xx}^b = \frac{\partial^2 w_b(x, z, t)}{\partial x^2}, \kappa_{xx}^s = \frac{\partial^2 w_s(x, z, t)}{\partial x^2}, \varepsilon_x^0 = \frac{\partial w_s}{\partial x} \tag{4.18}$$

in which $g(z) = 1 - \dfrac{df(z)}{dz}$.

4.3.2 Derivation of the Equations of Motion for Beams

The purpose of this section is to obtain the motion equations of a beam via extending the dynamic form of the principle of virtual work. In mechanical texts, the dynamic form of virtual work's principle is called Hamilton's principle. Based on this principle, the variation of the Lagrangian of a conservative continua is considered to be zero. Furthermore, this statement is free from transformations of the coordinate system. The mathematical form of Hamilton's principle can be written as

$$\delta \int_{t_0}^{t_1} L dt = 0, \tag{4.19}$$

where δ is the variation operator. In addition, L stands for the Lagrangian and can be stated as follows:

$$L = U - (K + V) \tag{4.20}$$

in which U, K and V denote strain energy, kinetic energy and work done by external loadings, respectively. The strain energy can be defined as

$$U = \int_V \boldsymbol{\sigma} : \varepsilon dV, \tag{4.21}$$

where $\boldsymbol{\sigma}$ refers to the Cauchy stress tensor. Also, V refers to the volume of the structure. On the other hand, the integral form of kinetic energy can be formulated in the following form:

$$K = \frac{1}{2} \int_V \rho \dot{\mathbf{u}}^2 dV \tag{4.22}$$

in which ρ and $\dot{\mathbf{u}}$ stand for mass density and first derivative of displacement field with respect to time, respectively. It is worth mentioning that work of external loadings cannot be generally formulated and possesses a unique form in each loading condition. Accordingly, the formulation of this term will be presented in the following sections.

4.3.2.1 Equations of Motion of Euler Bernoulli Beams

Here, we are about to derive the variation of strain energy and kinetic energy for an Euler–Bernoulli beam of length L, width b and thickness h. When Eq. (4.17) is extended and the effect of variation operator is imposed on the formula for linear elastic solids, the variation of strain energy can be written as follows:

$$\delta U = \int_V \sigma_{xx} \delta \varepsilon_{xx} dV. \tag{4.23}$$

Now, if Eq. (4.22) is substituted in Eq. (4.23), the above relation can be stated in the following form:

$$\delta U = \int_0^L (N \delta \varepsilon_{xx}^0 - M \delta \kappa_{xx}^0) dx \tag{4.24}$$

in which N and M can be computed via

$$[N, M] = \int_A [1, z] \sigma_{xx} dA. \tag{4.25}$$

Furthermore, the variation of kinetic energy can be expressed in the following form:

$$\delta K = \int_0^L \left[I_0 \left(\frac{\partial u}{\partial t} \frac{\partial \delta u}{\partial t} + \frac{\partial w}{\partial t} \frac{\partial \delta w}{\partial t} \right) - I_1 \left(\frac{\partial u}{\partial t} \frac{\partial^2 \delta w}{\partial x \partial t} + \frac{\partial \delta u}{\partial t} \frac{\partial^2 w}{\partial x \partial t} \right) + I_2 \frac{\partial^2 w}{\partial x \partial t} \frac{\partial^2 \delta w}{\partial x \partial t} \right] dx, \tag{4.26}$$

where mass moments of inertia can be calculated as

$$[I_0, I_1, I_2] = \int_A [1, z, z^2] \rho(z) dA. \tag{4.27}$$

Now, the Euler–Lagrange equations of an FG beam can be obtained by substituting Eqs. (4.24) and (4.26) in Eq. (4.19) and choosing the nontrivial response as

$$\frac{\partial N}{\partial x} = I_0 \frac{\partial^2 u}{\partial t^2} - I_1 \frac{\partial^3 w}{\partial x \partial t^2}, \tag{4.28}$$

$$\frac{\partial^2 M}{\partial x^2} = I_0 \frac{\partial^2 w}{\partial t^2} + I_1 \frac{\partial^3 u}{\partial x \partial t^2} - I_2 \frac{\partial^4 w}{\partial x^2 \partial t^2}. \tag{4.29}$$

4.3.2.2 Equations of Motion of Refined Sinusoidal Beams

Herein, the variation of strain energy and kinetic energy for a refined sinusoidal beam of geometry same as the one introduced in the previous section will be discussed. For this purpose, Eq. (4.21) is extended for a higher-order beam made from linear elastic materials. Thus, the variation of strain energy can be written as

$$\delta U = \int_V (\sigma_{xx} \delta \varepsilon_{xx} + \sigma_x \, \delta \varepsilon_x \,) dV. \tag{4.30}$$

Now, if Eq. (4.17) is substituted in Eq. (4.30), the strain energy can be reformulated as

$$\delta U = \int_0^L (N \delta \varepsilon_{xx}^0 - M^b \delta \kappa_{xx}^b - M^s \delta \kappa_{xx}^s + Q_x \, \delta \varepsilon_x^0 \,) dx \tag{4.31}$$

in which N, M^b, M^s and Q_x can be computed via

$$\left[N, M^b, M^s\right] = \int_A [1, z, f(z)]\,\sigma_{xx}dA, Q_x = \int_A g(z)\sigma_x\,dA. \tag{4.32}$$

On the other hand, the variation of kinetic energy can be written as follows:

$$\begin{aligned}
\delta K = \int_0^L \Bigg[& I_0\left(\dot{u}\delta\dot{u} + (\dot{w}_b + \dot{w}_s)\,\delta\left(\dot{w}_b + \dot{w}_s\right)\right) - I_1\left(\dot{u}\frac{\partial\delta\dot{w}_b}{\partial x} + \delta\dot{u}\frac{\partial\dot{w}_b}{\partial x}\right) \\
& - J_1\left(\dot{u}\frac{\partial\delta\dot{w}_s}{\partial x} + \delta\dot{u}\frac{\partial\dot{w}_s}{\partial x}\right) + I_2\frac{\partial\dot{w}_b}{\partial x}\frac{\partial\delta\dot{w}_b}{\partial x} + K_2\frac{\partial\dot{w}_s}{\partial x}\frac{\partial\delta\dot{w}_s}{\partial x} \\
& + J_2\left(\frac{\partial\dot{w}_b}{\partial x}\frac{\partial\delta\dot{w}_s}{\partial x} + \frac{\partial\delta\dot{w}_b}{\partial x}\frac{\partial\dot{w}_s}{\partial x}\right) \Bigg]dx,
\end{aligned} \tag{4.33}$$

where mass moments of inertia can be calculated as

$$[I_0, I_1, I_2, J_1, J_2, K_2] = \int_A \left[1, z, z^2, f(z), zf(z), f^2(z)\right]\rho(z)dA. \tag{4.34}$$

Now, the Euler–Lagrange equations of an FG sinusoidal beam are obtained by substituting Eqs. (4.31) and (4.33) in Eq. (4.19) as

$$\frac{\partial N}{\partial x} = I_0\ddot{u} - I_1\frac{\partial\ddot{w}_b}{\partial x} - J_1\frac{\partial\ddot{w}_s}{\partial x}, \tag{4.35}$$

$$\frac{\partial^2 M^b}{\partial x^2} = I_0(\ddot{w}_b + \ddot{w}_s) + I_1\frac{\partial\ddot{u}}{\partial x} - I_2\frac{\partial^2\ddot{w}_b}{\partial x^2} - J_2\frac{\partial^2\ddot{w}_s}{\partial x^2}, \tag{4.36}$$

$$\frac{\partial^2 M^s}{\partial x^2} + \frac{\partial Q_x}{\partial x} = I_0(\ddot{w}_b + \ddot{w}_s) + J_1\frac{\partial\ddot{u}}{\partial x} - J_2\frac{\partial^2\ddot{w}_b}{\partial x^2} - K_2\frac{\partial^2\ddot{w}_s}{\partial x^2}. \tag{4.37}$$

4.3.3 Constitutive Equations of FG Nanobeams

In this section, the size dependency will be considered for a composite FG nanobeam by applying a group of modifications on the stress–strain relationship of FG nanobeams. For this reason and also for the sake of completeness, the nonlocal strain gradient elasticity will be used. The main reason for this selection is that one can obtain the constitutive equations of the Eringen's nonlocal elasticity by omitting the length-scale parameter of the nonlocal strain gradient elasticity theory. In other words, it is easy to arrive at Eringen's theory from the stress–strain gradient elasticity theory. In what follows, the constitutive equations will be derived in terms of displacements and cross-sectional rigidities of the structure for both classical and refined higher-order nanobeams in the framework of separate parts to emphasize the differences between these two utilized beam hypotheses.

Whenever Eq. (2.21) is used and integrated over the thickness of the nanobeam, the stress–strain relationships can be modified in the following form for an Euler–Bernoulli beam element:

$$(1 - \mu^2\nabla^2)N = (1 - \lambda^2\nabla^2)\left(A_{xx}\frac{\partial u}{\partial x} - B_{xx}\frac{\partial^2 w}{\partial x^2}\right), \tag{4.38}$$

$$(1 - \mu^2\nabla^2)M = (1 - \lambda^2\nabla^2)\left(B_{xx}\frac{\partial u}{\partial x} - D_{xx}\frac{\partial^2 w}{\partial x^2}\right). \tag{4.39}$$

Also, via the same procedure, the stress–strain relationships of a refined beam can be formulated as follows:

$$(1 - \mu^2\nabla^2)N = (1 - \lambda^2\nabla^2)\left(A_{xx}\frac{\partial u}{\partial x} - B_{xx}\frac{\partial^2 w_b}{\partial x^2} - B_{xx}^s\frac{\partial^2 w_s}{\partial x^2}\right), \tag{4.40}$$

$$(1 - \mu^2\nabla^2)M^b = (1 - \lambda^2\nabla^2)\left(B_{xx}\frac{\partial u}{\partial x} - D_{xx}\frac{\partial^2 w_b}{\partial x^2} - D_{xx}^s\frac{\partial^2 w_s}{\partial x^2}\right), \tag{4.41}$$

$$(1 - \mu^2\nabla^2)M^s = (1 - \lambda^2\nabla^2)\left(B_{xx}^s\frac{\partial u}{\partial x} - D_{xx}^s\frac{\partial^2 w_b}{\partial x^2} - H_{xx}^s\frac{\partial^2 w_s}{\partial x^2}\right), \tag{4.42}$$

$$(1 - \mu^2\nabla^2)Q_x = (1 - \lambda^2\nabla^2)(A_x^s\,\frac{\partial w_s}{\partial x}), \tag{4.43}$$

where cross-sectional rigidities can be defined as

$$[A_{xx}, B_{xx}, B_{xx}^s, D_{xx}, D_{xx}^s, H_{xx}^s] = \int_A \left[1, z, f(z), z^2, zf(z), f^2(z)\right]E(z)dA, \tag{4.44}$$

$$A_x^s = \int_A g^2(z)G(z)dA. \tag{4.45}$$

It is worth mentioning that the differences between the Euler–Bernoulli and shear deformation beam models lie in either excluding or including the effects of the shear strain in the aforementioned models.

4.3.4 The Nonlocal Governing Equations of FG Nanobeams

Once the size-dependent constitutive relations, obtained in Section 4.3.3, are substituted in Euler–Lagrange equations, the nonlocal governing equations of the problem can be obtained. Whenever the governing equations of an Euler–Bernoulli nanobeam are required, Eqs. (4.38) and (4.39) are substituted in Eqs. (4.28) and (4.29), and the governing equations can be presented as follows:

$$\left(1 - \lambda^2\nabla^2\right)\left(A_{xx}\frac{\partial^2 u}{\partial x^2} - B_{xx}\frac{\partial^3 w}{\partial x^3}\right) + \left(1 - \mu^2\nabla^2\right)\left(-I_0\ddot{u} + I_1\frac{\partial\ddot{w}}{\partial x}\right) = 0, \tag{4.46}$$

$$\left(1 - \lambda^2\nabla^2\right)\left(B_{xx}\frac{\partial^3 u}{\partial x^3} - D_{xx}\frac{\partial^4 w}{\partial x^4}\right) + \left(1 - \mu^2\nabla^2\right)\left(-I_0\ddot{w} - I_1\frac{\partial\ddot{u}}{\partial x} + I_2\frac{\partial^2\ddot{w}}{\partial x^2}\right) = 0. \tag{4.47}$$

Via a similar procedure, the governing equations of sinusoidal nanobeams can be obtained by substituting Eqs. (4.40)–(4.43) in Eqs. (4.35)–(4.37) as

$$\left(1 - \lambda^2\nabla^2\right)\left(A_{xx}\frac{\partial^2 u}{\partial x^2} - B_{xx}\frac{\partial^3 w_b}{\partial x^3} - B_{xx}^s\frac{\partial^3 w_s}{\partial x^3}\right) + \left(1 - \mu^2\nabla^2\right)$$

$$\left(-I_0\ddot{u} + I_1\frac{\partial\ddot{w}_b}{\partial x} + J_1\frac{\partial\ddot{w}_s}{\partial x}\right) = 0, \tag{4.48}$$

$$\left(1 - \lambda^2\nabla^2\right)\left(B_{xx}\frac{\partial^3 u}{\partial x^3} - D_{xx}\frac{\partial^4 w_b}{\partial x^4} - D_{xx}^s\frac{\partial^4 w_s}{\partial x^4}\right) + \left(1 - \mu^2\nabla^2\right)$$

$$\left(-I_0\left(\ddot{w}_b + \ddot{w}_s\right) - I_1\frac{\partial\ddot{u}}{\partial x} + I_2\frac{\partial^2\ddot{w}_b}{\partial x^2} + J_2\frac{\partial^2\ddot{w}_s}{\partial x^2}\right) = 0, \tag{4.49}$$

$$\left(1 - \lambda^2\nabla^2\right)\left(B_{xx}^s\frac{\partial^3 u}{\partial x^3} - D_{xx}^s\frac{\partial^4 w_b}{\partial x^4} - H_{xx}^s\frac{\partial^4 w_s}{\partial x^4} + A_x^s\ \frac{\partial^2 w_s}{\partial x^2}\right) + \left(1 - \mu^2\nabla^2\right)$$

$$\left(-I_0\left(\ddot{w}_b + \ddot{w}_s\right) - J_1\frac{\partial \ddot{u}}{\partial x} + J_2\frac{\partial^2 \ddot{w}_b}{\partial x^2} + K_2\frac{\partial^2 \ddot{w}_s}{\partial x^2}\right) = 0. \tag{4.50}$$

4.3.5 Wave Solution for FG Nanobeams

Now, we are about to employ a solution method in order to solve the wave propagation problem of FG nanobeams. Due to our initial assumption, that is, utilization of infinitesimal strains, the governing equations obtained are linear equations. Thus, the system of equations can be easily solved on the basis of an analytical method.

Here, the well-known exponential solution functions, which were introduced in Chapter 1 will be presented for the displacement fields. By substituting the solution in the governing equations and solving an eigenvalue problem, the wave frequency and phase speed of the FG nanobeams can be obtained. In the following subsections, the introduced solution will be extended for both Euler–Bernoulli and higher-order nanobeams.

4.3.5.1 Solution of Euler Bernoulli FG Nanobeams

The solution functions of an Euler–Bernoulli nanobeam were presented in Chapter 1 in the framework of Eqs. (1.1) and (1.2). Now, once these solutions are substituted in Eqs. (4.42) and (4.43), the previously presented eigenvalue equation, which was expressed as Eq. (1.3), can be obtained. The corresponding arrays of the stiffness and mass matrices of the classical nanobeam can be written in the following forms:

$$k_{11} = -(1 + \lambda^2\beta^2)A_{xx}\beta^2, k_{12} = i(1 + \lambda^2\beta^2)B_{xx}\beta^3,$$
$$k_{22} = -(1 + \lambda^2\beta^2)D_{xx}\beta^4, \tag{4.51}$$

$$m_{11} = -(1 + \mu^2\beta^2)I_0, m_{12} = i\beta(1 + \mu^2\beta^2)I_1,$$
$$m_{22} = -(1 + \mu^2\beta^2)(I_0 + I_2\beta^2). \tag{4.52}$$

4.3.5.2 Solution of Refined Sinusoidal FG Nanobeams

As in the case of a Euler–Bernoulli nanobeam, herein, the eigenvalue equation will be extended for a refined sinusoidal nanobeam. The solution functions of this theorem can be found in Eqs. (1.5)–(1.7). Now, by substituting the aforementioned equations in Eqs. (4.44)–(4.46), an eigenvalue equation will be obtained. For this type of nanobeams, the corresponding mass and stiffness components are in the following forms:

$$k_{11} = -(1 + \lambda^2\beta^2)A_{xx}\beta^2, k_{12} = i(1 + \lambda^2\beta^2)B_{xx}\beta^3, k_{13} = i(1 + \lambda^2\beta^2)B_{xx}^s\beta^3,$$
$$k_{22} = -(1 + \lambda^2\beta^2)D_{xx}\beta^4, k_{23} = -(1 + \lambda^2\beta^2)D_{xx}^s\beta^4,$$
$$k_{33} = -(1 + \lambda^2\beta^2)(H_{xx}^s\beta^4 + A_x^s\ \beta^2), \tag{4.53}$$

$$m_{11} = -(1 + \mu^2\beta^2)I_0, m_{12} = i\beta(1 + \mu^2\beta^2)I_1, m_{13} = i\beta(1 + \mu^2\beta^2)J_1,$$
$$m_{22} = -(1 + \mu^2\beta^2)(I_0 + I_2\beta^2), m_{23} = -(1 + \mu^2\beta^2)(I_0 + J_2\beta^2),$$
$$m_{33} = -(1 + \mu^2\beta^2)(I_0 + K_2\beta^2). \tag{4.54}$$

4.3.6   Numerical Results and Discussion

This section is dedicated to numerically study the wave propagation problem of an FG nanobeam, shown in Figure 4.1, on the basis of nonlocal strain gradient elasticity theory. In this chapter, the effects of environment's temperature on the mechanical properties of the nanobeam are covered by using temperature-dependent material properties of alumina (Al_2O_3) and steel as the constituent materials of FG nanostructure [72]. On the other hand, a relative dimensionless size-dependent coefficient is presented for the sake of simplicity in the following form:

$$c = \frac{\lambda}{\mu}. \tag{4.55}$$

In Eq. (4.51), c is the scale factor which is used in the following illustrations. Indeed, the higher this coefficient, the higher is the system stiffness.

The validity of the presented model can be seen in Table 4.1 comparing the natural frequencies obtained from our modeling with those reported by Eltaher et al. [72]. It can be easily realized that our results are in remarkable agreement with those reported in Ref. [72].

Figure 4.2 is depicted to highlight the effect of length-scale parameter on the wave frequency of FG nanobeams. The mechanical behavior of the nanostructure can be divided into two completely different types. Indeed, wave frequency can be aggrandized continuously in the case of implementing Eringen's nonlocal theory ($\lambda = 0$), whereas wave frequency tends to ∞ whenever effects of strain gradient are included ($\lambda \neq 0$). It is clear that by increasing the value of length-scale parameter, the magnitude of wave frequency increases. This phenomenon can be justified when attention is paid to the stiffness-hardening effect which is covered within this theory. This increasing trend is more obvious in wave numbers

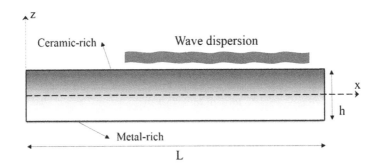

FIGURE 4.1

The geometry of an FG nanobeam subjected to the propagation of elastic waves.

TABLE 4.1

Comparison of the Natural Frequencies of FG Nanobeams for Various Gradient Indices and Nonlocal Parameters

μ	$p = 0.1$ Eltaher et al. [72]	This Study	$p = 0.2$ Eltaher et al. [72]	This Study	$p = 0.5$ Eltaher et al. [72]	This Study
0	9.2129	9.1887	7.8061	7.7377	7.0904	6.9885
1	8.7879	8.7663	7.4458	7.3820	6.7631	6.6672
2	8.4166	8.3972	7.1312	7.0712	6.4774	6.3865
3	8.0887	8.0712	6.8533	6.7966	6.2251	6.1386

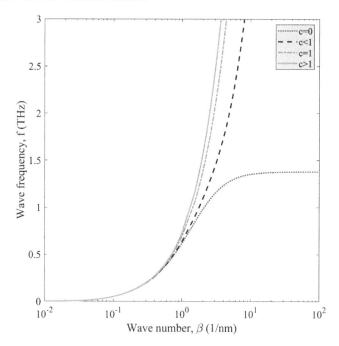

FIGURE 4.2
Variation of wave frequency against wave number for different length-scale parameters ($p = 2$, $\mu = 0.5$ nm).

higher than $\beta = 2$ (1/nm), meaning this hardening effect is not independent of wave number in wave dispersion problems.

On the other hand, in Figure 4.3, we try to show the differences of phase velocity curves of FG nanobeams for both classical and higher-order beams whenever wave number is changed. It can be figured out that Euler–Bernoulli beams possess greater wave responses in comparison with refined sinusoidal ones. As a matter of fact, classical beam theory presents a stiffer system. It is of significance to point out that differences of these theories can be well observed in a restricted range of wave numbers, nearly 0.5 (1/nm)$< \beta <$1 (1/nm). Therefore, it can be concluded that classical beam theory is not able to estimate the mechanical responses of an FG nanobeam in all wave numbers.

In Figure 4.4, the effect of length-scale parameter is covered again when the phase velocity curves of a flexural wave are plotted against wave number. According to the diagram, one can easily understand that phase velocity curves experience four kinds of behavior as length-scale parameter is varied. In the case of nonlocal elasticity ($\lambda = 0$), the path looks like a dome. In other words, phase velocity starts from a small amount, reaches its peak and eventually diminishes to zero. However, the path can be changed at higher wave numbers as nonlocal strain gradient elasticity is used ($\lambda \neq 0$). On the basis of the figure, the curves are shifted above at higher wave numbers when a greater value is assigned to the length-scale parameter. The phenomenon happens due to the enhanced stiffness of nanosize beams which is taken into consideration in nonlocal strain gradient elasticity and neglected in the nonlocal theory of Eringen. Moreover, one can obviously realize that there is no difference between Eringen's nonlocal theory and nonlocal strain gradient theory in small wave numbers.

Another illustration is presented in Figure 4.5 to show the coupled effects of length-scale parameter and gradient index on the phase speed of flexural waves dispersing inside

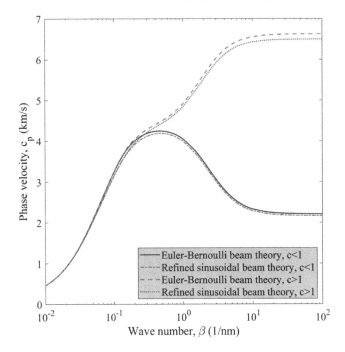

FIGURE 4.3
Comparison of the phase velocity curves of FG nanobeams via both Euler–Bernoulli and sinusoidal beam theories ($p = 2$).

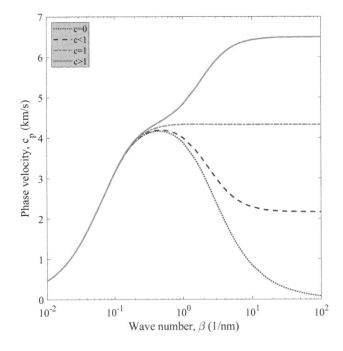

FIGURE 4.4
Variation of phase velocity against wave number for different length-scale parameters ($p = 2$, $\mu = 0.5$ nm).

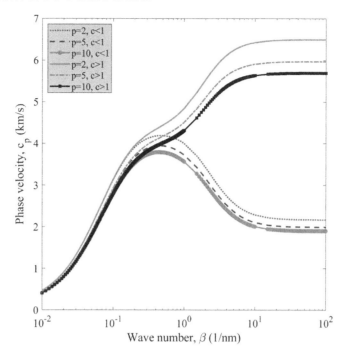

FIGURE 4.5
Variation of phase velocity against wave number for different gradient indices and relative ratios of size-dependent parameters.

a beam-type continuum. Clearly, the size-dependent behaviors of the nanostructure are the same as those observed in Figure 4.4. Indeed, the stiffness-hardening effect of length-scale coefficient can be well observed in this diagram. Besides, it can be found that phase velocity of FG nanobeams can be decreased whenever a greater material composition index is implemented. One should pay attention to the fact that phase velocity can be decreased further when gradient index is changed from 0.5 to 1 in comparison with the condition that this index is changed from 1 to 2. Thus, in conclusion, phase velocity has a two-step decrease as gradient index increases. At first, phase velocity suddenly drops (when p is changed from zero to a small nonzero value); thereafter, it gradually decreases as gradient index is increased.

Furthermore, Figure 4.6 shows a plot of the variation of wave frequency versus scale factor for various gradient indices in three desirable wave numbers. Based on the figure, the more the scale factor is, the greater is the wave frequency at all the wave numbers considered. This trend is mainly due to the fact that nanostructure's stiffness is amplified when a higher scale factor is chosen because higher scale factors correspond to bigger length-scale parameters. Moreover, using a higher gradient index results in obtaining a smaller mechanical response. This is due to the fact that by choosing a high gradient index, the volume of ceramic phase becomes higher and the elastic properties will become smaller compared to the situation in which the metal's volume fraction is greater than that of the ceramic. Also, it is clear that the effect of changing gradient index can be observed better in $\beta = 1$ (1/nm) compared with the cases in which wave number is greater.

As the final numerical example, Figure 4.7 is presented to highlight the effect of gradient index on the escape frequency of FG nanobeams when the effect of scale factor is covered.

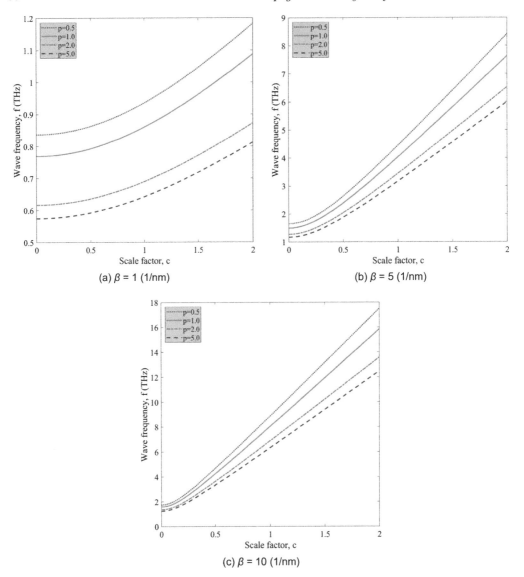

FIGURE 4.6
Variation of wave frequency versus scale factor for various values of gradient index and wave numbers.

It can be well observed that the escape frequency of propagated waves becomes lower as gradient index is raised. Moreover, the effect of gradient index, i.e., a two-step decrease in phase velocity, which was previously discussed in Figure 4.5 can be seen again. In other words, escape frequency enormously diminishes in the range of $p \in [0,1]$ and then follows a moderate decreasing path. In addition, the effect of adding scale factor can be clearly seen in this diagram. In other words, higher scale factors reveal higher escape frequencies. It is of high importance to point out that escape frequency does not depend on the nonlocal parameter at all. Actually, escape frequency can be reached when wave number reaches ∞, and at high wave numbers, the nonlocal parameter is not an effective coefficient anymore.

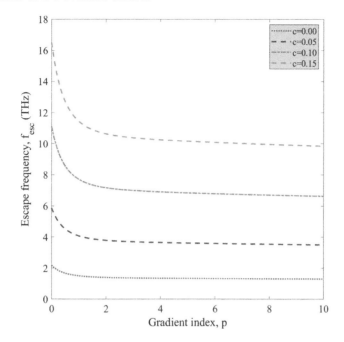

FIGURE 4.7
Variation of escape frequency against gradient index for various scale factors.

4.4 Analysis of FG Nanoplates

4.4.1 Kinematic Relations of Plates

This section is dedicated to deriving the displacement field of a plate for the purpose of obtaining the structure's equations of motion. As in the case of beams, in this section, the infinitesimal strain concept will be utilized to derive the nonzero strains of the structure. Our derivations will be performed on the basis of extending Eq. (4.11) for the displacement field of a plate. In what follows, both classical and refined shear deformation plate models will be introduced, and the introduced relations will be used to obtain the equations of motion.

4.4.1.1 Classical Plate Theory

Based on the classical theory of the plates, also known as Kirchhoff–Love plate theory, the cross-section of the plate experiences no distortion during the loading procedure and after it. Moreover, the normal line which is perpendicular to the cross-section of the plate before loading remains perpendicular after loading. Also, in this theory, the effects of shear deflection are not included. In fact, the important loading mode in this theorem is the bending one, and the effects of shear stress and strain are presumed to be small enough to be neglected. After above discussions, the displacement field of a classical plate can be formulated in the following form:

$$u_x(x, y, z, t) = u(x, y, t) - z\frac{\partial w(x, y, t)}{\partial x},$$

(4.56)

$$u_y(x, y, z, t) = v(x, y, t) - z\frac{\partial w(x, y, t)}{\partial y}, \tag{4.57}$$

$$u\ (x, y, z, t) = w(x, y, t), \tag{4.58}$$

where u and v denote displacement components of the mid-surface in axial and transverse directions, respectively. Also, w is the bending deflection of the plate. Substituting Eqs. (4.56)–(4.58) in Eq. (4.11), the nonzero strains of a classical plate can be written as follows:

$$\varepsilon_{xx} = \varepsilon_{xx}^0 - z\kappa_{xx}^0, \tag{4.59}$$

$$\varepsilon_{yy} = \varepsilon_{yy}^0 - z\kappa_{yy}^0, \tag{4.60}$$

$$\varepsilon_{xy} = \varepsilon_{xy}^0 - z\kappa_{xy}^0, \tag{4.61}$$

where

$$\varepsilon_{xx}^0 = \frac{\partial u(x, y, t)}{\partial x}, \varepsilon_{yy}^0 = \frac{\partial v(x, y, t)}{\partial y}, \varepsilon_{xy}^0 = \frac{\partial u(x, y, t)}{\partial y} + \frac{\partial v(x, y, t)}{\partial x},$$

$$\kappa_{xx}^0 = \frac{\partial^2 w(x, y, t)}{\partial x^2}, \kappa_{yy}^0 = \frac{\partial^2 w(x, y, t)}{\partial y^2}, \kappa_{xy}^0 = 2\frac{\partial^2 w(x, y, t)}{\partial x \partial y}. \tag{4.62}$$

4.4.1.2   Refined Sinusoidal Plate Theory

Now, this section is allocated to introduce and develop the displacement field of a refined higher-order plate model with a trigonometric shape function. As stated in the previous section, the classical theory of the plates is not able to handle the effects of shear deflection. To cover this deficiency, many shear deformation plate models were suggested by researchers to capture the effects of the variation of shear stress and strain across the thickness of the plate. In all these theories, two additional terms were introduced to consider the rotations of the plate's cross-section with respect to x- and y- axes. So, the total number of variants was changed from 3 to 5. In the refined plate models, the shear deformation is estimated using four variables which can lessen the computational cost of the program which is required for the analysis. The displacement field of a refined shear deformation plate is expressed in the following forms:

$$u_x(x, y, z, t) = u(x, y, t) - z\frac{\partial w_b(x, y, t)}{\partial x} - f(z)\frac{\partial w_s(x, y, t)}{\partial x}, \tag{4.63}$$

$$u_y(x, y, z, t) = v(x, y, t) - z\frac{\partial w_b(x, y, t)}{\partial y} - f(z)\frac{\partial w_s(x, y, t)}{\partial y}, \tag{4.64}$$

$$u\ (x, y, z, t) = w_b(x, y, t) + w_s(x, y, t), \tag{4.65}$$

where u and v denote displacement components of the mid-surface in axial and transverse directions, respectively. Also, w_b and w_s are the plate's bending and shear deflections, respectively. Again, by means of Eq. (4.11) and extending it for the displacement field expressed in Eqs. (4.63)–(4.65), the nonzero strains of a refined plate can be written in the following forms:

$$\varepsilon_{xx} = \varepsilon_{xx}^0 - z\kappa_{xx}^b - f(z)\kappa_{xx}^s, \tag{4.66}$$

$$\varepsilon_{yy} = \varepsilon_{yy}^0 - z\kappa_{yy}^b - f(z)\kappa_{yy}^s, \tag{4.67}$$

$$\varepsilon_{xy} = \varepsilon_{xy}^0 - z\kappa_{xy}^b - f(z)\kappa_{xy}^s, \tag{4.68}$$

$$\varepsilon_x = g(z)\varepsilon_x^0, g(z)\varepsilon_y = \varepsilon_y^0, \tag{4.69}$$

where

$$\varepsilon_{xx}^0 = \frac{\partial u(x,y,t)}{\partial x}, \varepsilon_{yy}^0 = \frac{\partial v(x,y,t)}{\partial y}, \varepsilon_{xy}^0 = \frac{\partial u(x,y,t)}{\partial y} + \frac{\partial v(x,y,t)}{\partial x},$$

$$\kappa_{xx}^b = \frac{\partial^2 w_b(x,y,t)}{\partial x^2}, \kappa_{yy}^b = \frac{\partial^2 w_b(x,y,t)}{\partial y^2}, \kappa_{xy}^b = 2\frac{\partial^2 w_b(x,y,t)}{\partial x \partial y},$$

$$\kappa_{xx}^s = \frac{\partial^2 w_s(x,y,t)}{\partial x^2}, \kappa_{yy}^s = \frac{\partial^2 w_s(x,y,t)}{\partial y^2}, \kappa_{xy}^s = 2\frac{\partial^2 w_s(x,y,t)}{\partial x \partial y},$$

$$\varepsilon_x^0 = \frac{\partial w_s(x,y,t)}{\partial x}, \varepsilon_y^0 = \frac{\partial w_s(x,y,t)}{\partial y}. \tag{4.70}$$

Also, the shape functions $f(z)$ and $g(z)$ are the same as those reported in the previous sections for the refined sinusoidal beams.

4.4.2 Derivation of the Equations of Motion for Plates

In this section, we will be using an energy-based approach to derive the equations of the plates' motion. For this purpose, Hamilton's principle will be used. The fundamentals of this principle were explained in Section 4.3.2, and the relations mentioned there will not be presented again in this section. In the following paragraphs, the aforementioned principle will be extended to both classical and refined plates in separate parts.

4.4.2.1 Equations of Motion for Classical Plates

This section discusses about obtaining the motion equations of a plate with length a, width b, and thickness h. By applying the variation operator in Eq. (4.19), the variation of the strain energy can be formulated in the following form:

$$\delta U = \int_V (\sigma_{xx}\delta\varepsilon_{xx} + \sigma_{yy}\delta\varepsilon_{yy} + \sigma_{xy}\delta\varepsilon_{xy})\,dV. \tag{4.71}$$

Now, by applying a set of mathematical simplifications and using the strains of the classical theory of the plates, Eq. (4.71) can be rewritten as

$$\delta U = \int_A \left(N_{xx}\delta\varepsilon_{xx}^0 - M_{xx}\delta\kappa_{xx}^0 + N_{yy}\delta\varepsilon_{yy}^0 - M_{yy}\delta\kappa_{yy}^0 + N_{xy}\delta\varepsilon_{xy}^0 - M_{xy}\delta\kappa_{xy}^0\right)dA, \tag{4.72}$$

where N_i and M_i components can be written in the following form:

$$[N_i, M_i] = \int_{-\frac{h}{2}}^{\frac{h}{2}} [1, z]\sigma_{ij}dz, (i,j = xx, yy, xy). \tag{4.73}$$

Moreover, the variation of the kinetic energy can be formulated as

$$\delta K = \int_A \left[I_0\left(\dot{u}\delta\dot{u} + \dot{v}\delta\dot{v} + \dot{w}\delta\dot{w}\right) - I_1\left(\dot{u}\frac{\partial\delta\dot{w}}{\partial x} + \delta\dot{u}\frac{\partial\dot{w}}{\partial x} + \dot{v}\frac{\partial\delta\dot{w}}{\partial y} + \delta\dot{v}\frac{\partial\dot{w}}{\partial y}\right) \right.$$
$$\left. + I_2\left(\frac{\partial\dot{w}}{\partial x}\frac{\partial\delta\dot{w}}{\partial x} + \frac{\partial\dot{w}}{\partial y}\frac{\partial\delta\dot{w}}{\partial y}\right) \right]dA \tag{4.74}$$

in which I_i components can be expressed in the following integral form:

$$[I_0, I_1, I_2] = \int_{-\frac{h}{2}}^{\frac{h}{2}} \left[1, z, z^2\right]\rho(z)dz. \tag{4.75}$$

Now, the Euler–Lagrange equations of the plate can be obtained by substituting for variations of strain energy and kinetic energy from Eqs. (4.72) and (4.74), respectively, in Eq. (4.19):

$$\frac{\partial N_{xx}}{\partial x} + \frac{\partial N_{xy}}{\partial y} = I_0 \ddot{u} - I_1 \frac{\partial \ddot{w}}{\partial x}, \tag{4.76}$$

$$\frac{\partial N_{xy}}{\partial x} + \frac{\partial N_{yy}}{\partial y} = I_0 \ddot{v} - I_1 \frac{\partial \ddot{w}}{\partial y}, \tag{4.77}$$

$$\frac{\partial^2 M_{xx}}{\partial x^2} + 2\frac{\partial^2 M_{xy}}{\partial x \partial y} + \frac{\partial^2 M_{yy}}{\partial y^2} = I_0 \ddot{w} + I_1 \left(\frac{\partial \ddot{u}}{\partial x} + \frac{\partial \ddot{v}}{\partial y} \right) - I_2 \left(\frac{\partial^2 \ddot{w}}{\partial x^2} + \frac{\partial^2 \ddot{w}}{\partial y^2} \right). \tag{4.78}$$

4.4.2.2   Equations of Motion for Refined Sinusoidal Plates

In this section, the motion equations for a plate will be derived by means of a refined sinusoidal plate hypothesis. The fundamentals of this section are the same as those of the previous section. Here, the variation of the strain energy can be written in the following form:

$$\delta U = \int_V \left(\sigma_{xx}\delta\varepsilon_{xx} + \sigma_{yy}\delta\varepsilon_{yy} + \sigma_{xy}\delta\varepsilon_{xy} + \sigma_x\,\delta\varepsilon_x + \sigma_y\,\delta\varepsilon_y \right) dV. \tag{4.79}$$

When the strains ε_{ij}s, obtained from Eqs. (4.66)–(4.69), are substituted in the above equation, the following formula can be written for the variation of the strain energy:

$$\delta U = \int_A \left(N_{xx}\delta\varepsilon_{xx}^0 - M_{xx}^b\delta\kappa_{xx}^b - M_{xx}^s\delta\kappa_{xx}^s + N_{yy}\delta\varepsilon_{yy}^0 - M_{yy}^b\delta\kappa_{yy}^b - M_{yy}^s\delta\kappa_{yy}^s + N_{xy}\delta\varepsilon_{xy}^0 \right.$$

$$\left. - M_{xy}^b\delta\kappa_{xy}^b - M_{xy}^s\delta\kappa_{xy}^s + Q_x\,\varepsilon_x^0 + Q_y\,\varepsilon_y^0 \right) dA, \tag{4.80}$$

where the stress resultants can be written as

$$\left[N_i, M_i^b, M_i^s \right] = \int_{-\frac{h}{2}}^{\frac{h}{2}} \left[1, z, f(z) \right] \sigma_i dz, \, (i = xx, yy, xy), \tag{4.81}$$

$$Q_j = \int_{-\frac{h}{2}}^{\frac{h}{2}} g(z)\sigma_j dz, \, (j = xz, yz). \tag{4.82}$$

By extending Eq. (4.74) for a refined shear deformation plate, the variation of the kinetic energy can be expressed as follows:

$$\delta K = \int_A \left[I_0 \left(\dot{u}\delta\dot{u} + \dot{v}\delta\dot{v} + (\dot{w}_b + \dot{w}_s)\delta(\dot{w}_b + \dot{w}_s) \right) - I_1 \left(\dot{u}\frac{\partial \delta\dot{w}_b}{\partial x} + \delta\dot{u}\frac{\partial \dot{w}_b}{\partial x} + \dot{v}\frac{\partial \delta\dot{w}_b}{\partial y} \right. \right.$$

$$+ \delta\dot{v}\frac{\partial \dot{w}_b}{\partial y} \right) - J_1 \left(\dot{u}\frac{\partial \delta\dot{w}_s}{\partial x} + \delta\dot{u}\frac{\partial \dot{w}_s}{\partial x} + \dot{v}\frac{\partial \delta\dot{w}_s}{\partial y} + \delta\dot{v}\frac{\partial \dot{w}_s}{\partial y} \right)$$

$$+ I_2 \left(\frac{\partial \dot{w}_b}{\partial x}\frac{\partial \delta\dot{w}_b}{\partial x} + \frac{\partial \dot{w}_b}{\partial y}\frac{\partial \delta\dot{w}_b}{\partial y} \right) + K_2 \left(\frac{\partial \dot{w}_s}{\partial x}\frac{\partial \delta\dot{w}_s}{\partial x} + \frac{\partial \dot{w}_s}{\partial y}\frac{\partial \delta\dot{w}_s}{\partial y} \right)$$

$$+ J_2 \left(\frac{\partial \dot{w}_b}{\partial x}\frac{\partial \delta\dot{w}_s}{\partial x} + \frac{\partial \dot{w}_s}{\partial x}\frac{\partial \delta\dot{w}_b}{\partial x} + \frac{\partial \dot{w}_b}{\partial y}\frac{\partial \delta\dot{w}_s}{\partial y} + \frac{\partial \dot{w}_s}{\partial y}\frac{\partial \delta\dot{w}_b}{\partial y} \right) \right] dA \tag{4.83}$$

in which

$$\left[I_0, I_1, J_1, I_2, J_2, K_2 \right] = \int_{-\frac{h}{2}}^{\frac{h}{2}} \left[1, z, f(z), z^2, zf(z), f^2(z) \right] \rho(z) dz. \tag{4.84}$$

Now, by substituting Eqs. (4.81) and (4.84) in Eq. (4.19) and selecting the nontrivial response, the Euler–Lagrange equations of a refined sinusoidal plate can be expressed as

$$\frac{\partial N_{xx}}{\partial x} + \frac{\partial N_{xy}}{\partial y} = I_0 \ddot{u} - I_1 \frac{\partial \ddot{w}_b}{\partial x} - J_1 \frac{\partial \ddot{w}_s}{\partial x}, \tag{4.85}$$

$$\frac{\partial N_{xy}}{\partial x} + \frac{\partial N_{yy}}{\partial y} = I_0 \ddot{v} - I_1 \frac{\partial \ddot{w}_b}{\partial y} - J_1 \frac{\partial \ddot{w}_s}{\partial y}, \tag{4.86}$$

$$\frac{\partial^2 M_{xx}^b}{\partial x^2} + 2\frac{\partial^2 M_{xy}^b}{\partial x \partial y} + \frac{\partial^2 M_{yy}^b}{\partial y^2} = I_0(\ddot{w}_b + \ddot{w}_s) + I_1\left(\frac{\partial \ddot{u}}{\partial x} + \frac{\partial \ddot{v}}{\partial y}\right) - I_2\left(\frac{\partial^2 \ddot{w}_b}{\partial x^2} + \frac{\partial^2 \ddot{w}_b}{\partial y^2}\right)$$
$$- J_2\left(\frac{\partial^2 \ddot{w}_s}{\partial x^2} + \frac{\partial^2 \ddot{w}_s}{\partial y^2}\right), \tag{4.87}$$

$$\frac{\partial^2 M_{xx}^s}{\partial x^2} + 2\frac{\partial^2 M_{xy}^s}{\partial x \partial y} + \frac{\partial^2 M_{yy}^s}{\partial y^2} + \frac{\partial Q_x}{\partial x} + \frac{\partial Q_y}{\partial y} = I_0(\ddot{w}_b + \ddot{w}_s) + J_1\left(\frac{\partial \ddot{u}}{\partial x} + \frac{\partial \ddot{v}}{\partial y}\right)$$
$$- J_2\left(\frac{\partial^2 \ddot{w}_b}{\partial x^2} + \frac{\partial^2 \ddot{w}_b}{\partial y^2}\right) - K_2\left(\frac{\partial^2 \ddot{w}_s}{\partial x^2} + \frac{\partial^2 \ddot{w}_s}{\partial y^2}\right). \tag{4.88}$$

4.4.3 Constitutive Equations of FG Nanoplates

In the previous sections on nanoplates' analysis, both classical and higher-order plate models were studied, and the equations of motion for a plate were obtained in the framework of an energy-based approach. The final step is to account for the effect of small size on the mechanical behaviors of FGM plates. As stated in the previous chapters, the effect of size dependency must be considered by making a set of modifications on the constitutive equations of a continuum. Also, nobody doubts that the theory of nonlocal strain gradient elasticity presented by Lim et al. [23] is more effective in analyzing the mechanical responses of tiny structures in comparison with the Eringen's nonlocal elasticity theory [11]. Henceforward, the constitutive equations of the FG nanoplates are modified in this section by means of the nonlocal strain gradient hypothesis.

Once the constitutive behaviors of a classical plate are determined, one can obtain the following relations by integrating Eq. (2.21) over the plate's thickness:

$$\left(1 - \mu^2 \nabla^2\right) \begin{bmatrix} N_{xx} \\ N_{yy} \\ N_{xy} \end{bmatrix} = \left(1 - \lambda^2 \nabla^2\right) \left(\begin{bmatrix} A_{11} & A_{12} & 0 \\ A_{12} & A_{22} & 0 \\ 0 & 0 & A_{66} \end{bmatrix} \begin{bmatrix} \frac{\partial u}{\partial x} \\ \frac{\partial v}{\partial y} \\ \frac{\partial u}{\partial y} + \frac{\partial v}{\partial x} \end{bmatrix} \right.$$
$$\left. + \begin{bmatrix} B_{11} & B_{12} & 0 \\ B_{12} & B_{22} & 0 \\ 0 & 0 & B_{66} \end{bmatrix} \begin{bmatrix} -\frac{\partial^2 w}{\partial x^2} \\ -\frac{\partial^2 w}{\partial y^2} \\ -2\frac{\partial^2 w}{\partial x \partial y} \end{bmatrix} \right), \tag{4.89}$$

$$\left(1 - \mu^2 \nabla^2\right) \begin{bmatrix} M_{xx} \\ M_{yy} \\ M_{xy} \end{bmatrix} = \left(1 - \lambda^2 \nabla^2\right) \left(\begin{bmatrix} B_{11} & B_{12} & 0 \\ B_{12} & B_{22} & 0 \\ 0 & 0 & B_{66} \end{bmatrix} \begin{bmatrix} \frac{\partial u}{\partial x} \\ \frac{\partial v}{\partial y} \\ \frac{\partial u}{\partial y} + \frac{\partial v}{\partial x} \end{bmatrix} \right.$$
$$\left. + \begin{bmatrix} D_{11} & D_{12} & 0 \\ D_{12} & D_{22} & 0 \\ 0 & 0 & D_{66} \end{bmatrix} \begin{bmatrix} -\frac{\partial^2 w}{\partial x^2} \\ -\frac{\partial^2 w}{\partial y^2} \\ -2\frac{\partial^2 w}{\partial x \partial y} \end{bmatrix} \right), \tag{4.90}$$

where the cross-sectional rigidities of the nanoplate can be expressed via

$$
\begin{bmatrix} A_{11} & B_{11} & D_{11} \\ A_{12} & B_{12} & D_{12} \\ A_{66} & B_{66} & D_{66} \end{bmatrix} = \int_{-\frac{h}{2}}^{\frac{h}{2}} \frac{E(z)}{1-\nu^2(z)} \begin{bmatrix} 1 & z & z^2 \end{bmatrix} \begin{bmatrix} 1 \\ \nu(z) \\ \frac{1-\nu(\ )}{2} \end{bmatrix} dz \tag{4.91}
$$

and

$$
[A_{22}, B_{22}, D_{22}] = [A_{11}, B_{11}, D_{11}]. \tag{4.92}
$$

Moreover, by following the same procedure for a refined sinusoidal plate model, the general constitutive equations can be expressed in the following forms:

$$
\left(1-\mu^2\nabla^2\right) \begin{bmatrix} N_{xx} \\ N_{yy} \\ N_{xy} \end{bmatrix} = \left(1-\lambda^2\nabla^2\right) \left(\begin{bmatrix} A_{11} & A_{12} & 0 \\ A_{12} & A_{22} & 0 \\ 0 & 0 & A_{66} \end{bmatrix} \begin{bmatrix} \frac{\partial u}{\partial x} \\ \frac{\partial v}{\partial y} \\ \frac{\partial u}{\partial y} + \frac{\partial v}{\partial x} \end{bmatrix} \right.
$$

$$
\left. + \begin{bmatrix} B_{11} & B_{12} & 0 \\ B_{12} & B_{22} & 0 \\ 0 & 0 & B_{66} \end{bmatrix} \begin{bmatrix} -\frac{\partial^2 w_b}{\partial x^2} \\ -\frac{\partial^2 w_b}{\partial y^2} \\ -2\frac{\partial^2 w_b}{\partial x \partial y} \end{bmatrix} + \begin{bmatrix} B_{11}^s & B_{12}^s & 0 \\ B_{12}^s & B_{22}^s & 0 \\ 0 & 0 & B_{66}^s \end{bmatrix} \begin{bmatrix} -\frac{\partial^2 w_s}{\partial x^2} \\ -\frac{\partial^2 w_s}{\partial y^2} \\ -2\frac{\partial^2 w_s}{\partial x \partial y} \end{bmatrix} \right),
$$

$$
\tag{4.93}
$$

$$
\left(1-\mu^2\nabla^2\right) \begin{bmatrix} M_{xx}^b \\ M_{yy}^b \\ M_{xy}^b \end{bmatrix} = \left(1-\lambda^2\nabla^2\right) \left(\begin{bmatrix} B_{11} & B_{12} & 0 \\ B_{12} & B_{22} & 0 \\ 0 & 0 & B_{66} \end{bmatrix} \begin{bmatrix} \frac{\partial u}{\partial x} \\ \frac{\partial v}{\partial y} \\ \frac{\partial u}{\partial y} + \frac{\partial v}{\partial x} \end{bmatrix} \right.
$$

$$
\left. + \begin{bmatrix} D_{11} & D_{12} & 0 \\ D_{12} & D_{22} & 0 \\ 0 & 0 & D_{66} \end{bmatrix} \begin{bmatrix} -\frac{\partial^2 w_b}{\partial x^2} \\ -\frac{\partial^2 w_b}{\partial y^2} \\ -2\frac{\partial^2 w_b}{\partial x \partial y} \end{bmatrix} + \begin{bmatrix} D_{11}^s & D_{12}^s & 0 \\ D_{12}^s & D_{22}^s & 0 \\ 0 & 0 & D_{66}^s \end{bmatrix} \begin{bmatrix} -\frac{\partial^2 w_s}{\partial x^2} \\ -\frac{\partial^2 w_s}{\partial y^2} \\ -2\frac{\partial^2 w_s}{\partial x \partial y} \end{bmatrix} \right),
$$

$$
\tag{4.94}
$$

$$
\left(1-\mu^2\nabla^2\right) \begin{bmatrix} M_{xx}^s \\ M_{yy}^s \\ M_{xy}^s \end{bmatrix} = \left(1-\lambda^2\nabla^2\right) \left(\begin{bmatrix} B_{11}^s & B_{12}^s & 0 \\ B_{12}^s & B_{22}^s & 0 \\ 0 & 0 & B_{66}^s \end{bmatrix} \begin{bmatrix} \frac{\partial u}{\partial x} \\ \frac{\partial v}{\partial y} \\ \frac{\partial u}{\partial y} + \frac{\partial v}{\partial x} \end{bmatrix} \right.
$$

$$
\left. + \begin{bmatrix} D_{11}^s & D_{12}^s & 0 \\ D_{12}^s & D_{22}^s & 0 \\ 0 & 0 & D_{66}^s \end{bmatrix} \begin{bmatrix} -\frac{\partial^2 w_b}{\partial x^2} \\ -\frac{\partial^2 w_b}{\partial y^2} \\ -2\frac{\partial^2 w_b}{\partial x \partial y} \end{bmatrix} + \begin{bmatrix} H_{11}^s & H_{12}^s & 0 \\ H_{12}^s & H_{22}^s & 0 \\ 0 & 0 & H_{66}^s \end{bmatrix} \begin{bmatrix} -\frac{\partial^2 w_s}{\partial x^2} \\ -\frac{\partial^2 w_s}{\partial y^2} \\ -2\frac{\partial^2 w_s}{\partial x \partial y} \end{bmatrix} \right),
$$

$$
\tag{4.95}
$$

$$
\left(1-\mu^2\nabla^2\right) \begin{bmatrix} Q_x \\ Q_y \end{bmatrix} = \left(1-\lambda^2\nabla^2\right) \begin{bmatrix} A_{44}^s & 0 \\ 0 & A_{55}^s \end{bmatrix} \begin{bmatrix} \frac{\partial w_s}{\partial x} \\ \frac{\partial w_s}{\partial y} \end{bmatrix}, \tag{4.96}
$$

where

$$
\begin{bmatrix}
A_{11} & B_{11} & D_{11} & B_{11}^s & D_{11}^s & H_{11}^s \\
A_{12} & B_{12} & D_{12} & B_{12}^s & D_{12}^s & H_{12}^s \\
A_{66} & B_{66} & D_{66} & B_{66}^s & D_{66}^s & H_{66}^s
\end{bmatrix}
= \int\limits_{-\frac{h}{2}}^{\frac{h}{2}} \frac{E(z)}{1-\nu^2(z)} \begin{bmatrix} 1 & z & z^2 & f(z) & zf(z) & f^2(z) \end{bmatrix}
$$

$$
\times \begin{bmatrix} 1 \\ \nu(z) \\ \frac{1-\nu()}{2} \end{bmatrix} dz,
\tag{4.97}
$$

$$
A_{44}^s = A_{55}^s = \int\limits_{-\frac{h}{2}}^{\frac{h}{2}} g^2(z) G(z) dz
\tag{4.98}
$$

and

$$
[A_{22}, B_{22}, D_{22}, B_{22}^s, D_{22}^s, H_{22}^s] = [A_{11}, B_{11}, D_{11}, B_{11}^s, D_{11}^s, H_{11}^s].
\tag{4.99}
$$

4.4.4 The Nonlocal Governing Equations of FG Nanoplates

In this section, the final partial differential equations which are required to solve the wave propagation problem of FG nanoplates will be obtained. Indeed, the final nonlocal governing equations can be obtained whenever the modified constitutive equations of nonlocal strain gradient theorem are substituted in the Euler–Lagrange equations of a plate. First, the governing equations will be developed for FGM classical nanoplates. For this purpose, Eqs. (4.89) and (4.90) must be substituted in Eqs. (4.76)–(4.78). Once the mathematical simplifications are performed, the nonlocal governing equations of a classical nanoplate made from FGM can be written in the following final forms:

$$
\left(1 - \lambda^2 \nabla^2\right) \left(A_{11} \frac{\partial^2 u}{\partial x^2} + (A_{12} + A_{66}) \frac{\partial^2 v}{\partial x \partial y} + A_{66} \frac{\partial^2 u}{\partial y^2} - B_{11} \frac{\partial^3 w}{\partial x^3} - (B_{12} + 2B_{66}) \frac{\partial^3 w}{\partial x \partial y^2} \right)
$$

$$
+ \left(1 - \mu^2 \nabla^2\right) \left(-I_0 \ddot{u} + I_1 \frac{\partial \ddot{w}}{\partial x} \right) = 0,
\tag{4.100}
$$

$$
\left(1 - \lambda^2 \nabla^2\right) \left(A_{22} \frac{\partial^2 v}{\partial y^2} + (A_{12} + A_{66}) \frac{\partial^2 u}{\partial x \partial y} + A_{66} \frac{\partial^2 v}{\partial x^2} - B_{22} \frac{\partial^3 w}{\partial y^3} - (B_{12} + 2B_{66}) \frac{\partial^3 w}{\partial x^2 \partial y} \right)
$$

$$
+ \left(1 - \mu^2 \nabla^2\right) \left(-I_0 \ddot{v} + I_1 \frac{\partial \ddot{w}}{\partial y} \right) = 0,
\tag{4.101}
$$

$$
\left(1 - \lambda^2 \nabla^2\right) \left(B_{11} \frac{\partial^3 u}{\partial x^3} + (B_{12} + 2B_{66}) \frac{\partial^3 u}{\partial x \partial y^2} + B_{22} \frac{\partial^3 v}{\partial y^3} + (B_{12} + 2B_{66}) \frac{\partial^3 v}{\partial x^2 \partial y} \right.
$$

$$
\left. - D_{11} \frac{\partial^4 w}{\partial x^4} - 2(D_{12} + 2D_{66}) \frac{\partial^4 w}{\partial x^2 \partial y^2} - D_{22} \frac{\partial^4 w}{\partial y^4} \right) + \left(1 - \mu^2 \nabla^2\right)
$$

$$
\times \left(-I_0 \ddot{w} - I_1 \left(\frac{\partial \ddot{u}}{\partial x} + \frac{\partial \ddot{v}}{\partial y} \right) + I_2 \left(\frac{\partial^2 \ddot{w}}{\partial x^2} + \frac{\partial^2 \ddot{w}}{\partial y^2} \right) \right) = 0.
\tag{4.102}
$$

Similarly, the final governing equations of FG nanoplates, considering the effects of shear stress and strain, can be obtained when Eqs. (4.93)–(4.96) are substituted in Eqs. (4.85)–(4.88):

$$\left(1 - \lambda^2 \nabla^2\right) \left(A_{11} \frac{\partial^2 u}{\partial x^2} + (A_{12} + A_{66}) \frac{\partial^2 v}{\partial x \partial y} + A_{66} \frac{\partial^2 u}{\partial y^2} - B_{11} \frac{\partial^3 w_b}{\partial x^3} - (B_{12} + 2B_{66}) \frac{\partial^3 w_b}{\partial x \partial y^2} \right.$$

$$\left. - B_{11}^s \frac{\partial^3 w_s}{\partial x^3} - (B_{12}^s + 2B_{66}^s) \frac{\partial^3 w_s}{\partial x \partial y^2} \right) + \left(1 - \mu^2 \nabla^2\right) \left(-I_0 \ddot{u} + I_1 \frac{\partial \ddot{w}_b}{\partial x} + J_1 \frac{\partial \ddot{w}_s}{\partial x} \right) = 0,$$

$$(4.103)$$

$$\left(1 - \lambda^2 \nabla^2\right) \left(A_{22} \frac{\partial^2 v}{\partial y^2} + (A_{12} + A_{66}) \frac{\partial^2 u}{\partial x \partial y} + A_{66} \frac{\partial^2 v}{\partial x^2} - B_{22} \frac{\partial^3 w_b}{\partial y^3} - (B_{12} + 2B_{66}) \frac{\partial^3 w_b}{\partial x^2 \partial y} \right.$$

$$\left. - B_{22}^s \frac{\partial^3 w_s}{\partial y^3} - (B_{12}^s + 2B_{66}^s) \frac{\partial^3 w_s}{\partial x^2 \partial y} \right) + \left(1 - \mu^2 \nabla^2\right) \left(-I_0 \ddot{v} + I_1 \frac{\partial \ddot{w}_b}{\partial y} + J_1 \frac{\partial \ddot{w}_s}{\partial y} \right) = 0,$$

$$(4.104)$$

$$\left(1 - \lambda^2 \nabla^2\right) \left(B_{11} \frac{\partial^3 u}{\partial x^3} + (B_{12} + 2B_{66}) \frac{\partial^3 u}{\partial x \partial y^2} + B_{22} \frac{\partial^3 v}{\partial y^3} + (B_{12} + 2B_{66}) \frac{\partial^3 v}{\partial x^2 \partial y} \right.$$

$$- D_{11} \frac{\partial^4 w_b}{\partial x^4} - 2(D_{12} + 2D_{66}) \frac{\partial^4 w_b}{\partial x^2 \partial y^2} - D_{22} \frac{\partial^4 w_b}{\partial y^4} - D_{11}^s \frac{\partial^4 w_s}{\partial x^4} - 2(D_{12}^s + 2D_{66}^s) \frac{\partial^4 w_s}{\partial x^2 \partial y^2}$$

$$\left. - D_{22}^s \frac{\partial^4 w_s}{\partial y^4} \right) + \left(1 - \mu^2 \nabla^2\right) \left(-I_0 (\ddot{w}_b + \ddot{w}_s) - I_1 \left(\frac{\partial \ddot{u}}{\partial x} + \frac{\partial \ddot{v}}{\partial y} \right) + I_2 \left(\frac{\partial^2 \ddot{w}_b}{\partial x^2} + \frac{\partial^2 \ddot{w}_b}{\partial y^2} \right) \right.$$

$$\left. + J_2 \left(\frac{\partial^2 \ddot{w}_s}{\partial x^2} + \frac{\partial^2 \ddot{w}_s}{\partial y^2} \right) \right) = 0,$$

$$(4.105)$$

$$\left(1 - \lambda^2 \nabla^2\right) \left(B_{11}^s \frac{\partial^3 u}{\partial x^3} + (B_{12}^s + 2B_{66}^s) \frac{\partial^3 u}{\partial x \partial y^2} + B_{22}^s \frac{\partial^3 v}{\partial y^3} + (B_{12}^s + 2B_{66}^s) \frac{\partial^3 v}{\partial x^2 \partial y} \right.$$

$$- D_{11}^s \frac{\partial^4 w_b}{\partial x^4} - 2(D_{12}^s + 2D_{66}^s) \frac{\partial^4 w_b}{\partial x^2 \partial y^2} - D_{22}^s \frac{\partial^4 w_b}{\partial y^4} - H_{11}^s \frac{\partial^4 w_s}{\partial x^4} - 2(H_{12}^s + 2H_{66}^s) \frac{\partial^4 w_s}{\partial x^2 \partial y^2}$$

$$\left. - H_{22}^s \frac{\partial^4 w_s}{\partial y^4} A_{44}^s \left(\frac{\partial^2 (w_b + w_s)}{\partial x^2} + \frac{\partial^2 (w_b + w_s)}{\partial y^2} \right) \right) + \left(1 - \mu^2 \nabla^2\right) \left(-I_0 (\ddot{w}_b + \ddot{w}_s) \right.$$

$$\left. - J_1 \left(\frac{\partial \ddot{u}}{\partial x} + \frac{\partial \ddot{v}}{\partial y} \right) + J_2 \left(\frac{\partial^2 \ddot{w}_b}{\partial x^2} + \frac{\partial^2 \ddot{w}_b}{\partial y^2} \right) + K_2 \left(\frac{\partial^2 \ddot{w}_s}{\partial x^2} + \frac{\partial^2 \ddot{w}_s}{\partial y^2} \right) \right) = 0.$$

$$(4.106)$$

4.4.5 Wave Solution for FG Nanoplates

Finally, it is time to solve the obtained governing equations to enrich the frequency and speed of the scattered waves in the nanoplate. As in the case of our previous problem, which was dedicated to solve for the equations of wave dispersion inside FG nanobeams, again the governing equations are considered to be linear based on the concept of the infinitesimal strains in continuum mechanics. The solution functions for both classical and higher-order plates were previously introduced in Section 1.3.2 of Chapter 1. In the following sentences, the details of the plate-type solution will be explained for both classical and shear deformation plate models in the framework of independent parts.

4.4.5.1 Solution of Classical FG Nanoplates

The formulations for a classical plate's solution are presented in Chapter 1. The solution functions for both spatial and temporal parts of displacement fields of the plate, namely u, v and w, were presented in Eqs. (1.9)–(1.11). Herein, we are about to show how one can derive the components of the stiffness and mass matrices of FG nanoplates to finish the procedure of solving the wave propagation problem of an FG nanoplate. By substituting Eqs. (1.9)–(1.11) in Eqs. (4.100)–(4.102), the components of the stiffness matrix can be expressed as

$$
\begin{aligned}
k_{11} &= -\left[1 + \lambda^2(\beta_1^2 + \beta_2^2)\right]\left(A_{11}\beta_1^2 + A_{66}\beta_1\beta_2\right), \\
k_{12} &= -\left[1 + \lambda^2(\beta_1^2 + \beta_2^2)\right]\left(A_{12} + A_{66}\right)\beta_1\beta_2, \\
k_{13} &= i\left[1 + \lambda^2(\beta_1^2 + \beta_2^2)\right]\left(B_{11}\beta_1^3 + (B_{12} + 2B_{66})\beta_1\beta_2^2\right), \\
k_{22} &= -\left[1 + \lambda^2(\beta_1^2 + \beta_2^2)\right]\left(A_{66}\beta_1^2 + A_{22}beta_2^2\right), \\
k_{23} &= i\left[1 + \lambda^2(\beta_1^2 + \beta_2^2)\right]\left(B_{22}\beta_2^3 + (B_{12} + 2B_{66})\beta_1^2\beta_2\right), \\
k_{33} &= -\left[1 + \lambda^2(\beta_1^2 + \beta_2^2)\right]\left(D_{11}\beta_1^4 + 2(D_{12} + 2D_{66})\beta_1^2\beta_2^2 + D_{22}\beta_2^4\right) \\
&\quad + \left[1 + \mu^2(\beta_1^2 + \beta_2^2)\right]N^T(\beta_1^2 + \beta_2^2).
\end{aligned}
\tag{4.107}
$$

Also, the components of the mass matrix can be written in the following forms:

$$
\begin{aligned}
m_{11} &= -\left[1 + \mu^2(\beta_1^2 + \beta_2^2)\right]I_0, \\
m_{12} &= 0, \quad m_{13} = i\left[1 + \mu^2(\beta_1^2 + \beta_2^2)\right]\beta_1 I_1, \\
m_{22} &= -\left[1 + \mu^2(\beta_1^2 + \beta_2^2)\right]I_0, \\
m_{23} &= i\left[1 + \mu^2(\beta_1^2 + \beta_2^2)\right]\beta_2 I_1, \\
m_{33} &= -\left[1 + \mu^2(\beta_1^2 + \beta_2^2)\right]\left(I_0 + I_2(\beta_1^2 + \beta_2^2)\right).
\end{aligned}
\tag{4.108}
$$

4.4.5.2 Solution of Refined Sinusoidal FG Nanoplates

In this section, the nonlocal governing equations of a refined sinusoidal FG nanoplate will be solved by means of the well-known analytical method which was introduced in Chapter 1. By substituting for u, v, w_b and w_s from Eqs. (1.12)–(1.15) in Eqs. (4.103)–(4.106), the components of the stiffness matrix of such nanoplates can be written as

$$
\begin{aligned}
k_{11} &= -\left[1 + \lambda^2(\beta_1^2 + \beta_2^2)\right]\left(A_{11}\beta_1^2 + A_{66}\beta_1\beta_2\right), \\
k_{12} &= -\left[1 + \lambda^2(\beta_1^2 + \beta_2^2)\right]\left(A_{12} + A_{66}\right)\beta_1\beta_2, \\
k_{13} &= i\left[1 + \lambda^2(\beta_1^2 + \beta_2^2)\right]\left(B_{11}\beta_1^3 + (B_{12} + 2B_{66})\beta_1\beta_2^2\right), \\
k_{14} &= i\left[1 + \lambda^2(\beta_1^2 + \beta_2^2)\right]\left(B_{11}^s\beta_1^3 + (B_{12}^s + 2B_{66}^s)\beta_1\beta_2^2\right), \\
k_{22} &= -\left[1 + \lambda^2(\beta_1^2 + \beta_2^2)\right]\left(A_{66}\beta_1^2 + A_{22}\beta_2^2\right), \\
k_{23} &= i\left[1 + \lambda^2(\beta_1^2 + \beta_2^2)\right]\left(B_{22}\beta_2^3 + (B_{12} + 2B_{66})\beta_1^2\beta_2\right), \\
k_{24} &= i\left[1 + \lambda^2(\beta_1^2 + \beta_2^2)\right]\left(B_{22}^s\beta_2^3 + (B_{12}^s + 2B_{66}^s)\beta_1^2\beta_2\right), \\
k_{33} &= -\left[1 + \lambda^2(\beta_1^2 + \beta_2^2)\right]\left(D_{11}\beta_1^4 + 2(D_{12} + 2D_{66})\beta_1^2\beta_2^2 + D_{22}\beta_2^4\right) \\
&\quad + \left[1 + \mu^2(\beta_1^2 + \beta_2^2)\right]N^T(\beta_1^2 + \beta_2^2), \\
k_{34} &= -\left[1 + \lambda^2(\beta_1^2 + \beta_2^2)\right]\left(D_{11}^s\beta_1^4 + 2(D_{12}^s + 2D_{66}^s)\beta_1^2\beta_2^2 + D_{22}^s\beta_2^4\right)
\end{aligned}
$$

$$+ \left[1 + \mu^2(\beta_1^2 + \beta_2^2)\right] N^T (\beta_1^2 + \beta_2^2),$$
$$k_{44} = -\left[1 + \lambda^2(\beta_1^2 + \beta_2^2)\right]\left(H_{11}^s \beta_1^4 + 2(H_{12}^s + 2H_{66}^s)\beta_1^2\beta_2^2 + H_{22}^s\beta_2^4 + A_{44}^s(\beta_1^2 + \beta_2^2)\right)$$
$$+ \left[1 + \mu^2(\beta_1^2 + \beta_2^2)\right] N^T (\beta_1^2 + \beta_2^2). \tag{4.109}$$

The components of the mass matrix are as follows:

$$m_{11} = -\left[1 + \mu^2(\beta_1^2 + \beta_2^2)\right] I_0,$$
$$m_{12} = 0, m_{13} = i\left[1 + \mu^2(\beta_1^2 + \beta_2^2)\right]\beta_1 I_1,$$
$$m_{14} = i\left[1 + \mu^2(\beta_1^2 + \beta_2^2)\right]\beta_1 J_1,$$
$$m_{22} = -\left[1 + \mu^2(\beta_1^2 + \beta_2^2)\right] I_0,$$
$$m_{23} = i\left[1 + \mu^2(\beta_1^2 + \beta_2^2)\right]\beta_2 I_1,$$
$$m_{24} = i\left[1 + \mu^2(\beta_1^2 + \beta_2^2)\right]\beta_2 J_1,$$
$$m_{33} = -\left[1 + \mu^2(\beta_1^2 + \beta_2^2)\right]\left(I_0 + I_2(\beta_1^2 + \beta_2^2)\right),$$
$$m_{34} = -\left[1 + \mu^2(\beta_1^2 + \beta_2^2)\right]\left(I_0 + J_2(\beta_1^2 + \beta_2^2)\right),$$
$$m_{44} = -\left[1 + \mu^2(\beta_1^2 + \beta_2^2)\right]\left(I_0 + K_2(\beta_1^2 + \beta_2^2)\right). \tag{4.110}$$

4.4.6   Numerical Results and Discussion

In this section, some illustrations will be studied to analyze the wave propagation behaviors of an FG nanoplate made of Al_2O_3 and steel. The geometry of the FG nanoplate can be seen in Figure 4.8. The material properties are assumed to be temperature dependent in order to obtain more accurate responses. The nonlocal strain gradient theory is applied in order to capture small-scale effects.

Herein, the validation of the presented methodology is performed in the framework of Table 4.2 presenting the natural frequency responses of FG nanoplates obtained from our modeling in comparison with those reported in a previously published paper of Natarajan et al. [88]. It is revealed that the presented model and solution can accurately predict the frequencies of FG nanoscale plates. On the other hand, it is worth mentioning that in all the presented results, axial and transverse wave numbers, β_1 and β_2, are assumed to be identical and equal with β.

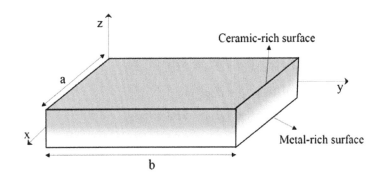

FIGURE 4.8
The geometry of an FG nanoplate subjected to the propagation of elastic waves.

TABLE 4.2

Comparison of the Natural Frequencies of FG Nanoplates for Various Nonlocal
Parameters and Length to Thickness Ratios

| | $a/h = 10$ | | $a/h = 20{=}0.2$ | |
μ	Natarajan et al. [88]	This Study	Natarajan et al. [88]	This Study
0	0.0441	0.043803	0.0113	0.011255
1	0.0403	0.040051	0.0103	0.010288
2	0.0374	0.037123	0.0096	0.009534
4	0.033	0.032791	0.0085	0.008418

In Figure 4.9, the propagation behaviors of FG nanoplates are monitored by drawing
the variation of wave frequency with wave number for various amounts of scale factors. It
is observable that the wave frequency of the nanoplate will experience a gradual increase
followed by a final stable situation whenever the nonlocal elasticity hypothesis is selected
(i.e., the scale factor is assumed to be zero). However, the wave frequency will tend to
infinity as a nonzero value is assigned to the scale factor. In this condition, the effects of the
strain gradient will be included, and because of the stiffness-hardening phenomenon which
happens in the nanostructure, the frequency of the propagated waves will be intensified.
The greater the used scale factor is, the higher will be the slope of the curve. Hence, both
aspects of size dependency are covered in this diagram.

In the next figure, the focus is on the effect of shear deflection on the dispersion curves
of FG nanoplates (see Figure 4.10). In this illustration, the variation of phase speed of
dispersed waves against wave number is plotted for nonlocal strain gradient nanosize plates

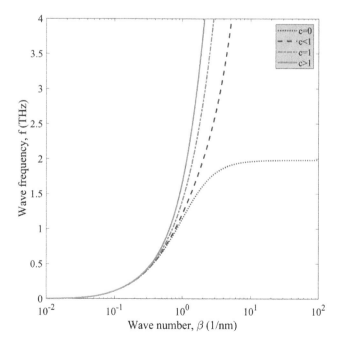

FIGURE 4.9

Variation of wave frequency of FG nanoplates versus wave number for different scale factors
($p = 1, \mu = 0.5$).

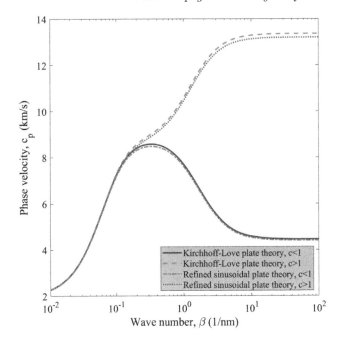

FIGURE 4.10

Comparison of phase velocity curves of FG nanoplates obtained using Kirchhoff–Love and refined sinusoidal plate theories ($p = 1$).

in two cases of $c < 1$ and $c > 1$ for both Kirchhoff–Love and higher-order FG nanoplates. The hardening impact of the scale factor as well as that of the shear deformation can be seen once again in this figure. It can be seen that the Kirchhoff–Love nanoplate theory estimates a response different from that estimated by the sinusoidal plate model because of its simplifying assumptions. However, the differences cannot be easily seen in the case of using small wave numbers. Indeed, the difference between these theories will be more observable at high wave numbers.

Furthermore, Figure 4.11 shows the phase velocity curves of FG nanoplates for the cases utilizing various size-dependent continuum mechanics theories. It is clear that the phase velocity of the composite nanoplate will increase with increase in the wave number until it reaches a maximum and after that, the phase velocity will decrease as the wave number continues its increasing trend when using nonlocal elasticity theory of Eringen. This trend is in complete compatibility with findings of Figure 4.9. In addition, similar to the frequency, the speed of the dispersed waves will experience a continuous increase as the scale factor is assumed to be a definite nonnegative value except zero. The major reason for this amplification trend is the hardening effect of the length-scale parameter on the stiffness of the nanostructure. Hence, it is natural to observe such an increment as the scale factor increases.

On the other hand, the small-scale effects are again included in another numerical example incorporated with the effects of the material distribution parameter. Indeed, Figure 4.12 depicts the dispersion curve of FG nanoplates for the cases using various gradient indices focusing on two general size-dependent cases, i.e., scale factors smaller than 1 and scale factors greater than 1 ($c < 1$ and $c > 1$). From this diagram, it is obvious that the phase speed will decrease when the gradient index is increased. As mentioned before, the total stiffness of the utilized FGM will be affected as the gradient index varies. In

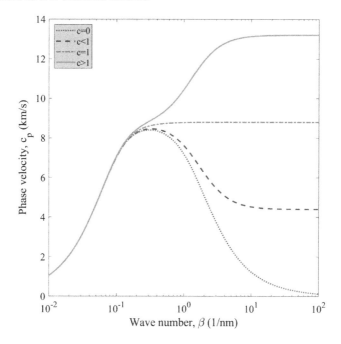

FIGURE 4.11
Variation of phase speed of propagated waves in FG nanoplates with wave number for different scale factors ($p = 1, \mu = 0.5$).

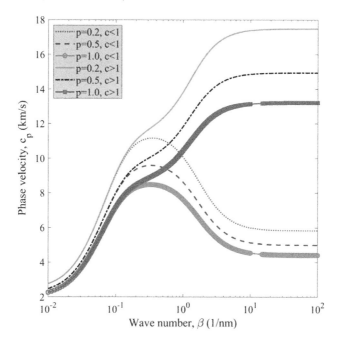

FIGURE 4.12
The effects of small scale parameter and gradient index on the phase velocity of FG nanoplates.

other words, the volume fraction of the ceramic-rich phase will be greater as the gradient index rises, and due to the lower stiffness of the ceramic compared with that of metal, the total stiffness will decrease in the case of using large material distribution parameters. Henceforward, it is natural to see a drop in the phase velocity curves whenever the gradient index is increased.

In Figure 4.13, coupled effects of gradient index and scale factor on the wave frequency of FG nanoplates are shown. In this graph, the effect of the selected wave number on the mechanical response of the waves scattered in the FG nanoplate is included too. The trends

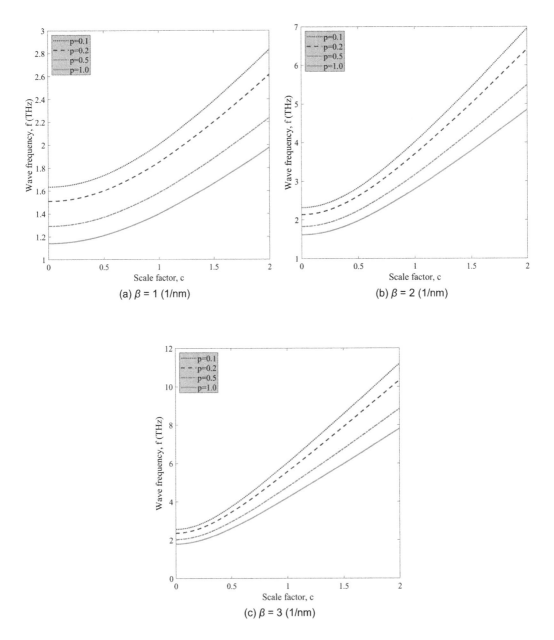

FIGURE 4.13
Variation of wave frequency of FG nanoplates versus scale factor for various gradient indices.

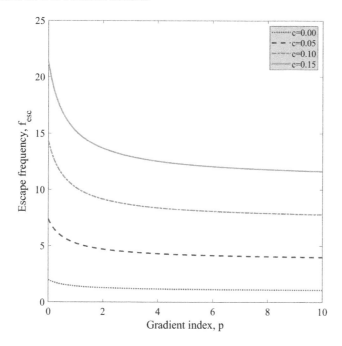

FIGURE 4.14
Variation of escape frequency of FG nanoplates with gradient index for different scale factors.

in this diagram have all been verified in the previous plots. Indeed, the intensifying effect of the scale factor on the mechanical response proves the stiffness-hardening role of the length-scale parameter shown in Figures 4.9–4.12. It is also shown that the frequencies of the dispersed waves will be amplified to a great extent when higher wave numbers are selected. This trend can be justified by observing Figure 4.9. Besides, it is clear that the wave frequency decreases if a higher value is assigned to the gradient index, which shows the lessening impact of the gradient index of the FGM which the nanoplate is made from.

Finally, emphasizing the crucial role of the gradient index, Figure 4.14 shows the variation of the escape frequency of the waves propagated through FG nanoplates with gradient index for various scale factors. It is clear that the escape frequency will follow a decreasing trend as the gradient index is increased. It is observable that beyond a certain value of gradient index, the variations in the escape frequency will be small enough to be ignored. The physical reason for the aforementioned phenomenon can be figured out by rereading the interpretations of previous figures. Also, the escape frequency increases as the scale factor is increased. The major reason for this trend can be understood from the explanations presented in previous paragraphs.

5

Porosity Effects on Wave Propagation Characteristics of Inhomogeneous Nanostructures

In this chapter, the effects of the existence of porosities in functionally graded materials (FGMs) will be considered during wave dispersion analysis of FG nanostructures. The equations of motion are derived in the same way as that discussed in the Chapter 4 for both beam-type and plate-type structures. However, the homogenization method used in this chapter will differ from that used in the previous chapter. In this chapter, two approaches will be discussed to formulate the effects of porosities on the stiffness of nanostructures. Also, a group of numerical illustrations will be presented in order to show the effect of porosity coefficient on the mechanical behaviors of nanobeams and nanoplates when an elastic wave is dispersed in these elements. After reading this chapter, one will be aware of the qualitative and quantitative effects of the existence of various types of porosities on the phase speed and wave frequency of the propagated waves inside a nanostructure.

5.1 Introduction

5.1.1 Porous FGM Structures

In the fabrication procedures, the theoretically designed materials cannot be produced due to various circumstances like complex fabrication process and low precision of the instruments that are used by companies to fabricate a designed material. Thus, it is natural to observe defects and gaps in the bulk of the materials produced by the companies. Due to this reality, it is of great importance to capture the effects of these destroying phenomena on the mechanical behaviors of elements made from any desired material. This issue must be given more attention in the case where an FGM would be utilized for analyzing a mechanical element. In fact, the FGMs require a complex production process due to their complicated mechanical properties which can vary along a particular direction. In some types of FGMs, the material properties are considered to be variable along more than one specific direction. Therefore, the probability of existence of porosities in these complicated materials is more than that in other usual materials. According to the above discussion, consideration of the effects of porosities in the FGMs is very important when either static or dynamic behaviors of an FG structure is to be probed. For this reason, many of the researchers tried to include the effects of porosities in their analyses on the mechanical behaviors of FGM beams, plates and shells. Wattanasakulpong and Ungbhakorn [89] analyzed both linear and nonlinear frequency behaviors of porous FGM beams with respect to the effects of porosities on the frequency variation of FG beams. They could solve the final governing equations by means of a powerful numerical method named differential transformation method (DTM). Also, the DTM was utilized in another research by Ebrahimi and Mokhtari [90] for analyzing the vibrational responses of rotating porous FG beams. Moreover, Ebrahimi and Zia [91] could show the effects of existence of porosities in an FGM on the nonlinear large amplitude frequency of the beams made from the aforementioned material. Atmane et al. [92] investigated the static deflection, stability and natural frequency characteristics of a porous FGM beam rested on an elastic substrate via an analytical approach by considering the effects of thickness stretching. Another endeavor related to the porous FGMs was performed by Ebrahimi et al. [93] dealing with the dynamic responses of smart porous FG beams in the framework of a higher-order beam model. Also, a quasi-3D shear deformation plate hypothesis was employed by Zenkour [94] to investigate the bending behaviors of exponentially FGM plates. Gupta and Talha [95] surveyed the bending and buckling problems of imperfect plates made from a porous FGM.

5.1.2   Porous FGM Nanostructures

As discussed in the previous section, the effects of the existence of porosities in a media must be considered by the designers, particularly if the material is an FGM. In Chapter 4, the advantages of FG nanostructures over macroscale FG elements were explained. There are many complicated processes during the fabrication procedures of the nanoscale FG continuous systems. Hence, the issue of the existence of porosities is of great importance in such tiny composites. For this reason, the effects of porosity on the mechanical responses of FG nanostructures were widely analyzed by many researchers. For example, Ebrahimi et al. [96,97] investigated the frequency responses of porous FG nanobeams in the framework of the nonlocal strain gradient hypothesis. They considered both straight and curved nanobeams in their analyses. Moreover, the vibrational characteristics of smart porous FG nanobeams were probed by Ebrahimi and Barati [98]. The wave dispersion problem of smart porous FG nanobeams was solved by Ebrahimi and Dabbagh [99]. Shahverdi and Barati [100] could study the nonlocal dynamic behaviors of porous FG nanoplates on the basis of a shear deformation theory. Barati [101] surveyed the forced vibration analysis of porous FG nanoplates based on stress–strain gradient nonlocal elasticity. Most recently, Eltaher et al. [102] could develop a new porosity-dependent homogenization scheme for the FGMs and showed its efficiency in a dynamic analysis of the FG nanobeams.

In what follows, two types of homogenization methods will be introduced for the FGMs, and these methods will be used to solve the wave propagation responses of FG nanostructures to show how the porosities can affect the wave dispersion characteristics of FG nanosize beams and plates.

5.2   Homogenization of Porous FGMs

5.2.1   Modified Power-Law Porous Model

In this model, the effects of existence of probable porosities will be added to the previously discussed power-law model. For this purpose, the FGM will be considered to be constructed of three phases instead of two. Indeed, the additional phase in comparison with the usual FGMs is the porosity phase. So, a volume fraction will be assigned to this phase. In the literature, it is common to denote the volume fraction of the porosities as α. It is clear that because this parameter is a relative term, it does not possess any dimension. In this model, the effect of porosity will be applied to both metal and ceramic phases of FGM. The equivalent material properties of the porous FGM can be expressed as

$$P_{eq}(z) = P_c \left(V_c - \frac{\alpha}{2} \right) + P_m \left(V_m - \frac{\alpha}{2} \right) \tag{5.1}$$

in which V_c and V_m are the volume fractions of the ceramic and metal phases. In the FGMs, the volume fractions of the constituents are assumed to be functions of the thickness direction. The volume fraction of the ceramic phase and the relation between the ceramic and metal phases can be seen in Eqs. (4.1) and (4.2), respectively. By doing mathematical simplifications, the equivalent material properties of the porous FGM can be presented in the following form:

$$P_{eq}(z) = (P_c - P_m)V_c + P_m - \frac{\alpha}{2}(P_c + P_m), \qquad (5.2)$$

where P can be any one of the material properties like Young's modulus E, shear modulus G, Poisson's ratio ν and mass density ρ. The effects of the existence of porosities on the equivalent Young's modulus of a porous FGM are illustrated in Figure 5.1. In this diagram, the influence of gradient index is covered, too. It can be found that the stiffness of the material decreases as the volume fraction of the porosities grows. Hence, it can be realized that the existence of porosity in the material can lessen the stiffness of the system. This trend will result in a reduction in the total stiffness of the continuous system, and due to this fact, the dynamic response of the system becomes smaller. This is due to the fact that the dynamic response of a continuous system has a direct relation with the equivalent stiffness of the system.

5.2.2 Coupled Elastic Kinetic Porous Model

In the previous section, the effects of porosities on the mechanical properties of the FGMs were studied on the basis of a modified model for the FGMs. However, it is important to know that the model presented can capture the effects of porosity on the various parameters in a discretized manner. In fact, the effects of the existence of the porosities on the equivalent material properties are studied individually. In other words, the destructive impacts of porosity on each variant are assumed to be functions of the studied variant itself, and there is no correlation between the different terms when the effects of porosities are studied. It is seen that the effect of the porosity on mechanical properties of FGMs cannot be summarized by defining a volume fraction for the porosities. Indeed, another crucial issue which plays an important role in determining the stiffness of the system is the correlation between the stiffness and density of the medium.

Actually, the effect of porosity on the density and elastic modulus of the FGM must be applied in different ways. It is demonstrated in one of the recent researches that the elastic modulus of a porous FGM must be modified by defining two density-dependent integrals which are computed by means of both true and apparent densities of the porous material [102]. In this modified model, the correlation between density and stiffness of the FGM, which was not included in the modified power-law model, will be covered. According to this model, the density and elastic modulus of a porous FGM can be expressed in the following forms:

$$\rho_{eq}(z) = (\rho_c - \rho_m)V_c + \rho_m - \frac{\alpha}{2}(\rho_c + \rho_m), \qquad (5.3)$$

$$E_{eq}(z) = (E_c - E_m)V_c + E_m - \frac{m_0 - m}{m_0}(E_c + E_m) \qquad (5.4)$$

in which m and m_0 are the density-based integrals which must be calculated using the apparent and true densities, respectively. These integrals can be formulated as

$$m_0 = \int_{-\frac{h}{2}}^{\frac{h}{2}} \rho_{eq}(z)dz, \alpha = 0, \qquad (5.5)$$

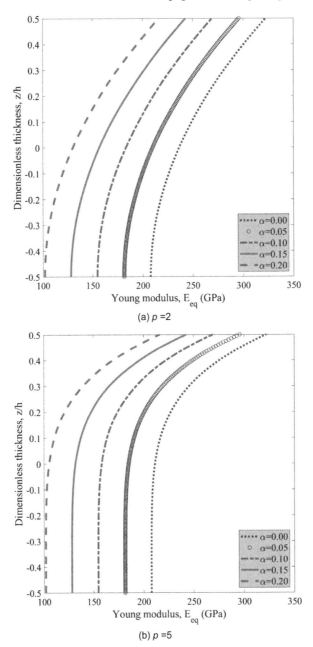

FIGURE 5.1
Effect of the volume fraction of porosities on the variation of the Young's modulus of the
FGMs across the thickness. The homogenization is performed by means of the modified
power-law porous model.

$$m = \int\limits_{-\frac{h}{2}}^{\frac{h}{2}} \rho_{eq}(z)dz, \alpha > 0. \tag{5.6}$$

The effect of utilizing the coupled elastic–kinetic model instead of the modified power-law
model is shown in the Figure 5.2. It can be seen that the curves of the modulus are similar

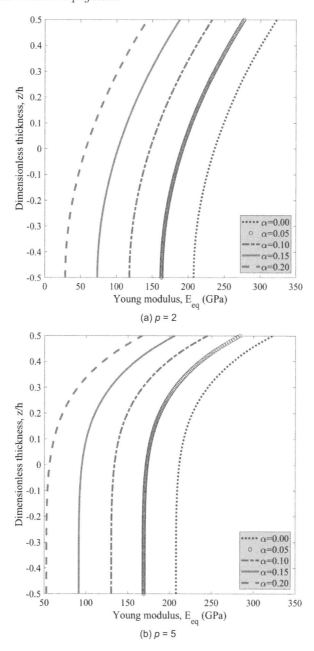

FIGURE 5.2
Effect of volume fraction of porosities on the variation of the Young's modulus of the FGMs across the thickness. The homogenization is performed by means of the coupled elastic–kinetic porous model.

to those illustrated in Figure 5.1 in the case of analyzing a nonporous FGM. Besides, again it can be realized that the elastic modulus possesses lower values when a smaller gradient index is chosen. The most crucial issue is the difference between the values reported by the modified power-law model and the coupled elastic–kinetic model in the case of using a porous FGM. Actually, the coupled elastic–kinetic model estimates a lower stiffness for the

FGM in comparison with the conventional modified power-law model. The major reason for the existence of such a remarkable difference is that, in the coupled elastic–kinetic model, the effects of both apparent and true densities of the material on the stiffness of the FGM are covered, whereas in the modified power-law model, such reality is neglected. In conclusion, it can be noted that considering the correlation between stiffness and density results in predicting a lower value for the elastic modulus of a porous FGM. Hence, it is better to employ the coupled elastic–kinetic model for the sake of caution.

5.3   Wave Propagation in Porous FG Nanostructures

In this section, the effects of porosity on the wave dispersion curves of a porous FG nanostructure will be studied. The formulations of the problem are the same as those presented in Chapter 4 for the refined shear deformation nonlocal strain gradient nanobeams and nanoplates. Thus, the derivation of the formulations of the problem will not be studied again. In fact, in this section, we will pay attention to only the results of the problem and try to highlight the changes that can be made in the dispersion curves of the nanobeams and nanoplates when a porous FGM is utilized.

5.3.1   Analysis of Porous FG Nanobeams

Based on the previous discussions, it is not possible to analyze the mechanical behaviors of the FG nanobeams without considering the probable porosities. In what follows, we will try to show the effects of porosity on the dispersion curves of FG nanobeams. In Figure 5.3,

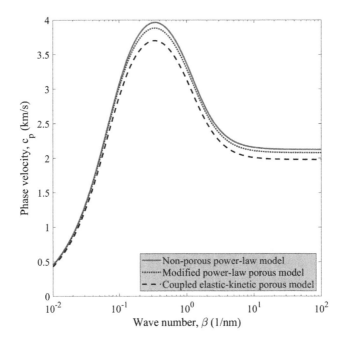

FIGURE 5.3
Comparison of phase velocity curves of both nonporous and porous FG nanobeams $(p = 2, c < 1)$.

an illustration is presented in order to compare the nonporous FG nanobeams with porous ones. It can be easily found that the previous findings, discussed in Figures 5.1 and 5.2, are shown in this diagram. Indeed, the nonporous ideal nanobeams present a higher range of phase speeds compared with the porous ones. This trend is natural because the existence of porosities in an FGM results in a reduction in the equivalent stiffness of the material, and this stiffness-softening phenomenon will result in a decrease in the dynamic responses of FG nanobeams. It is remarkable that the coupled elastic–kinetic porosity-dependent model can predict lower mechanical responses in comparison with the even porosity distribution which was presented in the modified power-law model. This phenomenon reveals that the coupled elastic–kinetic model, which is able to capture the correlation between the elastic and kinetic properties of the material, is a more reliable model for analyzing the wave propagation curves of the FG nanobeams because it provides more accurate responses.

Also, in this figure, the effects of tiny dimensions of the nanobeam are included, too. Actually, it can be realized that, because the ratio of length-scale parameter to nonlocal parameter, previously denoted as c, is less than, 1 the softening effect of the nonlocal parameter is stronger compared with that of the length-scale term. Due to this reason, the phase velocity experiences a gradual decrease at wave numbers higher than 0.3 (1/nm). However, it is worth mentioning that, because of the nonzero value which is assigned to the length-scale parameter, the final asymptotic magnitude of the phase speed reaches a constant nonzero value instead of converging to zero.

Furthermore, the coupled elastic–kinetic porous model is used in order to highlight the effects of adding the volume fraction of porosities on the wave frequency curves of FG nanobeams when $c < 1$. Clearly, it can be figured out that the wave frequency decreases as the volume fraction of the porosities grows. This decreasing effect comes from the decreasing effect of the porosity on the mechanical stiffness of the continuum. Besides, it must be considered that the effect of increasing the porosities' volume fraction can be better observed in the cases where high wave numbers are utilized. Also, it is evident that the wave frequency tends to infinity at high wave numbers because the effects of the length-scale parameter are covered in this (Figure 5.4).

As in the previous figure, the variation of phase velocity of a porous FG nanobeam against the wave number is illustrated in Figure 5.5 using the coupled elastic–kinetic homogenization method in order to present reliable results. It is clear that the phase velocity of the porous nanobeam will decrease when the volume fraction of the porosities increases. The main reason for this negative effect, caused by the existence of porosities, is explained in the previous paragraphs. Again, in this figure, it can be observed that the effect of the porosity on the phase speed of the dispersed waves depends on the wave number used. Indeed, the phase speed does not change remarkably on using various volume fractions of porosities at wave numbers smaller than 0.1 (1/nm).

5.3.2   Analysis of Porous FG Nanoplates

Herein, we will probe the wave propagation problem of a porous FG nanoplate. As in the case of FG nanobeams, similar diagrams will be presented and discussed for FG nanoplates. The phase velocity curves of both porous and nonporous nanoplates are shown in Figure 5.6. It is obvious that the waves scattered in the porous nanoplates travel faster in comparison with those dispersed in the nonporous nanoplates. The reason for this reality is mentioned in the previous section about FG nanobeams. Similarly, it is observable that more reliable dispersion curves for the nanoplates are obtained when the novel coupled elastic–kinetic model is chosen to estimate the behaviors of the nanostructure in the presence of porosities.

In addition, the effect of adding the amount of porosity in the media on the wave frequency behaviors of the FG nanoplates is shown in Figure 5.7. It can be seen that the wave

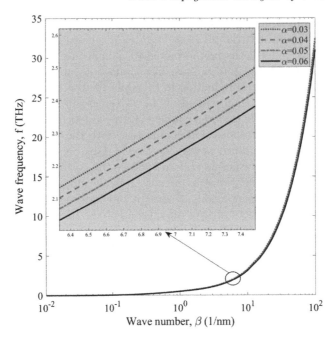

FIGURE 5.4

Comparison of frequency curves of porous FG nanobeams at various volume fractions of porosities ($p = 2, c < 1$).

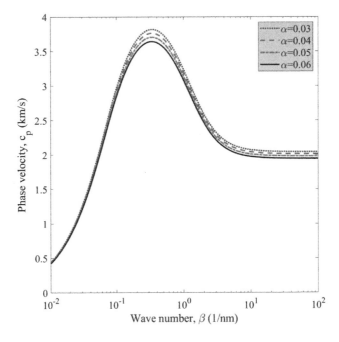

FIGURE 5.5

Comparison of phase velocity curves of porous FG nanobeams at various volume fractions of porosities ($p = 2, c < 1$).

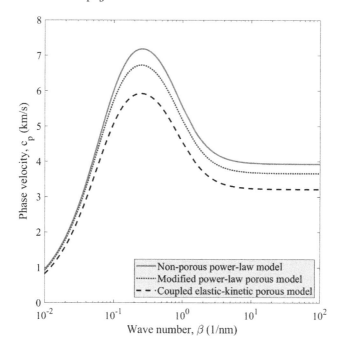

FIGURE 5.6
Comparison of phase velocity curves of nonporous and porous FG nanoplates ($p = 2, c < 1$).

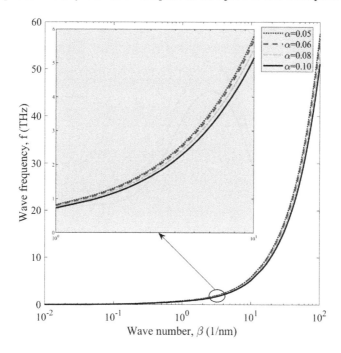

FIGURE 5.7
Comparison of frequency curves of porous FG nanoplates at various volume fractions of porosities ($p = 2, c < 1$).

frequency of the nanoplates decreases when using higher volume fractions for the porosities. Meanwhile, it is noticeable that the nanoplates cannot be affected by increasing the volume fraction of porosities when the axial and transverse wave numbers are assumed to be lower than 10 (1/nm). Also, it is clear that the estimations of the nonlocal strain gradient elasticity are included in this diagram. In other words, the main reason for the wave frequency to tend to infinity is the nonzero value that is assigned to the length-scale parameter. In fact, in the absence of the length-scale parameter (i.e., according to the nonlocal elasticity theory of Eringen), the wave frequency converges to a constant magnitude, and beyond that particular value, the wave frequency remains unchangeable. Thus, the graph presented is able to cover the stiffness-hardening effect of the length-scale parameter as well as the stiffness-softening impact of the nonlocal parameter.

Finally, the effects of various values of volume fraction of porosities on the phase velocity of the FG nanoplates are shown in Figure 5.8. It is clear that the phase speed of the propagated waves in an FG nanoplate shows a decreasing trend as the volume fraction of the porosities increases. The physical reasons for this behavior are explained in detail in the previous paragraphs. It is important to point out the fact that the effect of the existence of porosities on phase speed of the scattered waves is not the same at all wave numbers. As in the case of nanobeams, the phase velocity of the nanoplates at wave numbers lesser than a particular value (herein 0.05 (1/nm)) cannot be affected by making a change in the volume fraction of the porosities inside the nanostructure.

In conclusion, we wish to inform that the effects of probable porosities on the dispersion curves of the waves traveling inside FGM nanobeams and nanoplates are very important. Henceforward, such effects should be necessarily included when studying the wave propagation problems of nanostructures. Also, it is crucial to point out that it is

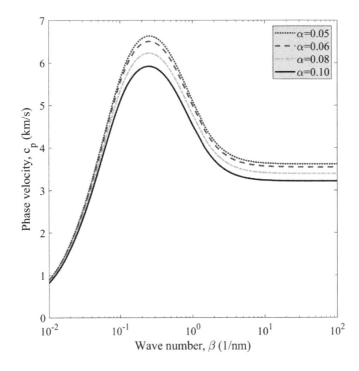

FIGURE 5.8

Comparison of phase velocity curves of porous FG nanoplates at various volume fractions of porosities ($p = 2, c < 1$).

better to employ the novel coupled elastic–kinetic porous model instead of the conventional modified power-law porous model for the FGMs because the newer model includes the effects of the mass density of the material with and without existence of porosities on the stiffness of the FGM. So, it is able to provide more reliable data on the propagation characteristics of FG nanostructures.

6

Wave Propagation Analysis of Smart Heterogeneous Pie oelectric Nanostructures

In this chapter, the wave propagation problem will be studied for a smart composite nanostructure with respect to the inherent electromechanical interactions of the material. The materials of both top and bottom edges of the nanostructure will be piezoelectric materials. The material's composition across the thickness of the nanobeam or nanoplate will be defined on the basis of the well-known power-law FG (functionally graded) model. Again, the motion equations will be derived by extending the Hamilton's principle to smart piezoelectric materials. The results show that the mechanical behaviors of the FG nanostructures can be deeply affected by changing the magnitude of the electric voltage. Indeed, the applied voltage can play a regressive role on the dynamic behaviors of the waves scattered in a smart nanostructure. In other words, it will be shown that the phase speed and wave frequency of the dispersed waves can be lessened whenever the applied electric voltage is increased. Moreover, it will be shown that this decreasing effect can be observed in a limited range of wave numbers. Also, we will try to illustrate the effect of the availability of porosities in the piezoelectric materials of the nanostructures on the wave dispersion curves of nanostructures.

6.1 Introduction

In recent technological attempts, the designers are moving as rapid as possible toward finding an element which can be utilized for a specific role with a multi-task functionality. This achievement can lower the cost of the engineering systems and also reduce the space required for producing a complicated system. So, it is beneficial to use a material with more than one capability in various subsections of a system. One of the most efficient types of smart multitasking materials is the piezoelectric material. This material can convert an electrical excitation to a mechanical output and vice versa. Actually, these smart materials experience an elastic strain when subjected to an electric voltage and generate an electric field on elongation of the structure. Such materials can be properly utilized as sensors, actuators and controllers in nano-electro-mechanical-systems (NEMSs). Moreover, the easy fabrication procedure and flexible design are crucial strong points of these materials. Due to the importance of these materials, many researchers found it necessary to analyze the mechanical responses of structures either made from or connected to piezoelectric materials under various loadings. For instance, Saravanos et al. [103] introduced a numerical discretized theory for the mechanical analysis of laminated composite piezoelectric plates. The investigation of the issue of controlling a multilayered plate with smart actuators was performed by Han and Lee [104] within the framework of a layer-wise study. Also, the effect of the position of the piezoelectric layer on the transverse frequency of the slender beams is included in an analysis done by Wang and Quek [105]. Oh et al. [106] combined a finite element nonlinear model with the layer-wise theory of the piezoelectric plates for probing the stability and frequency characteristics of composite plates in thermal environments. Wang et al. [107] studied the dynamic frequency behaviors of circular plates with two piezoelectric facesheets. They modeled the plate according to the classical theory of the plates. Also, Wang and Wang [108] presented a fundamental study of the control of the frequency of beams by means of piezoelectric actuators. They introduced an index for the controllability of the beam's frequency. In addition, He et al. [109] probed the effects of piezoelectric sensors and actuators on the dynamic responses of FG plates by extending the Kirchhoff–Love plate hypothesis to a finite- element-based approach. Some attempts were made by researchers to employ a control system for controlling the dynamic characteristics of composite plates based on both linear and nonlinear models [110,111]. The effects of shear deformation were covered by Gao and Shen [110] up to the first order on the basis of a finite- element-based method. They selected piezoelectric sensors and actuators to control the dynamic response of the plate. Liu et al. [112] studied both static bending and frequency responses of multilayered plates covered by piezosensors and actuators by implementing an interpolation-based numerical technique. The adaptive control of transient behaviors of cantilever beams with both piezoelectric sensors and actuators was investigated by Lin and Nien [113]. The effects of nonlinearities, generated from the existence of initial stress in the system, on the large deflection vibrational characteristics of piezoelectric composite plates were covered in a paper by Varelis and Saravanos [114]. The effects of bonding conditions between the structure and piezoelectric layers were covered and modeled using a spring by Bian et al. [115] for analyzing the mechanical responses of FGM beams. Moreover, Maurini et al. [116] performed the modal analysis of piezoelectric beams based on different numerical methods. The frequency behaviors of FG annular plates were studied by Ebrahimi and Rastgoo [117] in the case utilizing piezoelectric facesheets for the annular plate. The effects of initial geometrical imperfection on the vibrational behaviors of sandwich piezoelectric beams were covered by Azrar et al. [118]. The beam was presumed to be subjected to axial displacement and electric voltage. Also, Belouettar et al. [119] employed a control model for probing the nonlinear dynamic responses of three-layered beams consisting of an elastic core and two piezoelectric facesheets. Li and Shi [120] studied the vibrational responses of piezoelectric beams based on a numerical elasticity solution. An open-circuit condition was considered by Wu et al. [121] in order to analyze the natural frequency of the circular plates. Another nonlinear analysis was performed by Fu et al. [122] dealing with the static and dynamic characteristics of classical piezoelectric FG beams when a structure is subjected to a temperature gradient. Moreover, Elshafei and Alraiess [123] captured the effects of shear deformation when probing the mechanical behaviors of piezoelectric composite beams in the framework of a simulation study. Piezoelectric FGMs were utilized by Komijani et al. [124] to investigate the stability behaviors of beam-type elements based on a nonlinear finite element approach. On the other hand, Phung-Van et al. [125] performed an isogeometric analysis for investigating the mechanical characteristics of composite plates coupled with piezosensors and actuators. Barati et al. [126] could present a refined shear deformation model for the buckling responses of piezoelectric FG plates with respect to the porosities of the constituent material. Lately, the nonlinear thermoelectromechanical transient characteristics of piezoelectric FG plates were studied by Phung-Van et al. [127] by means of a numerical method.

In addition to the aforementioned studies, the increasing trend of using piezoelectric materials in nanosize devices has made it necessary to gather as much data as possible about the static and dynamic responses of piezoelectric nanostructures. Due to its importance, some of the researchers focused their studies on the investigation of the mechanical responses of piezoelectric tiny structures. Liu et al. [128] utilized the nonlocal theory of Eringen and combined it with the classical theory of plates to analyze the frequency behaviors of a rectangular biaxially loaded nanoplate. Hosseini-Hashemi et al. [129] probed the dynamic problem of a piezoelectric FG nanobeam in the framework of a nonlocal surface piezoelectricity hypothesis. The coupled effects of thermal environment and electric field on the buckling responses of FG nanobeams were covered by Ebrahimi and Salari [130]. Furthermore, Jandaghian and Rahmani [131] conducted an analytical investigation on the vibrational characteristics of smart piezoelectric FG nanoplates. Various types of the power-law model were employed by Ebrahimi and Salari [132] in order to study the natural frequency of the piezoelectric FG nanobeams with respect to the effect of the existence of thermal environment. Ebrahimi and Barati [133] solved the stability problem of a piezoelectric FG nanobeam utilizing a well-known shear deformation beam model. The wave propagation problem of a piezoelectric FG nanobeam, which is assumed to rotate around an axis parallel with its thickness direction, was studied by Ebrahimi and Dabbagh [134]. On the other hand, the wave dispersion analysis of piezoelectric FG nanoplates was performed by Ebrahimi and Dabbagh [135] based on a stress–strain gradient surface piezoelectricity theorem. In another valuable attempt, Ebrahimi and Barati [136] selected the nonlocal strain gradient elasticity for studying the vibration problem of FG piezoelectric nanoplates rested on a visco-Pasternak substrate with respect to the attenuation of the frequency. The combined effect of electric field, moisture concentration and temperature gradient on the nonlinear natural frequency of an piezoelectric FG nanobeam were covered by Ebrahimi et al. [137]. In one of the recent researches, a frequency analysis was performed by Ebrahimi and Barati [138] on the curved FG piezoelectric nanobeams by means of nonlocal strain gradient piezoelectricity.

6.2 Analysis of Piezoelectric FG Nanobeams

In this section, the motion equations of FG smart nanobeams will be derived. The previous governing equations which were used in the previous chapters cannot be employed here. Actually, the coupling of the electric field and elastic terms cannot be found in those equations, and due to this reality, the Hamilton's principle must be expanded again for the derivation of nanobeams' motion equations considering the effects of piezoelectric materials. For the sake of completeness, the motion equations of piezoelectric FG nanobeams will be derived for both Euler–Bernoulli and shear deformation nanobeams on the basis of nonlocal strain gradient hypothesis. At the end, the wave solutions that were introduced in Chapter 1 will be extended to the electric displacement to solve the problem analytically.

6.2.1 Euler Bernoulli Piezoelectric Nanobeams

In what follows, the nanobeam will be simulated based on the classical theory of the beams. The details of this theory were discussed in the Chapter 4. Hence, the derivation of the nonzero strains of the beam-type element will be dismissed, and the derivation procedure will start from the calculation of variations which need to be inserted in the definition of the Hamilton's principle.

6.2.1.1 Motion Equations of Piezoelectric Euler Bernoulli Beams

Before starting the derivation of motion equations, the Maxwell's equation must be satisfied. For this purpose, it is common to use a half-cosine half-linear expression for the electric potential of the piezoelectric material. The electric potential can be written as follows:

$$\bar{\Phi}(x, z, t) = -\cos(\xi z)\phi(x, t) + \frac{2z}{h}V \tag{6.1}$$

in which V is external applied electric voltage. Now, on the basis of Eq. (6.1), the electric field (E_x, E) and electric potential (ϕ) can be related to each other as

$$E_x = -\frac{\partial \bar{\Phi}}{\partial x} = \cos(\xi z)\frac{\partial \phi(x,t)}{\partial x}, \tag{6.2}$$

$$E = -\frac{\partial \bar{\Phi}}{\partial z} = -\xi \sin(\xi z)\phi(x,t) - \frac{2V}{h}. \tag{6.3}$$

Now, it is time to calculate the variation of strain energy, kinetic energy and work done by external loadings. It is worth mentioning that the strain energy, kinetic energy and work done by external loadings will be shown with symbols Π_S, Π_K and Π_W, respectively. The variation of strain energy can be calculated in an integral form as

$$\delta\Pi_S = \int_V \left(\sigma_{xx}\delta\varepsilon_{xx} - D_x\delta E_x - D \,\delta E \right)dV, \tag{6.4}$$

where D is the electric displacement. The above equation can be simplified by inserting the nonzero strains of the classical beams (see Eq. (4.12)) and electric fields (see Eqs. (6.2) and (6.3)) in it. Thus, this equation can be rewritten as follows:

$$\begin{aligned}
\delta\Pi_S = &\int_0^L \left(N\delta\varepsilon_{xx}^0 - M\delta\kappa_{xx}^0\right)dx \\
&+ \int_0^L \int_A \left(-D_x \cos(\xi z)\frac{\partial\delta\phi}{\partial x} + D \,\xi\sin(\xi z)\delta\phi\right)dAdx,
\end{aligned} \tag{6.5}$$

where axial force (N) and bending moment (M) can be computed based on the previously presented formulation in Eq. (4.25). Furthermore, the variation of work done by external forces can be formulated in the following form:

$$\delta\Pi_W = \int_0^L N^E \frac{\partial w}{\partial x}\frac{\partial\delta w}{\partial x}dx, \tag{6.6}$$

where N^E stands for the normal in-plane force generated due to electric voltage. This force can be computed as

$$N^E = \int_A \tilde{e}_{31}\frac{2V}{h}dA. \tag{6.7}$$

The variation of kinetic energy can be employed in the same form as expressed in Eq. (4.26); however, it is worth mentioning that this integral form should be shown with $\delta\Pi_K$ instead of δK. Now, the Euler–Lagrange equations of an Euler–Bernoulli smart beam can be obtained as follows by substituting Eqs. (6.5), (4.26) and (6.6) in Eq. (4.19) and using the nontrivial response:

$$\frac{\partial N}{\partial x} = I_0\ddot{u} - I_1\frac{\partial\ddot{w}}{\partial x}, \tag{6.8}$$

$$\frac{\partial^2 M}{\partial x^2} - N^E\frac{\partial^2 w}{\partial x^2} = I_0\ddot{w} + I_1\frac{\partial\ddot{u}}{\partial x} - I_2\frac{\partial^2\ddot{w}}{\partial x^2}, \tag{6.9}$$

$$\int_A \left(\cos(\xi z)\frac{\partial D_x}{\partial x} + \xi\sin(\xi z)D \right)dA = 0. \tag{6.10}$$

6.2.1.2 Nonlocal Strain Gradient Piezoelectricity for Euler Bernoulli Nanobeams

Next, the nonlocal strain gradient elasticity will be extended to piezoelectric nanobeams to capture the effect of the tiny scale of the nanostructure. The nonlocal strain gradient piezoelectricity theory can be expressed in the following relations for a classical piezoelectric nanobeam. To obtain the nonlocal constitutive equations of piezoelectric nanobeams, Eqs. (2.18) and (2.19) must be integrated over the thickness direction as follows:

$$(1 - \mu^2 \nabla^2)\sigma_{xx} = (1 - \lambda^2 \nabla^2)\left[\tilde{c}_{11}\varepsilon_{xx} - \tilde{e}_{31}E\right], \tag{6.11}$$

$$(1 - \mu^2 \nabla^2)D_x = (1 - \lambda^2 \nabla^2)\tilde{s}_{11}E_x, \tag{6.12}$$

$$(1 - \mu^2 \nabla^2)D = (1 - \lambda^2 \nabla^2)\left[\tilde{e}_{31}\varepsilon_{xx} + \tilde{s}_{33}E\right] \tag{6.13}$$

in which $\tilde{c}_{ij}$, $\tilde{e}_{ij}$, $\tilde{s}_{ij}$ are reduced coefficients of a piezoelectric FG nanobeam when it is subjected to a plane stress state [139]:

$$\tilde{c}_{11} = c_{11} - \frac{c_{13}^2}{c_{33}}, \tilde{e}_{31} = e_{31} - \frac{c_{13}e_{33}}{c_{33}}, \tilde{s}_{11} = s_{11}, \tilde{s}_{33} = s_{33} - \frac{e_{33}^2}{c_{33}}. \tag{6.14}$$

Thereafter, by integrating over the cross-sectional area of the nanobeam, the following relations can be obtained:

$$(1 - \mu^2 \nabla^2)N = (1 - \lambda^2 \nabla^2)\left[A_{xx}\frac{\partial u}{\partial x} - B_{xx}\frac{\partial^2 w}{\partial x^2} + A_{31}^e\phi\right] - N^E, \tag{6.15}$$

$$(1 - \mu^2 \nabla^2)M = (1 - \lambda^2 \nabla^2)\left[B_{xx}\frac{\partial u}{\partial x} - D_{xx}\frac{\partial^2 w}{\partial x^2} + E_{31}^e\phi\right] - M^E, \tag{6.16}$$

$$(1 - \mu^2 \nabla^2)\int_A D_x \cos(\xi z)dA = (1 - \lambda^2 \nabla^2)F_{11}^e\frac{\partial\phi}{\partial x}, \tag{6.17}$$

$$(1 - \mu^2 \nabla^2)\int_A D\,\xi\sin(\xi z)dA = (1 - \lambda^2 \nabla^2)\left[A_{31}^e\frac{\partial u}{\partial x} + E_{31}^e\frac{\partial^2 w}{\partial x^2} - F_{33}^e\phi\right], \tag{6.18}$$

where

$$[A_{31}^e, E_{31}^e] = \int_A [1, z]\tilde{e}_{31}\xi\sin(\xi z)dA, \tag{6.19}$$

$$A_{15}^e = \int_A \tilde{e}_{15}\cos(\xi z)dA, \tag{6.20}$$

$$[F_{11}^e, F_{33}^e] = \int_A \left[\tilde{s}_{11}\cos^2(\xi z), \tilde{s}_{33}\xi^2\sin^2(\xi z)\right]dA. \tag{6.21}$$

It is worth mentioning that the mechanical cross-sectional rigidities are

$$[A_{xx}, B_{xx}, D_{xx}] = \int_A \left[1, z, z^2\right]\tilde{c}_{11}dA. \tag{6.22}$$

Also, $M^E = -\int_A \tilde{e}_{31}\frac{2V}{h}zdA$ is the electric moment applied to the nanostructure.

6.2.1.3 Governing Equations of Piezoelectric Euler Bernoulli Nanobeams

Now, the nonlocal governing equations of smart piezoelectric nanobeams can be obtained. To obtain these equations, Eqs. (6.15)–(6.18) must be substituted in Eqs. (6.8)–(6.10). Through mathematical simplifications, the governing equations of a classical piezoelectric nanobeam can be expressed as follows:

$$\left(1 - \lambda^2 \nabla^2\right) \left[A_{xx} \frac{\partial^2 u}{\partial x^2} - B_{xx} \frac{\partial^3 w}{\partial x^3} + A_{31}^e \frac{\partial \phi}{\partial x} \right]$$
$$+ \left(1 - \mu^2 \nabla^2\right) \left[-I_0 \ddot{u} + I_1 \frac{\partial \ddot{w}}{\partial x} \right] = 0, \tag{6.23}$$

$$\left(1 - \lambda^2 \nabla^2\right) \left[B_{xx} \frac{\partial^3 u}{\partial x^3} - D_{xx} \frac{\partial^4 w}{\partial x^4} + E_{31}^e \frac{\partial^2 \phi}{\partial x^2} \right] + \left(1 - \mu^2 \nabla^2\right)$$
$$\times \left[-I_0 \ddot{w} - I_1 \frac{\partial \ddot{u}}{\partial x} + I_2 \frac{\partial^2 \ddot{w}}{\partial x^2} - N^E \frac{\partial^2 w}{\partial x^2} \right] = 0, \tag{6.24}$$

$$\left(1 - \lambda^2 \nabla^2\right) \left[A_{31}^e \frac{\partial u}{\partial x} - E_{31}^e \frac{\partial^2 w}{\partial x^2} + F_{11}^e \frac{\partial^2 \phi}{\partial x^2} - F_{33}^e \phi \right] = 0. \tag{6.25}$$

6.2.1.4 Wave Solution of the Euler Bernoulli Piezoelectric Nanobeams

In the following paragraphs, we are about to study the solution of wave propagation problem inside a piezoelectric FG nanobeam. The displacement fields can be considered to be equal to those reported in Eqs. (1.1) and (1.2). However, the electric potential is not covered yet, and we will probe it in this chapter. Henceforward, the analytical solution of the electric potential can be stated as follows:

$$\phi = \hat{\Phi} e^{i(\beta x - \omega t)}. \tag{6.26}$$

By substituting Eqs. (1.1), (1.2) and (6.26) in Eqs. (6.23)–(6.25), an eigenvalue problem similar to Eq. (1.3) will be obtained. However, the Δ vector in the present problem can be expressed as

$$\Delta = [U, W, \hat{\Phi}]^T. \tag{6.27}$$

Due to the mentioned differences between the present problem and those studied in the previous chapters, which were not dedicated to study the smart characteristics of the piezoelectric FG nanobeams, both stiffness and mass matrices must be rearranged. The components of the stiffness matrix can be calculated as

$$\begin{aligned}
k_{11} &= \left(1 + \lambda^2 \beta^2\right) A_{xx} \beta^2, \\
k_{12} &= i\left(1 + \lambda^2 \beta^2\right) B_{xx} \beta^3, \\
k_{13} &= i\left(1 + \lambda^2 \beta^2\right) \beta A_{31}^e, \\
k_{22} &= -\left(1 + \lambda^2 \beta^2\right) D_{xx} \beta^4 + \left(1 + \mu^2 \beta^2\right) \beta^2 N^E, \\
k_{23} &= -\left(1 + \lambda^2 \beta^2\right) \beta^2 E_{31}^e, \\
k_{33} &= -\left(1 + \lambda^2 \beta^2\right)(F_{33}^e + F_{11}^e \beta^2).
\end{aligned} \tag{6.28}$$

Also, the nonzero arrays of mass matrix can be assumed to be the same as those presented in Eq. (4.52).

6.2.2 Refined Sinusoidal Piezoelectric Nanobeams

In this section, the partial differential equations which must be solved to obtain the wave responses of a piezoelectric FG nanobeam will be derived on the basis of a refined sinusoidal beam model. The initial definitions of the nanobeam's displacement fields and their nonzero strains were previously discussed in the Chapter 4. Also, an expression for the electric potential was introduced and used to obtain the electric fields in Section 6.2.1. So, these relations will not be reviewed again.

6.2.2.1 Motion Equations of Piezoelectric Refined Shear Deformable Beams

In this section, we will obtain expressions for the motion equations of refined higher-order piezoelectric nanobeams. For this purpose, the variations of the nanobeam's energies must be again formulated on the basis of refined sinusoidal beam models. First of all, the variation of strain energy of the smart nanobeam can be written as

$$\delta\Pi_S = \int_V \left[\sigma_{xx}\delta\varepsilon_{xx} + \sigma_x\ \delta\varepsilon_x\ - D_x\delta E_x - D\ \delta E\ \right]dV. \tag{6.29}$$

By substituting Eqs. (4.17), (6.2) and (6.3) in Eq. (6.29), one can obtain the following simplified equation for the variation of strain energy of the beam:

$$\delta\Pi_S = \int_0^L \left(N\delta\varepsilon_{xx}^0 - M^b\delta\kappa_{xx}^b - M^s\delta\kappa_{xx}^s + Q_x\ \delta\varepsilon_x^0\ \right)dx$$
$$+ \int_0^L \int_A \left(- D_x\cos(\xi z)\frac{\partial\delta\phi}{\partial x} + D\ \xi\sin(\xi z)\delta\phi\right)dAdx \tag{6.30}$$

in which N, M^b, M^s and Q_x are computed in the same way as done in Eq. (4.32). Furthermore, the variation of work done by external forces can be formulated in the following form:

$$\delta\Pi_W = \int_0^L N^E \frac{\partial(w_b + w_s)}{\partial x} \frac{\partial\delta(w_b + w_s)}{\partial x}dx, \tag{6.31}$$

where N^E, that is, the electric force, can be expressed as

$$N^E = - \int_A \tilde{e}_{31}\frac{2V}{h}dA. \tag{6.32}$$

The variation of kinetic energy can be applied using the form given in Eq. (4.33); however, it is worth mentioning that this integral form should be shown with $\delta\Pi_K$ instead of δK. Now, the Euler–Lagrange equations of a refined sinusoidal smart beam can be obtained by substituting Eqs. (6.30), (4.33) and (6.31) in Eq. (4.19) and using the nontrivial response as follows:

$$\frac{\partial N}{\partial x} = I_0\ddot{u} - I_1\frac{\partial\ddot{w}_b}{\partial x} - J_1\frac{\partial\ddot{w}_s}{\partial x}, \tag{6.33}$$

$$\frac{\partial^2 M^b}{\partial x^2} - N^E\frac{\partial^2(w_b + w_s)}{\partial x^2} = I_0(\ddot{w}_b + \ddot{w}_s) + I_1\frac{\partial\ddot{u}}{\partial x} - I_2\frac{\partial^2\ddot{w}_b}{\partial x^2} - J_2\frac{\partial^2\ddot{w}_s}{\partial x^2}, \tag{6.34}$$

$$\frac{\partial^2 M^s}{\partial x^2} + \frac{\partial Q_x}{\partial x} - N^E\frac{\partial^2(w_b + w_s)}{\partial x^2} = I_0(\ddot{w}_b + \ddot{w}_s) + J_1\frac{\partial\ddot{u}}{\partial x} - J_2\frac{\partial^2\ddot{w}_b}{\partial x^2} - K_2\frac{\partial^2\ddot{w}_s}{\partial x^2}, \tag{6.35}$$

$$\int_A \left(\cos(\xi z)\frac{\partial D_x}{\partial x} + \xi\sin(\xi z)D\ \right)dA = 0. \tag{6.36}$$

6.2.2.2 Nonlocal Strain Gradient Piezoelectricity for Refined Sinusoidal Nanobeams

Now, it is time to obtain the constitutive equations of a piezoelectric nanobeam with respect to the small-scale softening and hardening effects. As in Section 6.2.1, these relations must be obtained by extending Eqs. (2.18) and (2.19) to refined shear deformable nanobeams. According to the aforementioned relations, the constitutive equations of a piezoelectric nanobeam can be written as

$$(1 - \mu^2 \nabla^2)\sigma_{xx} = (1 - \lambda^2 \nabla^2)[\tilde{c}_{11}\varepsilon_{xx} - \tilde{e}_{31}E\ ], \tag{6.37}$$

$$(1 - \mu^2 \nabla^2)\sigma_x = (1 - \lambda^2 \nabla^2)[\tilde{c}_{55}\varepsilon_x - \tilde{e}_{15}E_x], \tag{6.38}$$

$$(1 - \mu^2 \nabla^2)D_x = (1 - \lambda^2 \nabla^2)[\tilde{e}_{15}\varepsilon_x + \tilde{s}_{11}E_x], \tag{6.39}$$

$$(1 - \mu^2 \nabla^2)D = (1 - \lambda^2 \nabla^2)[\tilde{e}_{31}\varepsilon_{xx} + \tilde{s}_{33}E\ ]. \tag{6.40}$$

Thereafter, by integrating over the cross-sectional area of the Nanobeam, the following relations can be obtained:

$$(1 - \mu^2 \nabla^2)N = (1 - \lambda^2 \nabla^2)\left[A_{xx}\frac{\partial u}{\partial x} - B_{xx}\frac{\partial^2 w_b}{\partial x^2} - B^s_{xx}\frac{\partial^2 w_s}{\partial x^2} + A^e_{31}\phi\right] - N^E, \tag{6.41}$$

$$(1 - \mu^2 \nabla^2)M^b = (1 - \lambda^2 \nabla^2)\left[B_{xx}\frac{\partial u}{\partial x} - D_{xx}\frac{\partial^2 w_b}{\partial x^2} - D^s_{xx}\frac{\partial^2 w_s}{\partial x^2} + E^e_{31}\phi\right] - M^E_b, \tag{6.42}$$

$$(1 - \mu^2 \nabla^2)M^s = (1 - \lambda^2 \nabla^2)\left[B^s_{xx}\frac{\partial u}{\partial x} - D^s_{xx}\frac{\partial^2 w_b}{\partial x^2} - H^s_{xx}\frac{\partial^2 w_s}{\partial x^2} + F^e_{31}\phi\right] - M^E_s, \tag{6.43}$$

$$(1 - \mu^2 \nabla^2)Q_x = (1 - \lambda^2 \nabla^2)\left[A^s_x\frac{\partial w_s}{\partial x} - A^e_{15}\frac{\partial \phi}{\partial x}\right], \tag{6.44}$$

$$(1 - \mu^2 \nabla^2)\int_A D_x \cos(\xi z)dA = (1 - \lambda^2 \nabla^2)\left[E^e_{15}\frac{\partial w_s}{\partial x} + F^e_{11}\frac{\partial \phi}{\partial x}\right], \tag{6.45}$$

$$(1 - \mu^2 \nabla^2)\int_A D\ \xi \sin(\xi z)dA = (1 - \lambda^2 \nabla^2)\left[A^e_{31}\frac{\partial u}{\partial x} + E^e_{31}\frac{\partial^2 w_b}{\partial x^2} + F^e_{31}\frac{\partial^2 w_s}{\partial x^2} - F^e_{33}\phi\right], \tag{6.46}$$

where some of the electrical integrals used in the above relations are defined in Eqs. (6.19)–(6.21). However, some of them are not defined in those equations because they deal with the shear approximation shape function which is not available in the Euler–Bernoulli beam model. Thus, the rest of the electrical integrals can be expressed as

$$F^e_{31} = \int_A f(z)\tilde{e}_{31}\xi \sin(\xi z)dA, \tag{6.47}$$

$$E^e_{15} = \int_A g(z)\tilde{e}_{15} \cos(\xi z)dA. \tag{6.48}$$

Also, it is worth mentioning that the cross-sectional rigidities, which depend on the shape function of the shear deformation beam, can be calculated as

$$[B^s_{xx}, D^s_{xx}, H^s_{xx}] = \int_A [f(z), zf(z), f^2(z)]\tilde{c}_{11}dA. \tag{6.49}$$

Similarly, the electric resultants M_b^E and M_s^E can be expressed as

$$[M_b^E, M_s^E] = -\int_A [z, f(z)]\tilde{e}_{31}\frac{2V}{h}dA. \tag{6.50}$$

6.2.2.3 Governing Equations of Piezoelectric Refined Sinusoidal Nanobeams

In this section, we will go through the derivation of the final differential equations which govern the problem. Indeed, the nonlocal constitutive equations should be substituted in the Euler–Lagrange equations of the problem. In this case study, Eqs. (6.41)–(6.46) must be inserted in Eqs. (6.33)–(6.36). Finally, the governing equations are as follows:

$$\left(1 - \lambda^2\nabla^2\right)\left[A_{xx}\frac{\partial^2 u}{\partial x^2} - B_{xx}\frac{\partial^3 w_b}{\partial x^3} - B_{xx}^s\frac{\partial^3 w_s}{\partial x^3} + A_{31}^e\frac{\partial\phi}{\partial x}\right] + \left(1 - \mu^2\nabla^2\right)$$
$$\times\left[-I_0\ddot{u} + I_1\frac{\partial\ddot{w}_b}{\partial x} + J_1\frac{\partial\ddot{w}_s}{\partial x}\right] = 0, \tag{6.51}$$

$$\left(1 - \lambda^2\nabla^2\right)\left[B_{xx}\frac{\partial^3 u}{\partial x^3} - D_{xx}\frac{\partial^4 w_b}{\partial x^4} - D_{xx}^s\frac{\partial^4 w_s}{\partial x^4} + (E_{31}^e - A_{15}^e)\frac{\partial^2\phi}{\partial x^2}\right] + \left(1 - \mu^2\nabla^2\right)$$
$$\times\left[-I_0(\ddot{w}_b + \ddot{w}_s) - I_1\frac{\partial\ddot{u}}{\partial x} + I_2\frac{\partial^2\ddot{w}_b}{\partial x^2} + J_2\frac{\partial^2\ddot{w}_s}{\partial x^2} - N^E\frac{\partial^2(w_b + w_s)}{\partial x^2}\right] = 0, \tag{6.52}$$

$$\left(1 - \lambda^2\nabla^2\right)\left[B_{xx}^s\frac{\partial^3 u}{\partial x^3} - D_{xx}^s\frac{\partial^4 w_b}{\partial x^4} - H_{xx}^s\frac{\partial^4 w_s}{\partial x^4} + A_x^s\frac{\partial^2 w_s}{\partial x^2} + (F_{31}^e - E_{15}^e)\frac{\partial^2\phi}{\partial x^2}\right] + \left(1 - \mu^2\nabla^2\right)$$
$$\times\left[-I_0(\ddot{w}_b + \ddot{w}_s) - J_1\frac{\partial\ddot{u}}{\partial x} + J_2\frac{\partial^2\ddot{w}_b}{\partial x^2} + K_2\frac{\partial^2\ddot{w}_s}{\partial x^2} - N^E\frac{\partial^2(w_b + w_s)}{\partial x^2}\right] = 0, \tag{6.53}$$

$$\left(1 - \lambda^2\nabla^2\right)\left[A_{31}^e\frac{\partial u}{\partial x} - E_{31}^e\frac{\partial^2 w_b}{\partial x^2} - (F_{31}^e - E_{15}^e)\frac{\partial^2 w_s}{\partial x^2} + F_{11}^e\frac{\partial^2\phi}{\partial x^2} - F_{33}^e\phi\right] = 0. \tag{6.54}$$

6.2.2.4 Wave Solution of the Refined Piezoelectric Nanobeams

The wave dispersion solution for the electric potential was introduced in the previous section for the Euler–Bernoulli nanobeams. Herein, only the dimensions of the matrices vary compared to the previous situation due to the differences between the Euler–Bernoulli and higher-order nanobeams. Actually, the displacement vector (Δ) becomes

$$\Delta = [U, W_b, W_s, \hat{\Phi}]^T. \tag{6.55}$$

By substituting Eqs. (1.5)–(1.7) and (6.26) in Eqs. (6.51)–(6.54), an eigenvalue problem similar to Eq. (1.3) will be obtained. The components of the stiffness matrix can be calculated as

$$k_{11} = -\left(1 + \lambda^2\beta^2\right)A_{xx}\beta^2,$$
$$k_{12} = i\left(1 + \lambda^2\beta^2\right)B_{xx}\beta^3,$$
$$k_{13} = i\left(1 + \lambda^2\beta^2\right)B_{xx}^s\beta^3,$$
$$k_{14} = i\left(1 + \lambda^2\beta^2\right)\beta A_{31}^e,$$
$$k_{22} = -\left(1 + \lambda^2\beta^2\right)D_{xx}\beta^4 + (1 + \mu^2\beta^2)\beta^2 N^E,$$
$$k_{23} = -\left(1 + \lambda^2\beta^2\right)D_{xx}^s\beta^4 + (1 + \mu^2\beta^2)\beta^2 N^E,$$

$$k_{24} = -\left(1 + \lambda^2\beta^2\right)\beta^2 E_{31}^e,$$

$$k_{33} = -\left(1 + \lambda^2\beta^2\right)\left[H_{xx}^s\beta^4 + A_x^s\ \beta^2\right] + \left(1 + \mu^2\beta^2\right)\beta^2 N^E,$$

$$k_{34} = -\left(1 + \lambda^2\beta^2\right)\left(F_{31}^e - E_{15}^e\right)\beta^2,$$

$$k_{44} = -\left(1 + \lambda^2\beta^2\right)\left(F_{33}^e - F_{11}^e\beta^2\right). \tag{6.56}$$

Also, the nonzero arrays of mass matrix can be assumed to be the same as those presented in Eq. (4.54).

6.2.3   Numerical Results for Piezoelectric Nanobeams

In this section, the wave propagation curves of smart piezoelectric FG nanobeams will be illustrated in separate diagrams. It will be realized that the effects of electric voltage become clearer at low wave numbers. So, we try to use the phase speed curves instead of wave frequency ones because the dispersion curves of piezoelectric FG nanobeams can be better studied when the y-axis represents phase velocity instead of wave frequency. The effects of the existence of porosities in a smart FGM are included in the numerical examples of this section.

In Figure 6.1, the effect of changing the applied voltage on the phase velocity curves of the FG nanobeams is depicted. Before discussing about the behaviors of the smart nanostructure, it can be figured out that the nanobeam follows three paths as the scale factor varies. Indeed, the phase velocity of the nanobeam increases as the scale factor increases. It is clear that with respect to the relative value of scale factor in comparison with 1, the velocity curve can show a decreasing, stable or increasing trend after its initial rise when the scale factor is smaller than 1, equal to 1 or greater than 1, respectively. Furthermore, it is obvious that the electric voltage can result in a limited decrease in the phase velocity of the waves dispersed in a nanobeam whenever it is increased. It means, an increase in the voltage can make the nanostructure softer due to the heat generated from the voltage increment. It must be noticed that the wave frequency can be affected by the electric voltage only at wave numbers smaller than 0.1 (1/nm). Thus, it can be a good technique to tune the phase speed of the propagated waves in a piezoelectric continuum by making changes in the electric voltage.

The combined effects of electric voltage and gradient index of the smart FGM on the phase speed of the FG nanobeams are included in Figure 6.2 by plotting the variation of phase speed with applied voltage at various wave numbers. As explained in the interpretation of the previous diagram (see Figure 6.1), it can be realized that studying the effect of electric voltage cannot be useful at the wave numbers larger than 0.1 (1/nm). Henceforward, in Figure 6.2, two wave numbers smaller than 0.1 (1/nm) are used to observe the effects of voltage on the speed of the propagated waves. It is observable that the speed of the waves decreases continuously as the applied voltage increases. In other words, once again the softening effect of electric voltage on the nanostructure, which results in a reduction in the dynamic responses of the nanostructure, can be found. It is worth mentioning that the phase velocity of the nanobeam will be affected more by changing the magnitude of the applied voltage at the wave number of 0.02 (1/nm) than at the wave number of 0.05 (1/nm). It can be concluded that the smaller the wave number is, the greater is the effect of the electric voltage. Also, the stiffness of the FGM decreases whenever the gradient index of the FGM increases, so, it is natural that the phase velocity decreases in cases where higher gradient indices are used.

The coupled effects of electric voltage and length-scale parameter on the variation of phase velocity of smart piezoelectric FG nanobeams are illustrated in Figure 6.3. Clearly, the phase velocity at each desired value of nonlocal parameter can be increased by reducing the

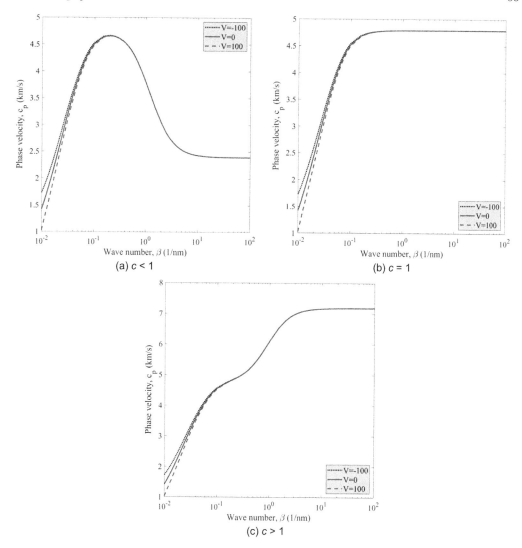

FIGURE 6.1

Variation of phase velocity with wave number at various amounts of applied voltage and scale factors $(p = 2)$.

amount of the applied voltage. The physical reason for this phenomenon is that the stiffness of the nanobeam can be decreased by increasing the magnitude of the electric voltage. On the other hand, it is observed that the dynamic response of the system diminishes when the value of the nonlocal parameter increases. Actually, this behavior can be justified by reviewing the softening effect of the nonlocal term on the equivalent stiffness of the nanostructure. In other words, the higher the value of the nonlocal parameter, the lower is the value of the system's stiffness. Therefore, increasing the nonlocal parameter can has the same effect as increasing the electric voltage.

Similarly, Figure 6.4 is plotted to investigate the effect of the length-scale parameter on the variation of the phase speed of the dispersed waves in an FG nanobeam for various applied voltages. It can be understood that the phase velocity can be decreased by increasing the value of the electric voltage. The physical reason for this behavior is thoroughly discussed

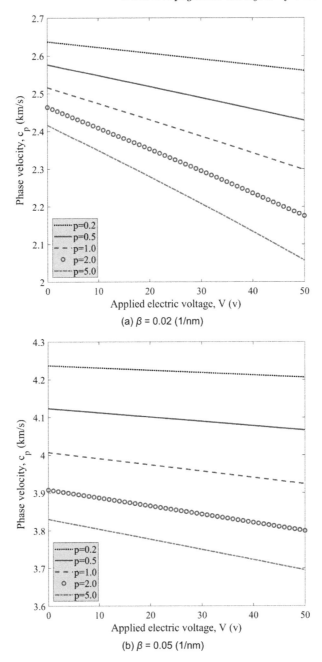

FIGURE 6.2
Variation of phase velocity with applied electric voltage at various gradient indices and wave
numbers $(c > 1)$.

in the previous illustrations. Also, the stiffness enhancement in tiny nanostructures which
is estimated from the nonlocal strain gradient elasticity can be seen here. In other words,
the phase speed increases when a higher value is assigned to the length-scale parameter.
It is observable that the effect of varying the length- scale parameter on the phase velocity
of the FG nanobeams is more than that of varying the nonlocal parameter. This trend can

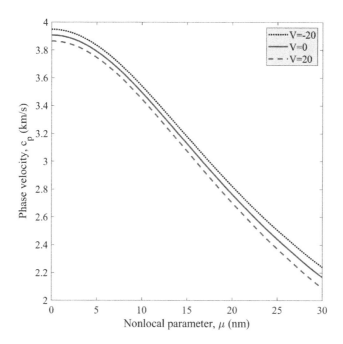

FIGURE 6.3

Variation of phase velocity with nonlocal parameter at various applied voltages ($\beta = 0.05$ (1/nm), $p = 2$, $\lambda = 1$ nm).

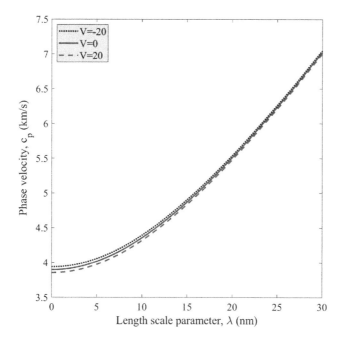

FIGURE 6.4

Variation of phase velocity with length-scale parameter at various applied voltages ($\beta = 0.05$ (1/nm), $p = 2$, $\mu = 1$ nm).

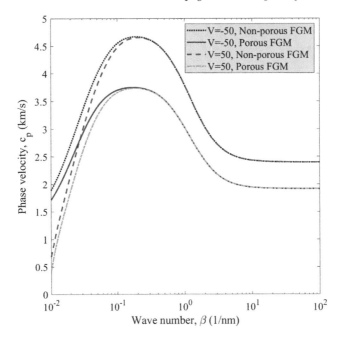

FIGURE 6.5
Variation of phase velocity with wave number at various applied voltages for both porous and nonporous smart piezoelectric FGMs ($c < 1$, $p = 2$).

be justified by observing Figure 6.1. It is obvious that at small wave numbers ($\beta < 0.1$ (1/nm)), the softening effect of the nonlocal parameter defeats the hardening effect of the length-scale term.

Finally, the effect of the existence of probable porosities on the phase velocity curves of the smart FG nanobeams can be observed in Figure 6.5. In this diagram, the variation of phase velocity with wave number is plotted for both positive and negative electric voltages. One can realize that the phase velocity of the piezoelectric nanobeam can be shifted up whenever a negative voltage is connected to the nanostructure. On the other hand, it is clear that the phase velocity's path drops down when the smart piezoelectric FGM is considered to be a porous one rather than a nonporous one. The reason for this phenomenon is completely explained in the Chapter 5; however, the reality will be again demonstrated that the stiffness of the FG nanostructure decreases in cases where some porosities can be found in the nanostructure. So, it is natural to see a reduction in the magnitude of the phase speed of the dispersed waves in a piezoelectric nanobeam when the utilized FGM is a porous one.

6.3　Analysis of FG Piezoelectric Nanoplates

In this section, the piezoelectrically affected wave propagation responses of smart FG nanoplates will be studied to show the effects of different variants on the wave dispersion curves of piezoelectric FG nanoplates. As in Section 6.2, the energy relations must be reformulated to obtain the governing equations of plates with respect to the effect of the piezoelectricity. Both Kirchhoff–Love and refined higher-order kinematic models of plates will be employed in what follows to derive the motion equations of smart piezoelectric nanoplates. Again, the most complicated form of the size-dependent continuum theories, which is commonly used for a wave dispersion problem, will be used to capture the effects of both nonlocal and length-scale terms.

6.3.1 Classical Piezoelectric Nanoplates

In this section, we will obtain the motion equations of classical nanoplates consisting of piezoelectric FGMs. Herein, the effects of shear deformation will not be covered to present the simplest forms of the governing equations for smart piezoelectric FG nanoplates. The small-scale effects will be included, and the obtained governing equations will be finally solved via an analytical method.

6.3.1.1 Motion Equations of Piezoelectric Kirchhoff Love Nanoplates

In the previous sections of this chapter, the equation needed for the electric potential was presented in such a way that the Maxwell's equation could be satisfied for a beam (see Section 6.2.1). However, there is a difference between the relations obtained in the aforementioned section and those required for a plate-type structure. Actually, the y dimension will be usually dismissed in a beam, whereas it must be included when a plate is to be analyzed. Therefore, the electric potential for a piezoelectric plate can be expressed as

$$\bar{\Phi}(x,y,z,t) = -\cos(\xi z)\phi(x,y,t) + \frac{2z}{h}V, \tag{6.57}$$

where V is the electric voltage. Now, on the basis of Eq. (6.57), the electric field (E_x, E_y, E) and electric potential (ϕ) can be related to each other as follows:

$$E_x = -\frac{\partial\bar{\Phi}}{\partial x} = \cos(\xi z)\frac{\partial\phi(x,y,t)}{\partial x} \tag{6.58}$$

$$E_y = -\frac{\partial\bar{\Phi}}{\partial x} = \cos(\xi z)\frac{\partial\phi(x,y,t)}{\partial y}, \tag{6.59}$$

$$E = -\frac{\partial\bar{\Phi}}{\partial z} = -\xi\sin(\xi z)\phi(x,y,t) - \frac{2V}{h}. \tag{6.60}$$

Then, the above relations can be utilized in the dynamic form of the principle of virtual work in order to obtain the equations of motion. The variation of the strain energy of a piezoelectric FG plate can be expressed as

$$\delta\Pi_S = \int_V \left(\sigma_{xx}\delta\varepsilon_{xx} + \sigma_{yy}\delta\varepsilon_{yy} + \sigma_{xy}\delta\varepsilon_{xy} - D_x\delta E_x - D_y\delta E_y - D \,\delta E \right)dV. \tag{6.61}$$

Equation (6.61) can be rewritten by substituting the nonzero strains of the Kirchhoff–Love theory (see Eqs. (4.59)–(4.61)) and electric field (see Eqs. (6.58)–(6.60)) in it in the following form:

$$\delta\Pi_S = \int_A \left(N_{xx}\delta\varepsilon_{xx}^0 - M_{xx}\delta\kappa_{xx}^0 + N_{yy}\delta\varepsilon_{yy}^0 - M_{yy}\delta\kappa_{yy}^0 + N_{xy}\delta\varepsilon_{xy}^0 - M_{xy}\delta\kappa_{xy}^0 \right)dA$$

$$+ \int_{-\frac{h}{2}}^{\frac{h}{2}} \left(-D_x\cos(\xi z)\frac{\partial\delta\phi}{\partial x} - D_y\cos(\xi z)\frac{\partial\delta\phi}{\partial y} + D \,\xi\sin(\xi z)\delta\phi \right)dAdz \tag{6.62}$$

in which N_{ij}, M_{ij} $(i, j = x, y)$ are stress resultants which were introduced in Eq. (4.73). As in the case of nanobeams, the electric force produced from the existence of applied voltage can be expressed in the framework of the variation of work done by extra forces as

$$\delta\Pi_W = \int_A N^E \left(\frac{\partial w}{\partial x} \frac{\partial \delta w}{\partial x} + \frac{\partial w}{\partial y} \frac{\partial \delta w}{\partial y} \right) dA \qquad (6.63)$$

in which N^E is the external in-plane electric force that will be applied to the structure. In this study, because of the fact that the homogenized equivalent material would be an isotropic material, we considered the electric force in both x and y directions to be

$$N^E = \int_{-\frac{h}{2}}^{\frac{h}{2}} \tilde{e}_{31} \frac{2V}{h} dz. \qquad (6.64)$$

It is worth mentioning that the variation of the kinetic energy of the plate will not be discussed in this section. Indeed, the variation of kinetic energy is just the same as that expressed in Eq. (4.74). Thus, the Euler–Lagrange equations of the plate can be obtained by inserting Eqs. (6.62), (6.63) and (4.74) in Eq. (4.19). The equations of motion of the plate can be written as

$$\frac{\partial N_{xx}}{\partial x} + \frac{\partial N_{xy}}{\partial y} = I_0 \ddot{u} - I_1 \frac{\partial \ddot{w}}{\partial x}, \qquad (6.65)$$

$$\frac{\partial N_{xy}}{\partial x} + \frac{\partial N_{yy}}{\partial y} = I_0 \ddot{v} - I_1 \frac{\partial \ddot{w}}{\partial y}, \qquad (6.66)$$

$$\frac{\partial^2 M_{xx}}{\partial x^2} + 2\frac{\partial^2 M_{xy}}{\partial x \partial y} + \frac{\partial^2 M_{yy}}{\partial y^2} - N^E \left(\frac{\partial^2 w}{\partial x^2} + \frac{\partial^2 w}{\partial y^2} \right) = I_0 \ddot{w} + I_1 \left(\frac{\partial \ddot{u}}{\partial x} + \frac{\partial \ddot{v}}{\partial y} \right) - I_2 \left(\frac{\partial^2 \ddot{w}}{\partial x^2} + \frac{\partial^2 \ddot{w}}{\partial y^2} \right), \qquad (6.67)$$

$$\int_A \left(\cos(\xi z) \frac{\partial D_x}{\partial x} + \cos(\xi z) \frac{\partial D_y}{\partial y} + \xi \sin(\xi z) D \right) dA = 0. \qquad (6.68)$$

6.3.1.2   Nonlocal Strain Gradient Piezoelectricity for Kirchhoff Love Nanoplates

In this section, the nonlocal strain gradient theorem will be extended to a piezoelectric nanoplate. The constitutive equations will be obtained for a simple classical plate for which shear stress and strain need not be considered. Based on this hypothesis and from Eqs. (2.18) and (2.19), the constitutive equations of a piezoelectric nanoplate can be expressed as follows:

$$\left(1 - \mu^2\nabla^2\right)\sigma_{xx} = \left(1 - \lambda^2\nabla^2\right)\left[\tilde{c}_{11}\varepsilon_{xx} + \tilde{c}_{12}\varepsilon_{yy} - \tilde{e}_{31}E\right], \qquad (6.69)$$

$$\left(1 - \mu^2\nabla^2\right)\sigma_{yy} = \left(1 - \lambda^2\nabla^2\right)\left[\tilde{c}_{12}\varepsilon_{xx} + \tilde{c}_{11}\varepsilon_{yy} - \tilde{e}_{31}E\right], \qquad (6.70)$$

$$\left(1 - \mu^2\nabla^2\right)\sigma_{xy} = \left(1 - \lambda^2\nabla^2\right)\tilde{c}_{66}\varepsilon_{xy}, \qquad (6.71)$$

$$\left(1 - \mu^2\nabla^2\right)D_x = \left(1 - \lambda^2\nabla^2\right)\tilde{s}_{11}E_x, \qquad (6.72)$$

$$\left(1 - \mu^2\nabla^2\right)D_y = \left(1 - \lambda^2\nabla^2\right)\tilde{s}_{11}E_y, \qquad (6.73)$$

$$\left(1 - \mu^2\nabla^2\right)D = \left(1 - \lambda^2\nabla^2\right)\left[\tilde{e}_{31}\varepsilon_{xx} + \tilde{e}_{31}\varepsilon_{yy} + \tilde{s}_{33}E\right], \qquad (6.74)$$

where $\tilde{c}_{ij}$, $\tilde{e}_{ij}$ and $\tilde{s}_{ij}$ are the reduced electromechanical coefficients of a piezoelectric FGM that were introduced in the Eq. (6.14). By integrating the above equations over the thickness of the nanoplate, the following resultant relations can be obtained:

$$
\left(1 - \mu^2 \nabla^2\right) \begin{bmatrix} N_{xx} \\ N_{yy} \\ N_{xy} \end{bmatrix} = \left(1 - \lambda^2 \nabla^2\right) \left(\begin{bmatrix} A_{11} & A_{12} & 0 \\ A_{12} & A_{22} & 0 \\ 0 & 0 & A_{66} \end{bmatrix} \begin{bmatrix} \frac{\partial u}{\partial x} \\ \frac{\partial v}{\partial y} \\ \frac{\partial u}{\partial y} + \frac{\partial v}{\partial x} \end{bmatrix} \right.
$$
$$
\left. + \begin{bmatrix} B_{11} & B_{12} & 0 \\ B_{12} & B_{22} & 0 \\ 0 & 0 & B_{66} \end{bmatrix} \begin{bmatrix} -\frac{\partial^2 w}{\partial x^2} \\ -\frac{\partial^2 w}{\partial y^2} \\ -2\frac{\partial^2 w}{\partial x \partial y} \end{bmatrix} + \begin{bmatrix} A_{31}^e \\ A_{31}^e \\ 0 \end{bmatrix} \phi - \begin{bmatrix} N^E \\ N^E \\ 0 \end{bmatrix} \right), \quad (6.75)
$$

$$
\left(1 - \mu^2 \nabla^2\right) \begin{bmatrix} M_{xx} \\ M_{yy} \\ M_{xy} \end{bmatrix} = \left(1 - \lambda^2 \nabla^2\right) \left(\begin{bmatrix} B_{11} & B_{12} & 0 \\ B_{12} & B_{22} & 0 \\ 0 & 0 & B_{66} \end{bmatrix} \begin{bmatrix} \frac{\partial u}{\partial x} \\ \frac{\partial v}{\partial y} \\ \frac{\partial u}{\partial y} + \frac{\partial v}{\partial x} \end{bmatrix} \right.
$$
$$
\left. + \begin{bmatrix} D_{11} & D_{12} & 0 \\ D_{12} & D_{22} & 0 \\ 0 & 0 & D_{66} \end{bmatrix} \begin{bmatrix} -\frac{\partial^2 w}{\partial x^2} \\ -\frac{\partial^2 w}{\partial y^2} \\ -2\frac{\partial^2 w}{\partial x \partial y} \end{bmatrix} + \begin{bmatrix} E_{31}^e \\ E_{31}^e \\ 0 \end{bmatrix} \phi - \begin{bmatrix} M^E \\ M^E \\ 0 \end{bmatrix} \right), \quad (6.76)
$$

$$
\left(1 - \mu^2 \nabla^2\right) \int_A \begin{bmatrix} D_x \\ D_y \end{bmatrix} \cos(\xi z) dz = \left(1 - \lambda^2 \nabla^2\right) F_{11}^e \begin{bmatrix} \frac{\partial \phi}{\partial x} \\ \frac{\partial \phi}{\partial y} \end{bmatrix}, \quad (6.77)
$$

$$
\left(1 - \mu^2 \nabla^2\right) \int_A D \, \xi \sin(\xi z) dz = \left(1 - \lambda^2 \nabla^2\right) \left(A_{31}^e \left(\frac{\partial u}{\partial x} + \frac{\partial v}{\partial y} \right) - E_{31}^e \left(\frac{\partial^2 w}{\partial x^2} + \frac{\partial^2 w}{\partial y^2} \right) - F_{33}^e \phi \right)
$$
$$
(6.78)
$$

in which the cross-sectional rigidities can be calculated as

$$
\begin{bmatrix} A_{11} & B_{11} & D_{11} \\ A_{12} & B_{12} & D_{12} \\ A_{66} & B_{66} & D_{66} \end{bmatrix} = \int_{-\frac{h}{2}}^{\frac{h}{2}} \begin{bmatrix} 1 & z & z^2 \end{bmatrix} \begin{bmatrix} \tilde{c}_{11} \\ \tilde{c}_{12} \\ \tilde{c}_{66} \end{bmatrix} dz, \quad (6.79)
$$

$$
[A_{31}^e, E_{31}^e] = \int_{-\frac{h}{2}}^{\frac{h}{2}} [1, z] \tilde{e}_{31} \xi \sin(\xi z) dz, \quad (6.80)
$$

$$
[F_{11}^e, F_{33}^e] = \int_{-\frac{h}{2}}^{\frac{h}{2}} [\tilde{s}_{11} \cos^2(\xi z), \tilde{s}_{33} \xi^2 \sin^2(\xi z)] dz \quad (6.81)
$$

and

$$
\begin{bmatrix} A_{22}, B_{22}, D_{22} \end{bmatrix} = \begin{bmatrix} A_{11}, B_{11}, D_{11} \end{bmatrix}. \quad (6.82)
$$

Furthermore, the electrical resultants N^E and M^E in Eqs. (6.75) and (6.76) can be expressed as

$$
N^E = - \int_{-\frac{h}{2}}^{\frac{h}{2}} \tilde{e}_{31} \frac{2V}{h} dz, \quad (6.83)
$$

$$M^E = -\int_{-\frac{h}{2}}^{\frac{h}{2}} \tilde{e}_{31} \frac{2V}{h} z dz. \tag{6.84}$$

6.3.1.3 Governing Equations of Piezoelectric Kirchhoff Love Nanoplates

In this section, the constitutive equations of the nanoplate will be used for deriving the nonlocal governing equations of a smart nanoplate. In fact, it is necessary to substitute Eqs. (6.75)–(6.78) in Eqs. (6.65)–(6.68). By this substitution, the partial differential equations of the problem can be expressed in the following forms:

$$\left(1 - \lambda^2 \nabla^2\right) \left(A_{11} \frac{\partial^2 u}{\partial x^2} + (A_{12} + A_{66}) \frac{\partial^2 v}{\partial x \partial y} + A_{66} \frac{\partial^2 u}{\partial y^2} - B_{11} \frac{\partial^3 w}{\partial x^3} \right.$$
$$\left. - (B_{12} + 2B_{66}) \frac{\partial^3 w}{\partial x \partial y^2} + A^e_{31} \frac{\partial \phi}{\partial x} \right) + \left(1 - \mu^2 \nabla^2\right) \left(-I_0 \ddot{u} + I_1 \frac{\partial \ddot{w}}{\partial x} \right) = 0, \tag{6.85}$$

$$\left(1 - \lambda^2 \nabla^2\right) \left(A_{22} \frac{\partial^2 v}{\partial y^2} + (A_{12} + A_{66}) \frac{\partial^2 u}{\partial x \partial y} + A_{66} \frac{\partial^2 v}{\partial x^2} - B_{22} \frac{\partial^3 w}{\partial y^3} \right.$$
$$\left. - (B_{12} + 2B_{66}) \frac{\partial^3 w}{\partial x^2 \partial y} + A^e_{31} \frac{\partial \phi}{\partial y} \right) + \left(1 - \mu^2 \nabla^2\right) \left(-I_0 \ddot{v} + I_1 \frac{\partial \ddot{w}}{\partial y} \right) = 0, \tag{6.86}$$

$$\left(1 - \lambda^2 \nabla^2\right) \left(B_{11} \frac{\partial^3 u}{\partial x^3} + (B_{12} + 2B_{66}) \frac{\partial^3 u}{\partial x \partial y^2} + B_{22} \frac{\partial^3 v}{\partial y^3} \right.$$
$$+ (B_{12} + 2B_{66}) \frac{\partial^3 v}{\partial x^2 \partial y} - D_{11} \frac{\partial^4 w}{\partial x^4} - 2(D_{12} + 2D_{66}) \frac{\partial^4 w}{\partial x^2 \partial y^2}$$
$$\left. - D_{22} \frac{\partial^4 w}{\partial y^4} + E^e_{31} \left(\frac{\partial^2 \phi}{\partial x^2} + \frac{\partial^2 \phi}{\partial y^2} \right) \right) + \left(1 - \mu^2 \nabla^2\right) \left(-I_0 \ddot{w} - I_1 \left(\frac{\partial \ddot{u}}{\partial x} + \frac{\partial \ddot{v}}{\partial y} \right) \right)$$
$$+ I_2 \left(\frac{\partial^2 \ddot{w}}{\partial x^2} + \frac{\partial^2 \ddot{w}}{\partial y^2} \right) - N^E \left(\frac{\partial^2 w}{\partial x^2} + \frac{\partial^2 w}{\partial y^2} \right) = 0, \tag{6.87}$$

$$\left(1 - \lambda^2 \nabla^2\right) \left(A^e_{31} \left(\frac{\partial u}{\partial x} + \frac{\partial v}{\partial y} \right) - E^e_{31} \left(\frac{\partial^2 w}{\partial x^2} + \frac{\partial^2 w}{\partial y^2} \right) + F^e_{11} \left(\frac{\partial^2 \phi}{\partial x^2} + \frac{\partial^2 \phi}{\partial y^2} \right) - F^e_{33} \phi \right) = 0. \tag{6.88}$$

6.3.1.4 Wave Solution for the Kirchhoff Love Piezoelectric Nanoplates

In this section, we will obtain a wave propagation solution for the smart piezoelectric FG nanoplates. In Section 6.2, the solution of the electric potential was considered to be dependent on the x coordinate and time t. However, in this part, we need the response of a nanoplate. Thus, the electric potential must be considered to be a function of both x and y coordinates as well as time (t). The electric potential of a piezoelectric nanoplate can be expressed as

$$\phi = \hat{\Phi} e^{i(\beta_1 x + \beta_2 y - \omega t)}. \tag{6.89}$$

Now, the final eigenvalue equation of the problem can be obtained by inserting Eqs. (1.9)–(1.11) and (6.89) in Eqs. (6.85)–(6.88). By doing this operation, we will obtain Eq. (1.3) again; however, the amplitude vector of the obtained equation must be considered to be

$$\Delta = [U, V, W, \hat{\Phi}]^T. \tag{6.90}$$

The components of the stiffness matrix of the aforementioned equation can be written in the following forms:

$$k_{11} = -\left(1 + \lambda^2\left(\beta_1^2 + \beta_2^2\right)\right)\left[A_{11}\beta_1^2 + A_{66}\beta_1\beta_2\right],$$

$$k_{12} = -\left(1 + \lambda^2\left(\beta_1^2 + \beta_2^2\right)\right)\left[A_{12} + A_{66}\right]\beta_1\beta_2,$$

$$k_{13} = i\left(1 + \lambda^2\left(\beta_1^2 + \beta_2^2\right)\right)\left[B_{11}\beta_1^3 + \left(B_{12} + 2B_{66}\right)\beta_1\beta_2^2\right],$$

$$k_{14} = \left(1 + \lambda^2\left(\beta_1^2 + \beta_2^2\right)\right)i\beta_1 A_{31}^e,$$

$$k_{22} = -\left(1 + \lambda^2\left(\beta_1^2 + \beta_2^2\right)\right)\left[A_{66}\beta_1^2 + A_{22}\beta_2^2\right],$$

$$k_{23} = i\left(1 + \lambda^2\left(\beta_1^2 + \beta_2^2\right)\right)\left[B_{22}\beta_2^3 + \left(B_{12} + 2B_{66}\right)\beta_1^2\beta_2\right],$$

$$k_{24} = \left(1 + \lambda^2\left(\beta_1^2 + \beta_2^2\right)\right)i\beta_2 A_{31}^e,$$

$$k_{33} = -\left(1 + \lambda^2\left(\beta_1^2 + \beta_2^2\right)\right)\left[D_{11}\beta_1^4 + 2\left(D_{12} + 2D_{66}\right)\beta_1^2\beta_2^2 + D_{22}\beta_2^4\right]$$
$$+ \left(1 + \mu^2\left(\beta_1^2 + \beta_2^2\right)\right)N^E\left(\beta_1^2 + \beta_2^2\right),$$

$$k_{34} = -\left(1 + \lambda^2\left(\beta_1^2 + \beta_2^2\right)\right)E_{31}^e\left(\beta_1^2 + \beta_2^2\right),$$

$$k_{44} = -\left(1 + \lambda^2\left(\beta_1^2 + \beta_2^2\right)\right)\left[F_{11}^e\left(\beta_1^2 + \beta_2^2\right) + F_{33}^e\right]. \tag{6.91}$$

The nonzero components of the mass matrix are the same as the arrays in Eq. (4.110).

6.3.2 Refined Sinusoidal Piezoelectric Nanoplates

Now, it is time to complete the wave dispersion formulations of piezoelectric FG nanobeams with respect to a refined higher-order plate model. In Section 6.3.1, the problem was expressed in mathematical language without considering the effect of shear deformation. To complete our investigations of the wave propagation phenomenon in the piezoelectric nanoplates, the problem's formulation will be introduced in subsequent sections of this chapter.

6.3.2.1 Equations of Motion of Piezoelectric Refined Sinusoidal Nanoplates

The equations of electric potential and the electric field components for a piezoelectric nanoplate were fully derived in the previous sections. Herein, these will not be reviewed again, and volunteers can review these by taking a brief look at Eqs. (6.57)–(6.60). Based on the refined shear deformation plate models, the variation of the strain energy of a plate can be expressed as

$$\delta\Pi_S = \int_V \left(\sigma_{xx}\delta\varepsilon_{xx} + \sigma_{yy}\delta\varepsilon_{yy} + \sigma_{xy}\delta\varepsilon_{xy} + \sigma_x\ \delta\varepsilon_x\ + \sigma_y\ \delta\varepsilon_y\right.$$
$$\left. - D_x\delta E_x - D_y\delta E_y - D\ \delta E\ \right)dV. \tag{6.92}$$

Equation (6.92) can be rewritten by inserting the nonzero strains of the refined plate model in it and using the electric fields obtained through Eqs. (6.58)–(6.60). The simplified form of the variation of strain energy can be expressed as follows:

$$\delta\Pi_S = \int_A \left(N_{xx}\delta\varepsilon_{xx}^0 - M_{xx}^b\delta\kappa_{xx}^b - M_{xx}^s\delta\kappa_{xx}^s + N_{yy}\delta\varepsilon_{yy}^0 - M_{yy}^b\delta\kappa_{yy}^b\right.$$
$$\left. - M_{yy}^s\delta\kappa_{yy}^s + N_{xy}\delta\varepsilon_{xy}^0 - M_{xy}^b\delta\kappa_{xy}^b - M_{xy}^s\delta\kappa_{xy}^s + Q_x\ \varepsilon_x^0\ + Q_y\ \varepsilon_y^0\right)dA$$
$$+ \int_{-\frac{h}{2}}^{\frac{h}{2}} \left(-D_x\cos(\xi z)\frac{\partial\delta\phi}{\partial x} - D_y\cos(\xi z)\frac{\partial\delta\phi}{\partial y} + D\ \xi\sin(\xi z)\delta\phi\right)dAdz, \tag{6.93}$$

where the stress resultants N_i, M_i^b, M_i^s, Q_j ($i = xx, yy, xy$ and $j = xz, yz$) can be assumed to be the same as the integrals in Eqs. (4.82) and (4.83). Now, it is time to formulate the variation of work done by external electrical loading for refined shear deformation plates. The variation of work done by external loading can be expressed as

$$\delta\Pi_W = \int_A N^E \left(\frac{\partial(w_b + w_s)}{\partial x} \frac{\partial\delta(w_b + w_s)}{\partial x} + \frac{\partial(w_b + w_s)}{\partial y} \frac{\partial\delta(w_b + w_s)}{\partial y} \right) dA. \tag{6.94}$$

It is worth mentioning that the variation of the kinetic energy of the piezoelectric plate is not different from that of an elastic one. Thus, the variation of the kinetic energy can be assumed to be as expressed in Eq. (4.84). By substituting Eqs. (6.93), (6.94) and (4.84) in Eq. (4.19), the Euler–Lagrange equations of the refined sinusoidal plates can be written as follows:

$$\frac{\partial N_{xx}}{\partial x} + \frac{\partial N_{xy}}{\partial y} = I_0 \frac{\partial^2 u}{\partial t^2} - I_1 \frac{\partial^3 w_b}{\partial x \partial t^2} - J_1 \frac{\partial^3 w_s}{\partial x \partial t^2}, \tag{6.95}$$

$$\frac{\partial N_{xy}}{\partial x} + \frac{\partial N_{yy}}{\partial y} = I_0 \frac{\partial^2 v}{\partial t^2} - I_1 \frac{\partial^3 w_b}{\partial y \partial t^2} - J_1 \frac{\partial^3 w_s}{\partial y \partial t^2}, \tag{6.96}$$

$$\frac{\partial^2 M_{xx}^b}{\partial x^2} + 2\frac{\partial^2 M_{xy}^b}{\partial x \partial y} + \frac{\partial^2 M_{yy}^b}{\partial y^2} - N^E \left(\frac{\partial^2(w_b + w_s)}{\partial x^2} + \frac{\partial^2(w_b + w_s)}{\partial y^2} \right)$$
$$= I_0 \frac{\partial^2(w_b + w_s)}{\partial t^2} + I_1 \left(\frac{\partial^3 u}{\partial x \partial t^2} + \frac{\partial^3 v}{\partial y \partial t^2} \right) - I_2 \left(\frac{\partial^4 w_b}{\partial x^2 \partial t^2} + \frac{\partial^4 w_b}{\partial y^2 \partial t^2} \right)$$
$$- J_2 \left(\frac{\partial^4 w_s}{\partial x^2 \partial t^2} + \frac{\partial^4 w_s}{\partial y^2 \partial t^2} \right), \tag{6.97}$$

$$\frac{\partial^2 M_{xx}^s}{\partial x^2} + 2\frac{\partial^2 M_{xy}^s}{\partial x \partial y} + \frac{\partial^2 M_{yy}^s}{\partial y^2} + \frac{\partial Q_x}{\partial x} + \frac{\partial Q_y}{\partial y} - N^E \left(\frac{\partial^2(w_b + w_s)}{\partial x^2} + \frac{\partial^2(w_b + w_s)}{\partial y^2} \right)$$
$$= I_0 \frac{\partial^2(w_b + w_s)}{\partial t^2} + J_1 \left(\frac{\partial^3 u}{\partial x \partial t^2} + \frac{\partial^3 v}{\partial y \partial t^2} \right) - J_2 \left(\frac{\partial^4 w_b}{\partial x^2 \partial t^2} + \frac{\partial^4 w_b}{\partial y^2 \partial t^2} \right)$$
$$- K_2 \left(\frac{\partial^4 w_s}{\partial x^2 \partial t^2} + \frac{\partial^4 w_s}{\partial y^2 \partial t^2} \right), \tag{6.98}$$

$$\int_A \left(\cos(\xi z)\frac{\partial D_x}{\partial x} + \cos(\xi z)\frac{\partial D_y}{\partial y} + \xi \sin(\xi z)D \right) dA = 0. \tag{6.99}$$

6.3.2.2 Nonlocal Strain Gradient Piezoelectricity for Refined Sinusoidal Nanoplates

In this section, the constitutive equations of a smart shear deformable nanoplate will be obtained to cover the lack of a shear-deformation-based model for the wave dispersion characteristics of piezoelectric FG nanoplates. As done in previous sections and by extending Eqs. (2.18) and (2.19) to refined sinusoidal nanoplates, the constitutive equations of a refined piezoelectric nanoplate can be expressed as

$$(1 - \mu^2\nabla^2)\sigma_{xx} = (1 - \lambda^2\nabla^2) \left[\tilde{c}_{11}\varepsilon_{xx} + \tilde{c}_{12}\varepsilon_{yy} - \tilde{e}_{31}E \right], \tag{6.100}$$

$$(1 - \mu^2\nabla^2)\sigma_{yy} = (1 - \lambda^2\nabla^2) \left[\tilde{c}_{12}\varepsilon_{xx} + \tilde{c}_{11}\varepsilon_{yy} - \tilde{e}_{31}E \right], \tag{6.101}$$

$$(1 - \mu^2\nabla^2)\sigma_{xy} = (1 - \lambda^2\nabla^2)\tilde{c}_{66}\varepsilon_{xy}, \tag{6.102}$$

$$\left(1 - \mu^2\nabla^2\right)\sigma_x \ = \left(1 - \lambda^2\nabla^2\right)\left[\tilde{c}_{55}\varepsilon_x \ - \tilde{e}_{15}E_x\right], \tag{6.103}$$

$$\left(1 - \mu^2\nabla^2\right)\sigma_y \ = \left(1 - \lambda^2\nabla^2\right)\left[\tilde{c}_{55}\varepsilon_y \ - \tilde{e}_{15}E_y\right], \tag{6.104}$$

$$\left(1 - \mu^2\nabla^2\right)D_x = \left(1 - \lambda^2\nabla^2\right)\left[\tilde{e}_{15}\varepsilon_x \ + \tilde{s}_{11}E_x\right], \tag{6.105}$$

$$\left(1 - \mu^2\nabla^2\right)D_y = \left(1 - \lambda^2\nabla^2\right)\left[\tilde{e}_{15}\varepsilon_y \ + \tilde{s}_{11}E_y\right], \tag{6.106}$$

$$\left(1 - \mu^2\nabla^2\right)D \ = \left(1 - \lambda^2\nabla^2\right)\left[\tilde{e}_{31}\varepsilon_{xx} + \tilde{e}_{31}\varepsilon_{yy} + \tilde{s}_{33}E \ \right]. \tag{6.107}$$

By integrating the above equations over the thickness of the nanoplate, the equations required to obtain the governing equations of the nanoplate can be written in the following forms:

$$
\left(1 - \mu^2\nabla^2\right)\begin{bmatrix} N_{xx} \\ N_{yy} \\ N_{xy} \end{bmatrix} = \left(1 - \lambda^2\nabla^2\right)\left(\begin{bmatrix} A_{11} & A_{12} & 0 \\ A_{12} & A_{22} & 0 \\ 0 & 0 & A_{66} \end{bmatrix} \begin{bmatrix} \frac{\partial u}{\partial x} \\ \frac{\partial v}{\partial y} \\ \frac{\partial u}{\partial y} + \frac{\partial v}{\partial x} \end{bmatrix} \right.
$$
$$
+ \begin{bmatrix} B_{11} & B_{12} & 0 \\ B_{12} & B_{22} & 0 \\ 0 & 0 & B_{66} \end{bmatrix} \begin{bmatrix} -\frac{\partial^2 w_b}{\partial x^2} \\ -\frac{\partial^2 w_b}{\partial y^2} \\ -2\frac{\partial^2 w_b}{\partial x\partial y} \end{bmatrix}
$$
$$
\left. + \begin{bmatrix} B_{11}^s & B_{12}^s & 0 \\ B_{12}^s & B_{22}^s & 0 \\ 0 & 0 & B_{66}^s \end{bmatrix} \begin{bmatrix} -\frac{\partial^2 w_s}{\partial x^2} \\ -\frac{\partial^2 w_s}{\partial y^2} \\ -2\frac{\partial^2 w_s}{\partial x\partial y} \end{bmatrix} + \begin{bmatrix} A_{31}^e \\ A_{31}^e \\ 0 \end{bmatrix}\phi - \begin{bmatrix} N^E \\ N^E \\ 0 \end{bmatrix} \right), \tag{6.108}
$$

$$
\left(1 - \mu^2\nabla^2\right)\begin{bmatrix} M_{xx}^b \\ M_{yy}^b \\ M_{xy}^b \end{bmatrix} = \left(1 - \lambda^2\nabla^2\right)\left(\begin{bmatrix} B_{11} & B_{12} & 0 \\ B_{12} & B_{22} & 0 \\ 0 & 0 & B_{66} \end{bmatrix} \begin{bmatrix} \frac{\partial u}{\partial x} \\ \frac{\partial v}{\partial y} \\ \frac{\partial u}{\partial y} + \frac{\partial v}{\partial x} \end{bmatrix} \right.
$$
$$
+ \begin{bmatrix} D_{11} & D_{12} & 0 \\ D_{12} & D_{22} & 0 \\ 0 & 0 & D_{66} \end{bmatrix} \begin{bmatrix} -\frac{\partial^2 w_b}{\partial x^2} \\ -\frac{\partial^2 w_b}{\partial y^2} \\ -2\frac{\partial^2 w_b}{\partial x\partial y} \end{bmatrix}
$$
$$
\left. + \begin{bmatrix} D_{11}^s & D_{12}^s & 0 \\ D_{12}^s & D_{22}^s & 0 \\ 0 & 0 & D_{66}^s \end{bmatrix} \begin{bmatrix} -\frac{\partial^2 w_s}{\partial x^2} \\ -\frac{\partial^2 w_s}{\partial y^2} \\ -2\frac{\partial^2 w_s}{\partial x\partial y} \end{bmatrix} + \begin{bmatrix} E_{31}^e \\ E_{31}^e \\ 0 \end{bmatrix}\phi - \begin{bmatrix} M_b^E \\ M_b^E \\ 0 \end{bmatrix} \right), \tag{6.109}
$$

$$
\left(1 - \mu^2\nabla^2\right)\begin{bmatrix} M_{xx}^s \\ M_{yy}^s \\ M_{xy}^s \end{bmatrix} = \left(1 - \lambda^2\nabla^2\right)\left(\begin{bmatrix} B_{11}^s & B_{12}^s & 0 \\ B_{12}^s & B_{22}^s & 0 \\ 0 & 0 & B_{66}^s \end{bmatrix} \begin{bmatrix} \frac{\partial u}{\partial x} \\ \frac{\partial v}{\partial y} \\ \frac{\partial u}{\partial y} + \frac{\partial v}{\partial x} \end{bmatrix} \right.
$$
$$
+ \begin{bmatrix} D_{11}^s & D_{12}^s & 0 \\ D_{12}^s & D_{22}^s & 0 \\ 0 & 0 & D_{66}^s \end{bmatrix} \begin{bmatrix} -\frac{\partial^2 w_b}{\partial x^2} \\ -\frac{\partial^2 w_b}{\partial y^2} \\ -2\frac{\partial^2 w_b}{\partial x\partial y} \end{bmatrix}
$$
$$
\left. + \begin{bmatrix} H_{11}^s & H_{12}^s & 0 \\ H_{12}^s & H_{22}^s & 0 \\ 0 & 0 & H_{66}^s \end{bmatrix} \begin{bmatrix} -\frac{\partial^2 w_s}{\partial x^2} \\ -\frac{\partial^2 w_s}{\partial y^2} \\ -2\frac{\partial^2 w_s}{\partial x\partial y} \end{bmatrix} + \begin{bmatrix} F_{31}^e \\ F_{31}^e \\ 0 \end{bmatrix}\phi - \begin{bmatrix} M_s^E \\ M_s^E \\ 0 \end{bmatrix} \right), \tag{6.110}
$$

$$\left(1 - \mu^2\nabla^2\right) \begin{bmatrix} Q_x \\ Q_y \end{bmatrix} = \left(1 - \lambda^2\nabla^2\right) \left(\begin{bmatrix} A_{44}^s & 0 \\ 0 & A_{55}^s \end{bmatrix} \begin{bmatrix} \frac{\partial w_s}{\partial x} \\ \frac{\partial w_s}{\partial y} \end{bmatrix} - A_{15}^e \begin{bmatrix} \frac{\partial \phi}{\partial x} \\ \frac{\partial \phi}{\partial y} \end{bmatrix} \right), \tag{6.111}$$

$$\left(1 - \mu^2\nabla^2\right) \int_A \begin{bmatrix} D_x \\ D_y \end{bmatrix} \cos(\xi z) dz = \left(1 - \lambda^2\nabla^2\right) \left(E_{15}^e \begin{bmatrix} \frac{\partial w_s}{\partial x} \\ \frac{\partial w_s}{\partial y} \end{bmatrix} + F_{11}^e \begin{bmatrix} \frac{\partial \phi}{\partial x} \\ \frac{\partial \phi}{\partial y} \end{bmatrix} \right), \tag{6.112}$$

$$\left(1 - \mu^2\nabla^2\right) \int_A D\ \xi\sin(\xi z) dz = \left(1 - \lambda^2\nabla^2\right) \left(A_{31}^e \left(\frac{\partial u}{\partial x} + \frac{\partial v}{\partial y} \right) \right.$$
$$\left. - E_{31}^e \left(\frac{\partial^2 w_b}{\partial x^2} + \frac{\partial^2 w_b}{\partial y^2} \right) - F_{31}^e \left(\frac{\partial^2 w_s}{\partial x^2} + \frac{\partial^2 w_s}{\partial y^2} \right) - F_{33}^e \phi \right). \tag{6.113}$$

In the above equations, some of the cross-sectional rigidities are similar to those introduced in Eqs. (6.79)–(6.82). The rest of the electromechanical resultants that are dependent on the shear approximation function of the refined plate theory can be evaluated as follows:

$$\begin{bmatrix} B_{11}^s & D_{11}^s & H_{11}^s \\ B_{12}^s & D_{12}^s & H_{12}^s \\ B_{66}^s & D_{66}^s & H_{66}^s \end{bmatrix} = \int_{-\frac{h}{2}}^{\frac{h}{2}} \begin{bmatrix} f(z) & zf(z) & f^2(z) \end{bmatrix} \begin{bmatrix} \tilde{c}_{11} \\ \tilde{c}_{12} \\ \tilde{c}_{66} \end{bmatrix} dz, \tag{6.114}$$

$$F_{31}^e = \int_{-\frac{h}{2}}^{\frac{h}{2}} f(z)\tilde{e}_{31}\xi\sin(\xi z) dz, \tag{6.115}$$

$$[A_{15}^e, E_{15}^e] = \int_{-\frac{h}{2}}^{\frac{h}{2}} \tilde{e}_{15}\cos(\xi z)[1, g(z)] dz. \tag{6.116}$$

It is worth mentioning that the electrical resultants of refined nanoplates can be written in the following form:

$$[N^E, M_b^E, M_s^E] = -\int_{-\frac{h}{2}}^{\frac{h}{2}} \tilde{e}_{31}\frac{2V}{h}[1, z, f(z)] dz. \tag{6.117}$$

6.3.2.3   Governing Equations of Piezoelectric Refined Sinusoidal Nanoplates

In this section, the final differential equations that control the wave dispersion responses of a piezoelectric nanoplate will be obtained. By substituting Eqs. (6.108)–(6.113) in Eqs. (6.95)–(6.99), the following equations can be obtained for the refined sinusoidal piezoelectric nanoplate:

$$\left(1 - \lambda^2\nabla^2\right) \left(A_{11}\frac{\partial^2 u}{\partial x^2} + (A_{12} + A_{66})\frac{\partial^2 v}{\partial x\partial y} + A_{66}\frac{\partial^2 u}{\partial y^2} - B_{11}\frac{\partial^3 w_b}{\partial x^3} \right.$$
$$\left. - (B_{12} + 2B_{66})\frac{\partial^3 w_b}{\partial x\partial y^2} - B_{11}^s\frac{\partial^3 w_s}{\partial x^3} - (B_{12}^s + 2B_{66}^s)\frac{\partial^3 w_s}{\partial x\partial y^2} + A_{31}^e\frac{\partial \phi}{\partial x} \right)$$
$$+ \left(1 - \mu^2\nabla^2\right) \left(-I_0\frac{\partial^2 u}{\partial t^2} + I_1\frac{\partial^3 w_b}{\partial x\partial t^2} + J_1\frac{\partial^3 w_s}{\partial x\partial t^2} \right) = 0, \tag{6.118}$$

$$(1 - \lambda^2 \nabla^2)\left(A_{22}\frac{\partial^2 v}{\partial y^2} + (A_{12} + A_{66})\frac{\partial^2 u}{\partial x \partial y} + A_{66}\frac{\partial^2 v}{\partial x^2} - B_{22}\frac{\partial^3 w_b}{\partial y^3}\right.$$

$$\left. - (B_{12} + 2B_{66})\frac{\partial^3 w_b}{\partial x^2 \partial y} - B_{22}^s\frac{\partial^3 w_s}{\partial y^3} - (B_{12}^s + 2B_{66}^s)\frac{\partial^3 w_s}{\partial x^2 \partial y} + A_{31}^e\frac{\partial \phi}{\partial y}\right)$$

$$+ (1 - \mu^2 \nabla^2)\left(-I_0\frac{\partial^2 v}{\partial t^2} + I_1\frac{\partial^3 w_b}{\partial y \partial t^2} + J_1\frac{\partial^3 w_s}{\partial y \partial t^2}\right) = 0, \tag{6.119}$$

$$(1 - \lambda^2 \nabla^2)\left(B_{11}\frac{\partial^3 u}{\partial x^3} + (B_{12} + 2B_{66})\frac{\partial^3 u}{\partial x \partial y^2} + B_{22}\frac{\partial^3 v}{\partial y^3} + (B_{12} + 2B_{66})\frac{\partial^3 v}{\partial x^2 \partial y}\right.$$

$$- D_{11}\frac{\partial^4 w_b}{\partial x^4} - 2(D_{12} + 2D_{66})\frac{\partial^4 w_b}{\partial x^2 \partial y^2} - D_{22}\frac{\partial^4 w_b}{\partial y^4} - D_{11}^s\frac{\partial^4 w_s}{\partial x^4} - 2(D_{12}^s + 2D_{66}^s)$$

$$\left. \times \frac{\partial^4 w_s}{\partial x^2 \partial y^2} - D_{22}^s\frac{\partial^4 w_s}{\partial y^4} + (E_{31}^e - A_{15}^e)\left(\frac{\partial^2 \phi}{\partial x^2} + \frac{\partial^2 \phi}{\partial y^2}\right)\right) + (1 - \mu^2 \nabla^2)$$

$$\times \left(-I_0\frac{\partial^2(w_b + w_s)}{\partial t^2} - I_1\left(\frac{\partial^3 u}{\partial x \partial t^2} + \frac{\partial^3 v}{\partial y \partial t^2}\right) + I_2\left(\frac{\partial^4 w_b}{\partial x^2 \partial t^2} + \frac{\partial^4 w_b}{\partial y^2 \partial t^2}\right)\right.$$

$$\left. + J_2\left(\frac{\partial^4 w_s}{\partial x^2 \partial t^2} + \frac{\partial^4 w_s}{\partial y^2 \partial t^2}\right) - N^E\left(\frac{\partial^2(w_b + w_s)}{\partial x^2} + \frac{\partial^2(w_b + w_s)}{\partial y^2}\right)\right) = 0, \tag{6.120}$$

$$(1 - \lambda^2 \nabla^2)\left(B_{11}^s\frac{\partial^3 u}{\partial x^3} + (B_{12}^s + 2B_{66}^s)\frac{\partial^3 u}{\partial x \partial y^2} + B_{22}^s\frac{\partial^3 v}{\partial y^3} + (B_{12}^s + 2B_{66}^s)\frac{\partial^3 v}{\partial x^2 \partial y} - D_{11}^s\frac{\partial^4 w_b}{\partial x^4}\right.$$

$$- 2(D_{12}^s + 2D_{66}^s)\frac{\partial^4 w_b}{\partial x^2 \partial y^2} - D_{22}^s\frac{\partial^4 w_b}{\partial y^4} - H_{11}^s\frac{\partial^4 w_s}{\partial x^4} - 2(H_{12}^s + 2H_{66}^s)\frac{\partial^4 w_s}{\partial x^2 \partial y^2}$$

$$\left. - H_{22}^s\frac{\partial^4 w_s}{\partial y^4} A_{44}^s\left(\frac{\partial^2(w_b + w_s)}{\partial x^2} + \frac{\partial^2(w_b + w_s)}{\partial y^2}\right) + (F_{31}^e - A_{15}^e)\left(\frac{\partial^2 \phi}{\partial x^2} + \frac{\partial^2 \phi}{\partial y^2}\right)\right)$$

$$+ (1 - \mu^2 \nabla^2)\left(-I_0\frac{\partial^2(w_b + w_s)}{\partial t^2} - J_1\left(\frac{\partial^3 u}{\partial x \partial t^2} + \frac{\partial^3 v}{\partial y \partial t^2}\right) + J_2\left(\frac{\partial^4 w_b}{\partial x^2 \partial t^2} + \frac{\partial^4 w_b}{\partial y^2 \partial t^2}\right)\right.$$

$$\left. + K_2\left(\frac{\partial^4 w_s}{\partial x^2 \partial t^2} + \frac{\partial^4 w_s}{\partial y^2 \partial t^2}\right) - N^E\left(\frac{\partial^2(w_b + w_s)}{\partial x^2} + \frac{\partial^2(w_b + w_s)}{\partial y^2}\right)\right) = 0, \tag{6.121}$$

$$(1 - \lambda^2 \nabla^2)\left(A_{31}^e\left(\frac{\partial u}{\partial x} + \frac{\partial v}{\partial y}\right) - E_{31}^e\left(\frac{\partial^2 w_b}{\partial x^2} + \frac{\partial^2 w_b}{\partial y^2}\right) - (F_{31}^e - E_{15}^e)\right.$$

$$\left. \times \left(\frac{\partial^2 w_s}{\partial x^2} + \frac{\partial^2 w_s}{\partial y^2}\right) + F_{11}^e\left(\frac{\partial^2 \phi}{\partial x^2} + \frac{\partial^2 \phi}{\partial y^2}\right) - F_{33}^e\phi\right) = 0. \tag{6.122}$$

6.3.2.4 Wave Solution for the Refined Sinusoidal Piezoelectric Nanoplates

Finally, it is time to solve the governing equations of a refined shear deformable smart nanostructure in order to obtain the phase velocity values of piezoelectric nanoplates. For this purpose, the electric potential obtained via Eq. (6.89) and Eqs. (1.12)–(1.15) will be used and substituted in Eqs. (6.118)–(6.122) to obtain an eigenvalue equation similar to

Eq. (1.3). It must be mentioned that the only difference in this situation in comparison with the case using a classical piezoelectric nanoplate is that the amplitude vector of the refined nanoplates differs from that introduced in Eq. (6.90). The amplitude vector of refined piezoelectric nanoplate can be written as

$$\Delta = \left[U, V, W_b, W_s, \hat{\Phi} \right]^T. \tag{6.123}$$

The components of the stiffness matrix of the refined piezoelectric nanoplates can be expressed as

$$k_{11} = - \left(1 + \lambda^2 (\beta_1^2 + \beta_2^2)\right) \left[A_{11}\beta_1^2 + A_{66}\beta_1\beta_2 \right],$$

$$k_{12} = - \left(1 + \lambda^2 (\beta_1^2 + \beta_2^2)\right) \left[A_{12} + A_{66} \right] \beta_1\beta_2,$$

$$k_{13} = i \left(1 + \lambda^2 (\beta_1^2 + \beta_2^2)\right) \left[B_{11}\beta_1^3 + (B_{12} + 2B_{66})\beta_1\beta_2^2 \right],$$

$$k_{14} = i \left(1 + \lambda^2 (\beta_1^2 + \beta_2^2)\right) \left[B_{11}^s\beta_1^3 + (B_{12}^s + 2B_{66}^s)\beta_1\beta_2^2 \right],$$

$$k_{15} = \left(1 + \lambda^2 (\beta_1^2 + \beta_2^2)\right) i\beta_1 A_{31}^e,$$

$$k_{22} = - \left(1 + \lambda^2 (\beta_1^2 + \beta_2^2)\right) \left[A_{66}\beta_1^2 + A_{22}\beta_2^2 \right],$$

$$k_{23} = i \left(1 + \lambda^2 (\beta_1^2 + \beta_2^2)\right) \left[B_{22}\beta_2^3 + (B_{12} + 2B_{66})\beta_1^2\beta_2 \right],$$

$$k_{24} = i \left(1 + \lambda^2 (\beta_1^2 + \beta_2^2)\right) \left[B_{22}^s\beta_2^3 + (B_{12}^s + 2B_{66}^s)\beta_1^2\beta_2 \right],$$

$$k_{25} = \left(1 + \lambda^2 (\beta_1^2 + \beta_2^2)\right) i\beta_2 A_{31}^e,$$

$$k_{33} = - \left(1 + \lambda^2 (\beta_1^2 + \beta_2^2)\right) \left[D_{11}\beta_1^4 + 2(D_{12} + 2D_{66})\beta_1^2\beta_2^2 + D_{22}\beta_2^4 \right]$$
$$\quad + \left(1 + \mu^2 (\beta_1^2 + \beta_2^2)\right) N^E (\beta_1^2 + \beta_2^2),$$

$$k_{34} = - \left(1 + \lambda^2 (\beta_1^2 + \beta_2^2)\right) \left[D_{11}^s\beta_1^4 + 2(D_{12}^s + 2D_{66}^s)\beta_1^2\beta_2^2 + D_{22}^s\beta_2^4 \right]$$
$$\quad + \left(1 + \mu^2 (\beta_1^2 + \beta_2^2)\right) N^E (\beta_1^2 + \beta_2^2),$$

$$k_{35} = - \left(1 + \lambda^2 (\beta_1^2 + \beta_2^2)\right) \left(E_{31}^e - A_{15}^e \right) (\beta_1^2 + \beta_2^2),$$

$$k_{44} = - \left(1 + \lambda^2 (\beta_1^2 + \beta_2^2)\right) \left[H_{11}^s\beta_1^4 + 2(H_{12}^s + 2H_{66}^s)\beta_1^2\beta_2^2 + H_{22}^s\beta_2^4 \right.$$
$$\quad \left. + A_{44}^s(\beta_1^2 + \beta_2^2) \right] + \left(1 + \mu^2 (\beta_1^2 + \beta_2^2)\right) N^E (\beta_1^2 + \beta_2^2),$$

$$k_{45} = - \left(1 + \lambda^2 (\beta_1^2 + \beta_2^2)\right) \left(F_{31}^e - A_{15}^e \right) (\beta_1^2 + \beta_2^2),$$

$$k_{55} = - \left(1 + \lambda^2 (\beta_1^2 + \beta_2^2)\right) \left[F_{11}^e (\beta_1^2 + \beta_2^2) + F_{33}^e \right]. \tag{6.124}$$

It is worth mentioning that the nonzero components of the mass matrix are defined in Eq. (4.110).

6.3.3   Numerical Results for Piezoelectric Nanoplates

In this section, a group of numerical examples will be presented to show the effects of different terms on the piezoelectrically affected wave dispersion curves of smart FG nanoplates. As done in the section on the results for nanobeams (Section 6.2.3), in this section, illustrations will be presented for phase speeds of the propagated waves. It will be seen that the effect of wave number is more observable in the wave propagation plots of a piezoelectric nanoplate compared with a nanobeam of the same material. Indeed, in the analysis of a nanoplate, two wave numbers are involved, namely the axial one and the transverse one. Therefore, it is natural to see that the wave number can play a more observable role in determining the wave dispersion paths of FG nanoplates compared with nanobeams. On the other hand, the effect of the existence of porosities in the smart FGM, which is considered in constructing the nanostructure, on the wave responses of FG nanoplates is shown in a graphical example.

Figure 6.6 is presented in order to emphasize the crucial role of the scale factor on phase velocity curves of smart piezoelectric FG nanoplates when subjected to different types of electric voltages. It is clear that at all values of the scale factor, the phase velocity increases gradually at first, and after this initial rise, the behavior of the waves can be classified into three types. The first type is seen in the case where the softening effect of the nonlocal parameter surmounts the hardening effect of the length-scale parameter. In this case, the phase velocity decreases in a continuous manner. In the next case, the phase speed remains constant because there is a similarity between the effects of couple stresses and those of nonlocal parameters ($c = 1$). Finally, the phase speed increases a second time when the scale factor exceeds one ($c > 1$). In this case, the effects of couple stresses on the surface of the nanostructure are assumed to be more powerful than those of the nonlocal strains. However, in all these cases, the phase velocity of the smart nanoplates can be decreased by increasing the voltage intensity. Actually, the increment of the applied voltage can generate external heat in the medium, which leads to a reduction in the stiffness of the nanostructure, and due to this reality, the dynamic response of the system decreases gradually. One should

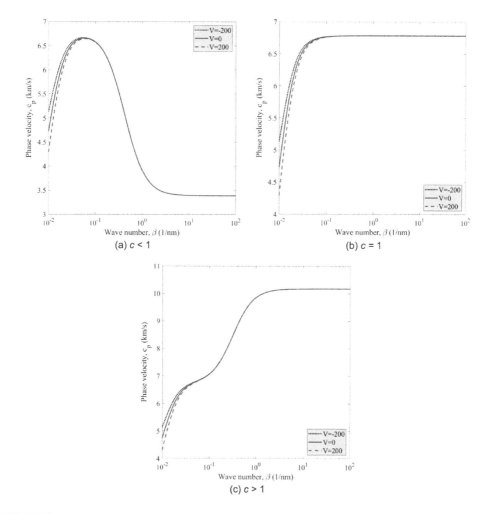

FIGURE 6.6
Variation of phase velocity with wave number at various amounts of applied voltage and scale factors ($p = 2$).

pay attention to the fact that changing the applied voltage cannot be a useful approach for changing the phase velocity of the piezoelectric nanoplates. Clearly, the wave speed can be influenced by the electric voltage at the wave numbers less than 0.05 (1/nm).

Moreover, the significance of using various wave numbers in determining the effect of efficiency of the applied electric voltage on the variations of phase speeds of the waves scattered in a piezoelectric nanoplate can be seen in Figure 6.7. It is obvious that the phase velocity cannot be greatly influenced by changing the electric voltage at even wave

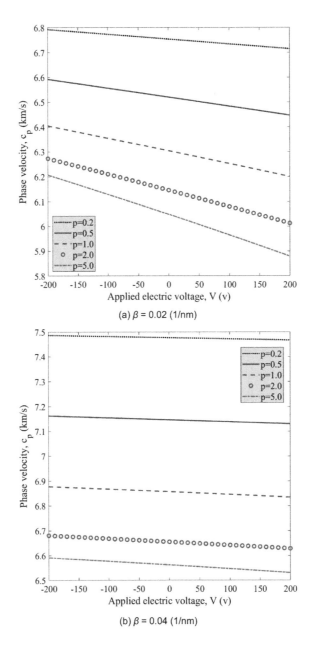

FIGURE 6.7
Variation of phase velocity with applied electric voltage at various gradient indices and wave numbers ($c > 1$).

numbers close to 0.05 (1/nm). This observation reveals that it is of great importance to select suitable wave numbers when one intends to control the speed of the dispersed wave through a piezoelectric medium. As in the case of FG nanobeams, here it can be figured out that the phase speed of the wave can be decreased when a higher gradient index is used. The physical reason for this is that by increasing this index, the material's behavior will be closer to the part that has smaller stiffness. Thus, it is not strange to observe such a reducing impact on increasing the gradient index.

The effect of nonlocal parameter on the phase velocity of the smart piezoelectric FG nanoplates at a particular desired wave number is depicted in Figure 6.8. It is clear that the effect of the nonlocal parameter is a gradual decreasing effect as estimated before. One should pay attention to the fact that this decreasing effect can be better observed at high values of the nonlocal parameter (i.e., small scale factor). It is also clear that the phase velocity of the nanoplate can be changed by tuning the applied electric voltage. Just as in our previous observations, here it can be again seen that the dynamic response of the system can be reduced by intensifying the magnitude of the applied voltage. The reason for this fact is the decreasing effect of the heat generated from increasing of the voltage on the stiffness of the nanostructure.

On the other hand, the effect of length-scale parameter on the phase velocity of the nanoplate at various applied electric voltages can be observed in Figure 6.9. Clearly, it can be seen that the phase speed of the waves scattered in a medium can be increased by increasing the length-scale parameter of the nanostructure (i.e., large scale factor). Also, it is observable that the effect of changing the electric voltage cannot be seen in a tangible manner. Indeed, the difference in various parameters at different applied voltages can be seen by enlarging the diagram. The main reason for this phenomenon is that in plate-type structures, two directions must be covered, and this is difficult when compared with beams which have only one axial direction.

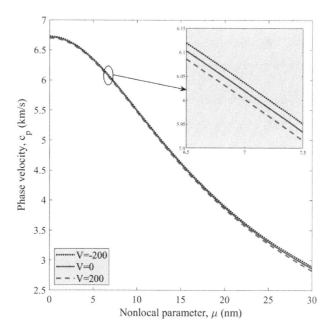

FIGURE 6.8

Variation of phase velocity with nonlocal parameter at various amounts of applied electric voltage ($\beta = 0.05$ (1/nm), $p = 2$, $\lambda = 1$ nm).

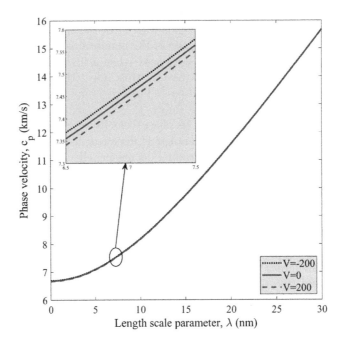

FIGURE 6.9

Variation of phase velocity with length-scale parameter at various amounts of applied electric voltage ($\beta = 0.05$ (1/nm), $p = 2$, $\mu = 1$ nm).

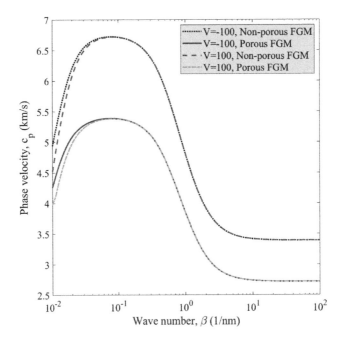

FIGURE 6.10

Variation of phase velocity with wave number at various amounts of applied electric voltage in both porous and nonporous FGMs ($c < 1$, $p = 2$).

Finally, we will discuss the effect of the existence of porosities in a smart piezoelectric FGM, which is chosen for the fabrication of a nanoplate, on the wave dispersion curves of the FG nanoplate. As investigated and discussed in Chapter 5, the existence of porosities in the constituent material of a nanostructure can result in a decrease in the stiffness of the continuous system because of the defects that can be found in the porous materials in comparison with the nonporous ones. On the other hand, it can be easily understood that the dynamic response of a system (e.g., phase velocity of the propagated waves) depends on the stiffness of that studied system. Thus, it is natural to see that the phase velocity of the nanostructure shifts downward in the case of porous smart materials. This fact is illustrated in Figure 6.10 plotting the variations of phase velocity with wave number for various types of porous and nonporous smart FGMs. It can be seen that the phase velocity is the lowest for porous nanoplates with positive applied electric voltage. It is worth mentioning that the effect of existence of porosities in a smart material can be seen at all wave numbers. However, the effect of changing the voltage can be observed only at wave numbers less than 0.05 (1/nm).

7

Wave Dispersion Characteristics of Magnetostrictive Nanostructures

In this chapter, we will derive the equations of motion of wave propagation in nanosize beams and plates considering the effects of magnetostrictive materials to demonstrate wave dispersion behaviors of nanostructures consisting of smart materials. The derivation of the equations of motion is again conducted by means of the well-known Hamilton's principle. The effect of the magnetostrictive property of the constituent material of the nanostructures will be formulated by introducing a velocity feedback control system. Illustrations reveal that the effects of the velocity feedback gain can be dramatically changed by varying the wave number. It will be shown that the mechanical response of the system will be intensified when the velocity feedback gain of the feedback control system is amplified.

7.1   Magnetostriction and Magnetostrictive Materials

Magnetostriction is a phenomenon which can happen in the most of the ferromagnetic materials which are sensitive to variations in magnetic field [140]. In such materials, the small regions consisting of randomly oriented grains will change their orientation and acquire a shape different from the initial one whenever they are subjected to a variable magnetic field. This phenomenon was discovered by the well-known English physicist James Joule in the nineteenth century (1842) [141]. He observed an elongation in iron, which is a ferromagnetic material, by placing it in an environment of varying magnetic field. In the case reported by Joule, negative magnetostriction phenomenon was observed. However, magnetostrictive materials can show both positive and negative magnetostriction. Actually, in magnetostrictive materials, the reorientation of the magnetic regions will result in the generation of strain in the material which can be either a tension strain or a compression one. In other words, positive magnetostriction causes extension, whereas negative magnetostriction causes compression. One should pay attention to the fact that during the process of reorientation of the ferromagnetic regions of the smart magnetostrictive material in routine operating conditions, the total volume of the material can be presumed to be unchangeable because of the reduction that happens in the cross-section of the material structure. Besides, the reorientation process will continued up to the saturation point of the material. The saturation point is the point at which all the regions are oriented in the same direction and no region can be found with an orientation different from that of the others [141].

During the magnetostriction process, different strain–magnetic field relations can be observed in the material. The first step is a small increase in the magnetic field which produces very small strains depending on the homogeneity of the material structure. Next, the most interesting stage of the magnetostriction for engineers occurs, which is the increasing of the magnetic field in a way that produces a linear strain–magnetic field relation in the magnetostrictive medium. This region is of great interest because the material's behavior can be controlled in it, and due to this reality, most of the smart devices made from magnetostrictive materials will be designed to work within this region [141]. Afterward, the final increasing of the magnetic field happens which is generally not able to produce notable strains in the material because of the reorientation of most of the ferromagnetic regions in the previous step. In this step, a nonlinear behavior can be seen from the magnetostrictive material. Because of the strains being negligible in this step, the magnetostrictive material reaches its saturation point at the end of this step.

On the other hand, the magnetostrictive materials have different behaviors and complexities which must be discussed here for the sake of completeness. In fact, magnetostrictive materials will be subjected to various operating conditions (e.g., varying magnetic field), and because of these changes in the working conditions, these materials exhibit various behaviors. One of the most significant effects which can be observed in magnetostrictive materials is the **Joule effect** which corresponds to the elongation (positive magnetostriction) or contraction (negative magnetostriction) of any desired structure when it works under a variable magnetic field. This aspect of magnetostrictive materials makes such materials suitable choices in the cases where the use of an actuator is required. It is worth mentioning that the strain produced in a magnetostrictive device can be omitted when the magnetic field is set to zero. Due to this field-dependent characteristic of the strain in magnetostrictive materials, actuators made from these smart materials are capable of being used for various purposes in modern systems [141]. Furthermore, **Villari effect** states that in the case where an initial prestress is applied on a magnetostrictive material, the smart material will experience a variation in the magnetic flux density, and this flux density gradient results in magnetic flux flow. This effect is similar to the Joule effect from the reversibility point of view. In addition, this effect enables engineers to select magnetostrictive materials to construct sensors for different applications [141]. As stated in the previous paragraphs, the Joule effect is related to the axial displacement of the magnetostrictive material. Similarly, **Wiedemann effect** denotes that variations in the magnetic field applied to the magnetostrictive material can generate torsional displacements which causes shear strains, either positive or negative, in magnetostrictive media. The physical explanations for this effect are similar to those related to the Joule effect: they are obtained by replacing the axial strain with the shear one [141]. On the other hand, the volume of some of the magnetostrictive materials can be changed when the material is subjected to a magnetic field. This effect is called as **Barret effect** in literature [141]. In some magnetostrictive materials, Barret effect is small enough to be assumed as a negligible effect. For instance, nickel will experience a change of about 10^{-7} in its volume when it works under a magnetic field of 80 kA/m [141].

In conclusion, it is proven that magnetostrictive materials are good candidates for use in sensor and actuator devices in complex systems because of their different behaviors when various external excitations are applied to these materials. Therefore, the control procedure is of great importance in such materials. In the following section, we will discuss about the most effective way to control magnetostrictive structures in a systematic manner.

7.2   Velocity Feedback Control System

Herein, we will introduce a feedback control system to govern the magnetic field, which affects the behaviors of any desired continuous system made from magnetostrictive materials. Using a closed-loop velocity control system, the magnetic field is considered to be a function of the coil current [142]. Hence, the magnetic field applied on a continuous element can be expressed as

$$H(\mathbf{x}, t) = k_c I(\mathbf{x}, t), \tag{7.1}$$

where $\mathbf{x}$ is the vector that includes the dependent spatial coordinates that are required for an accurate modeling of the continuous system. For instance, this vector includes x variable in the Cartesian coordinate system for a beam, whereas it includes y as well as x for a plate in the same coordinate system. Also, t is the time variable to cover the temporal distribution, and k_c is the coil constant. It is worth mentioning that the coil current, $I(\mathbf{x}, t)$, can be considered as a function of the velocity of the system in the following form:

$$I(\mathbf{x}, t) = c(t) \frac{\partial u\ (\mathbf{x}, t)}{\partial t} \tag{7.2}$$

in which u is the deflection of the system which is employed to compute the flexural speed of the continuous system. In addition, $c(t)$ is the control gain, which is a time-dependent variable. Now, it is time to present the formulation of the coil constant as a multivariable function of the involved variants. This term is related to some variables such as the width of the coil (b_c), the coil's radius (r_c) and the number of coils (n_c). The coil constant can therefore be expressed in the following form:

$$k_c = \frac{n_c}{\sqrt{b_c^2 + 4r_c^2}}. \tag{7.3}$$

7.3 Constitutive Equations of Magnetostrictive Nanostructures

In this section, the constitutive equations of smart magnetostrictive solids will be presented to cover the effects of magnetic field on the stress–strain relationships of such smart materials. Following the constitutive equations presented by researchers who worked on the mechanical responses of magnetostrictive structures, the constitutive equation of the smart magnetostrictive isotropic solids can be written in the following index form [142–144]:

$$\sigma_{ij} = C_{ijkl}\varepsilon_{kl} - e_{ijk}H_k \tag{7.4}$$

in which σ_{ij}, C_{ijkl}, ε_{kl}, e_{ijk} and H_k correspond to the components of the second-order stress tensor, fourth-order elasticity tensor, second-order strain tensor, third-order magnetostrictive coupling tensor and magnetic field vector, respectively. According to the fundamentals of the continuum mechanics, the above constitutive equation can be rewritten for the plane stress situation in the following expanded form:

$$
\begin{bmatrix} \sigma_{xx} \\ \sigma_{yy} \\ \tau_{xy} \\ \tau_x \\ \tau_y \end{bmatrix} =
\begin{bmatrix} C_{11} & C_{12} & 0 & 0 & 0 \\ C_{12} & C_{22} & 0 & 0 & 0 \\ 0 & 0 & C_{66} & 0 & 0 \\ 0 & 0 & 0 & C_{44} & 0 \\ 0 & 0 & 0 & 0 & C_{55} \end{bmatrix}
\begin{bmatrix} \varepsilon_{xx} \\ \varepsilon_{yy} \\ \varepsilon_{xy} \\ \varepsilon_x \\ \varepsilon_y \end{bmatrix} -
\begin{bmatrix} 0 & 0 & e_{31} & 0 & 0 \\ 0 & 0 & e_{32} & 0 & 0 \\ 0 & 0 & e_{34} & 0 & 0 \\ 0 & 0 & 0 & 0 & 0 \\ 0 & 0 & 0 & 0 & 0 \end{bmatrix}
\begin{bmatrix} 0 \\ 0 \\ H \\ 0 \\ 0 \end{bmatrix} \tag{7.5}
$$

in which C_{ij}'s can be obtained using those reported in Chapter 4 for both beam-type and plate-type elements. In addition, the magnetostrictive coupling coefficients e_{ij}'s can be defined as follows:

$$e_{31} = \tilde{e}_{31} \cos^2 \theta + \tilde{e}_{32} \sin^2 \theta,$$
$$e_{32} = \tilde{e}_{31} \sin^2 \theta + \tilde{e}_{32} \cos^2 \theta,$$
$$e_{34} = \left(\tilde{e}_{31} - \tilde{e}_{32}\right) \sin \theta \cos \theta \tag{7.6}$$

Now, the obtained constitutive equation must be transformed to the nanoscale in order to apply them for the dynamic problem of a nanostructure. It was demonstrated in Chapter 2 that the nonlocal strain gradient elasticity is the best alternative for considering the effects of size dependency of the nanosize elements. Due to this fact, the constitutive equation, given in Eq. (7.5), will be combined with the initial form of the nonlocal strain gradient elasticity theory (see Eq. (2.17)) to achieve the nonlocal strain gradient constitutive equations of magnetostrictive solids.

In the case of investigating Euler–Bernoulli nanobeams, the nonlocal strain gradient constitutive equations can be expressed in the following form:

$$\left(1 - \mu^2 \nabla^2\right) \sigma_{xx} = \left(1 - \lambda^2 \nabla^2\right) \left[C_{11} \varepsilon_{xx} - e_{31} H \right]. \tag{7.7}$$

Integrating the above equation over the cross-sectional area of the nanobeam, the stress resultants can be obtained in the following forms:

$$\left(1 - \mu^2 \nabla^2\right) N = \left(1 - \lambda^2 \nabla^2\right) \left[A_{xx} \frac{\partial u}{\partial x} - B_{xx} \frac{\partial^2 w}{\partial x^2} - \Xi_{31} H \right], \tag{7.8}$$

$$\left(1 - \mu^2 \nabla^2\right) M = \left(1 - \lambda^2 \nabla^2\right) \left[B_{xx} \frac{\partial u}{\partial x} - D_{xx} \frac{\partial^2 w}{\partial x^2} - \Pi_{31} H \right]. \tag{7.9}$$

Similarly, the constitutive equations of refined higher-order nanobeams can be obtained as follows when the effects of shear deflection are included:

$$\begin{aligned}
\left(1 - \mu^2 \nabla^2\right) \sigma_{xx} &= \left(1 - \lambda^2 \nabla^2\right) \left[C_{11} \varepsilon_{xx} - e_{31} H \right], \\
\left(1 - \mu^2 \nabla^2\right) \sigma_x &= \left(1 - \lambda^2 \nabla^2\right) C_{44} \varepsilon_x .
\end{aligned} \tag{7.10}$$

Now, on integrating over the nanobeam's cross-section, one can obtain the following resultants for refined shear deformable nanosize beams:

$$\left(1 - \mu^2 \nabla^2\right) N = \left(1 - \lambda^2 \nabla^2\right) \left[A_{xx} \frac{\partial u}{\partial x} - B_{xx} \frac{\partial^2 w_b}{\partial x^2} - B_{xx}^s \frac{\partial^2 w_s}{\partial x^2} - \Xi_{31} H \right], \tag{7.11}$$

$$\left(1 - \mu^2 \nabla^2\right) M^b = \left(1 - \lambda^2 \nabla^2\right) \left[B_{xx} \frac{\partial u}{\partial x} - D_{xx} \frac{\partial^2 w_b}{\partial x^2} - D_{xx}^s \frac{\partial^2 w_s}{\partial x^2} - \Pi_{31} H \right], \tag{7.12}$$

$$\left(1 - \mu^2 \nabla^2\right) M^s = \left(1 - \lambda^2 \nabla^2\right) \left[D_{xx} \frac{\partial u}{\partial x} - D_{xx}^s \frac{\partial^2 w_b}{\partial x^2} - H_{xx}^s \frac{\partial^2 w_s}{\partial x^2} - \Gamma_{31} H \right], \tag{7.13}$$

$$\left(1 - \mu^2 \nabla^2\right) Q_x = \left(1 - \lambda^2 \nabla^2\right) A_x^s \frac{\partial w_s}{\partial x}. \tag{7.14}$$

In the above equations, the cross-sectional rigidities can be achieved using the following formulae:

$$[A_{xx}, B_{xx}, B_{xx}^s, D_{xx}, D_{xx}^s, H_{xx}^s] = \int_A C_{11} \left[1, z, f(z), z^2, zf(z), f^2(z)\right] dA, \tag{7.15}$$

$$A_x^s = \int_A C_{44} g^2(z) dA. \tag{7.16}$$

Also, the magnetostrictive constants can be obtained from

$$[\Xi_{31}, \Pi_{31}, \Gamma_{31}] = \int_A e_{31} \left[1, z, f(z)\right] dA. \tag{7.17}$$

Next, we will derive the constitutive equations of smart magnetostrictive nanoplates. The constitutive equation for magnetostrictive plates is presented as Eq. (7.5). Ignoring shear deformation, the constitutive equations for Kirchhoff–Love nanoplates can be expressed as follows:

$$
\left(1 - \mu^2 \nabla^2\right)
\begin{bmatrix} \sigma_{xx} \\ \sigma_{yy} \\ \tau_{xy} \end{bmatrix}
= \left(1 - \lambda^2 \nabla^2\right)
\left(
\begin{bmatrix} C_{11} & C_{12} & 0 \\ C_{12} & C_{22} & 0 \\ 0 & 0 & C_{66} \end{bmatrix}
\begin{bmatrix} \varepsilon_{xx} \\ \varepsilon_{yy} \\ \varepsilon_{xy} \end{bmatrix}
-
\begin{bmatrix} 0 & 0 & e_{31} \\ 0 & 0 & e_{32} \\ 0 & 0 & e_{34} \end{bmatrix}
\begin{bmatrix} 0 \\ 0 \\ H \end{bmatrix}
\right).
$$
(7.18)

Now, the stress resultants of such nanosize plates can be calculated on integrating the aforementioned formula over the thickness of the nanosize plate when this expression is multiplied by 1 and z as follows:

$$
\left(1 - \mu^2 \nabla^2\right)
\begin{bmatrix} N_{xx} \\ N_{yy} \\ N_{xy} \end{bmatrix}
= \left(1 - \lambda^2 \nabla^2\right)
\left(
\begin{bmatrix} A_{11} & A_{12} & 0 \\ A_{12} & A_{22} & 0 \\ 0 & 0 & A_{66} \end{bmatrix}
\begin{bmatrix} \frac{\partial u}{\partial x} \\ \frac{\partial v}{\partial y} \\ \frac{\partial u}{\partial y} + \frac{\partial v}{\partial x} \end{bmatrix}
\right.
$$
$$
\left.
+
\begin{bmatrix} B_{11} & B_{12} & 0 \\ B_{12} & B_{22} & 0 \\ 0 & 0 & B_{66} \end{bmatrix}
\begin{bmatrix} -\frac{\partial^2 w}{\partial x^2} \\ -\frac{\partial^2 w}{\partial y^2} \\ -2\frac{\partial^2 w}{\partial x \partial y} \end{bmatrix}
-
\begin{bmatrix} \Xi_{31} \\ \Xi_{32} \\ \Xi_{34} \end{bmatrix} H
\right),
$$
(7.19)

$$
\left(1 - \mu^2 \nabla^2\right)
\begin{bmatrix} M_{xx} \\ M_{yy} \\ M_{xy} \end{bmatrix}
= \left(1 - \lambda^2 \nabla^2\right)
\left(
\begin{bmatrix} B_{11} & B_{12} & 0 \\ B_{12} & B_{22} & 0 \\ 0 & 0 & B_{66} \end{bmatrix}
\begin{bmatrix} \frac{\partial u}{\partial x} \\ \frac{\partial v}{\partial y} \\ \frac{\partial u}{\partial y} + \frac{\partial v}{\partial x} \end{bmatrix}
\right.
$$
$$
\left.
+
\begin{bmatrix} D_{11} & D_{12} & 0 \\ D_{12} & D_{22} & 0 \\ 0 & 0 & D_{66} \end{bmatrix}
\begin{bmatrix} -\frac{\partial^2 w}{\partial x^2} \\ -\frac{\partial^2 w}{\partial y^2} \\ -2\frac{\partial^2 w}{\partial x \partial y} \end{bmatrix}
-
\begin{bmatrix} \Pi_{31} \\ \Pi_{32} \\ \Pi_{34} \end{bmatrix} H
\right).
$$
(7.20)

Through a similar procedure, the nonlocal strain gradient constitutive equations of smart magnetostrictive nanoplates with respect to the effects of shear deflection can be presented in the following form:

$$
\left(1 - \mu^2 \nabla^2\right)
\begin{bmatrix} \sigma_{xx} \\ \sigma_{yy} \\ \tau_{xy} \\ \tau_x \\ \tau_y \end{bmatrix}
= \left(1 - \lambda^2 \nabla^2\right)
\left(
\begin{bmatrix}
C_{11} & C_{12} & 0 & 0 & 0 \\
C_{12} & C_{22} & 0 & 0 & 0 \\
0 & 0 & C_{66} & 0 & 0 \\
0 & 0 & 0 & C_{44} & 0 \\
0 & 0 & 0 & 0 & C_{55}
\end{bmatrix}
\begin{bmatrix} \varepsilon_{xx} \\ \varepsilon_{yy} \\ \varepsilon_{xy} \\ \varepsilon_x \\ \varepsilon_y \end{bmatrix}
\right.
$$
$$
\left.
-
\begin{bmatrix}
0 & 0 & e_{31} & 0 & 0 \\
0 & 0 & e_{32} & 0 & 0 \\
0 & 0 & e_{34} & 0 & 0 \\
0 & 0 & 0 & 0 & 0 \\
0 & 0 & 0 & 0 & 0
\end{bmatrix}
\begin{bmatrix} 0 \\ 0 \\ H \\ 0 \\ 0 \end{bmatrix}
\right).
$$
(7.21)

Henceforward, the constitutive equations of smart shear deformable nanoplates can be expressed as follows in terms of the nanoplates' displacement field:

$$\left(1 - \mu^2 \nabla^2\right) \begin{bmatrix} N_{xx} \\ N_{yy} \\ N_{xy} \end{bmatrix} = \left(1 - \lambda^2 \nabla^2\right) \left(\begin{bmatrix} A_{11} & A_{12} & 0 \\ A_{12} & A_{22} & 0 \\ 0 & 0 & A_{66} \end{bmatrix} \begin{bmatrix} \frac{\partial u}{\partial x} \\ \frac{\partial v}{\partial y} \\ \frac{\partial u}{\partial y} + \frac{\partial v}{\partial x} \end{bmatrix} \right.$$
$$+ \begin{bmatrix} B_{11} & B_{12} & 0 \\ B_{12} & B_{22} & 0 \\ 0 & 0 & B_{66} \end{bmatrix} \begin{bmatrix} -\frac{\partial^2 w_b}{\partial x^2} \\ -\frac{\partial^2 w_b}{\partial y^2} \\ -2\frac{\partial^2 w_b}{\partial x \partial y} \end{bmatrix} + \begin{bmatrix} B_{11}^s & B_{12}^s & 0 \\ B_{12}^s & B_{22}^s & 0 \\ 0 & 0 & B_{66}^s \end{bmatrix} \begin{bmatrix} -\frac{\partial^2 w_s}{\partial x^2} \\ -\frac{\partial^2 w_s}{\partial y^2} \\ -2\frac{\partial^2 w_s}{\partial x \partial y} \end{bmatrix} - \begin{bmatrix} \Xi_{31} \\ \Xi_{32} \\ \Xi_{34} \end{bmatrix} H \right),$$

$$(7.22)$$

$$\left(1 - \mu^2 \nabla^2\right) \begin{bmatrix} M_{xx}^b \\ M_{yy}^b \\ M_{xy}^b \end{bmatrix} = \left(1 - \lambda^2 \nabla^2\right) \left(\begin{bmatrix} B_{11} & B_{12} & 0 \\ B_{12} & B_{22} & 0 \\ 0 & 0 & B_{66} \end{bmatrix} \begin{bmatrix} \frac{\partial u}{\partial x} \\ \frac{\partial v}{\partial y} \\ \frac{\partial u}{\partial y} + \frac{\partial v}{\partial x} \end{bmatrix} \right.$$
$$+ \begin{bmatrix} D_{11} & D_{12} & 0 \\ D_{12} & D_{22} & 0 \\ 0 & 0 & D_{66} \end{bmatrix} \begin{bmatrix} -\frac{\partial^2 w_b}{\partial x^2} \\ -\frac{\partial^2 w_b}{\partial y^2} \\ -2\frac{\partial^2 w_b}{\partial x \partial y} \end{bmatrix} + \begin{bmatrix} D_{11}^s & D_{12}^s & 0 \\ D_{12}^s & D_{22}^s & 0 \\ 0 & 0 & D_{66}^s \end{bmatrix} \begin{bmatrix} -\frac{\partial^2 w_s}{\partial x^2} \\ -\frac{\partial^2 w_s}{\partial y^2} \\ -2\frac{\partial^2 w_s}{\partial x \partial y} \end{bmatrix} - \begin{bmatrix} \Pi_{31} \\ \Pi_{32} \\ \Pi_{34} \end{bmatrix} H \right),$$

$$(7.23)$$

$$\left(1 - \mu^2 \nabla^2\right) \begin{bmatrix} M_{xx}^s \\ M_{yy}^s \\ M_{xy}^s \end{bmatrix} = \left(1 - \lambda^2 \nabla^2\right) \left(\begin{bmatrix} B_{11}^s & B_{12}^s & 0 \\ B_{12}^s & B_{22}^s & 0 \\ 0 & 0 & B_{66}^s \end{bmatrix} \begin{bmatrix} \frac{\partial u}{\partial x} \\ \frac{\partial v}{\partial y} \\ \frac{\partial u}{\partial y} + \frac{\partial v}{\partial x} \end{bmatrix} \right.$$
$$+ \begin{bmatrix} D_{11}^s & D_{12}^s & 0 \\ D_{12}^s & D_{22}^s & 0 \\ 0 & 0 & D_{66}^s \end{bmatrix} \begin{bmatrix} -\frac{\partial^2 w_b}{\partial x^2} \\ -\frac{\partial^2 w_b}{\partial y^2} \\ -2\frac{\partial^2 w_b}{\partial x \partial y} \end{bmatrix} + \begin{bmatrix} H_{11}^s & H_{12}^s & 0 \\ H_{12}^s & H_{22}^s & 0 \\ 0 & 0 & H_{66}^s \end{bmatrix} \begin{bmatrix} -\frac{\partial^2 w_s}{\partial x^2} \\ -\frac{\partial^2 w_s}{\partial y^2} \\ -2\frac{\partial^2 w_s}{\partial x \partial y} \end{bmatrix} - \begin{bmatrix} \Gamma_{31} \\ \Gamma_{32} \\ \Gamma_{34} \end{bmatrix} H \right),$$

$$(7.24)$$

$$\left(1 - \mu^2 \nabla^2\right) \begin{bmatrix} Q_x \\ Q_y \end{bmatrix} = \left(1 - \lambda^2 \nabla^2\right) \begin{bmatrix} A_{44}^s & 0 \\ 0 & A_{55}^s \end{bmatrix} \begin{bmatrix} \frac{\partial w_s}{\partial x} \\ \frac{\partial w_s}{\partial y} \end{bmatrix}, \qquad (7.25)$$

where cross-sectional rigidities and magnetostrictive constants can be calculated utilizing the following formulations:

$$\begin{bmatrix} A_{11} & B_{11} & D_{11} & B_{11}^s & D_{11}^s & H_{11}^s \\ A_{12} & B_{12} & D_{12} & B_{12}^s & D_{12}^s & H_{12}^s \\ A_{66} & B_{66} & D_{66} & B_{66}^s & D_{66}^s & H_{66}^s \end{bmatrix} = \int_{-\frac{h}{2}}^{\frac{h}{2}} \begin{bmatrix} 1 & z & z^2 & f(z) & zf(z) & f^2(z) \end{bmatrix} \begin{bmatrix} C_{11} \\ C_{12} \\ C_{66} \end{bmatrix} dz,$$

$$(7.26)$$

$$[A_{44}^s, A_{55}^s] = \int_{-\frac{h}{2}}^{\frac{h}{2}} g^2(z)[C_{44}, C_{55}]dz, \qquad (7.27)$$

$$\begin{bmatrix} \Xi_{31} & \Pi_{31} & \Gamma_{31} \\ \Xi_{32} & \Pi_{32} & \Gamma_{32} \\ \Xi_{34} & \Pi_{34} & \Gamma_{34} \end{bmatrix} = \int_{-\frac{h}{2}}^{\frac{h}{2}} \begin{bmatrix} 1 & z & f(z) \end{bmatrix} \begin{bmatrix} e_{31} \\ e_{32} \\ e_{34} \end{bmatrix} dz. \qquad (7.28)$$

It must be mentioned that as in previous chapters, the rigidities with subscript 22, meaning $\square_{22}$'s in which $\square$ can be either A, B, B^s, D, D^s or H^s, are identical with their $\square_{11}$ component.

7.4 Derivation of the Governing Equations

In this section, the procedure required to derive the governing equations of motion of magnetostrictive beams and plates will be discussed in the framework of both classical and higher-order shear deformation kinematic hypotheses. First of all, it must be declared that the Euler–Lagrange equations of smart magnetostrictive nanostructures are the same as those of simple elastic beams and plates and differences will appear in the constitutive equations. Thus, we can take the Euler–Lagrange equations introduced in Chapter 4 as the equations of motion in the macroscale and modify these equations based on the constitutive equations derived in the previous section to obtain the governing equations of smart nanostructures in terms of displacement field and considering magnetostriction. For this purpose, the following two completely independent sections are allocated to deriving the governing equations of nanosize beams and plates.

7.4.1 Governing Equations of Magnetostrictive Nanobeams

In this section, we aim to insert the constitutive equations of nanobeams in the Euler–Lagrange equations of the same nanostructures to obtain the governing equations of smart magnetostrictive nanobeams. First, Eqs. (7.11)–(7.14) must be substituted in Eqs. (4.28) and (4.29) to derive the governing equations of magnetostrictive nanobeams. Once the this is completed, the equations of motion of smart nanobeams on the basis of the classical beam hypothesis can be written as follows:

$$
\left(1 - \lambda^2 \nabla^2\right) \left[A_{xx} \frac{\partial^2 u}{\partial x^2} - B_{xx} \frac{\partial^3 w}{\partial x^3} - \Xi_{31} \frac{\partial H_z}{\partial x} \right] + \left(1 - \mu^2 \nabla^2\right) \left[-I_0 \frac{\partial^2 u}{\partial t^2} + I_1 \frac{\partial^3 w}{\partial x \partial t^2} \right] = 0,
$$
(7.29)

$$
\left(1 - \lambda^2 \nabla^2\right) \left[B_{xx} \frac{\partial^3 u}{\partial x^3} - D_{xx} \frac{\partial^4 w}{\partial x^4} - \Pi_{31} \frac{\partial^2 H_z}{\partial x^2} \right] + \left(1 - \mu^2 \nabla^2\right)
$$
$$
\left[-I_0 \frac{\partial^2 w}{\partial t^2} - I_1 \frac{\partial^3 u}{\partial x \partial t^2} + I_2 \frac{\partial^4 w}{\partial x^2 \partial t^2} \right] = 0.
$$
(7.30)

Also, once Eqs. (7.11)–(7.14) are inserted in Eqs. (4.35)–(4.37), the governing differential equations of refined sinusoidal smart magnetostrictive nanobeams are obtained. After the aforementioned substitution, the governing equations can be expressed as

$$
\left(1 - \lambda^2 \nabla^2\right) \left[A_{xx} \frac{\partial^2 u}{\partial x^2} - B_{xx} \frac{\partial^3 w_b}{\partial x^3} - B_{xx}^s \frac{\partial^3 w_s}{\partial x^3} - \Xi_{31} \frac{\partial H_z}{\partial x} \right] + \left(1 - \mu^2 \nabla^2\right)
$$
$$
\left[-I_0 \frac{\partial^2 u}{\partial t^2} + I_1 \frac{\partial^3 w_b}{\partial x \partial t^2} + J_1 \frac{\partial^3 w_s}{\partial x \partial t^2} \right] = 0,
$$
(7.31)

$$
\left(1 - \lambda^2 \nabla^2\right) \left[B_{xx} \frac{\partial^3 u}{\partial x^3} - D_{xx} \frac{\partial^4 w_b}{\partial x^4} - D_{xx}^s \frac{\partial^4 w_s}{\partial x^4} - \Pi_{31} \frac{\partial^2 H_z}{\partial x^2} \right] + \left(1 - \mu^2 \nabla^2\right)
$$
$$
\left[-I_0 \frac{\partial^2 (w_b + w_s)}{\partial t^2} - I_1 \frac{\partial^3 u}{\partial x \partial t^2} + I_2 \frac{\partial^4 w_b}{\partial x^2 \partial t^2} + J_2 \frac{\partial^4 w_s}{\partial x^2 \partial t^2} \right] = 0,
$$
(7.32)

$$
\left(1 - \lambda^2 \nabla^2\right) \left[B_{xx}^s \frac{\partial^3 u}{\partial x^3} - D_{xx}^s \frac{\partial^4 w_b}{\partial x^4} - H_{xx}^s \frac{\partial^4 w_s}{\partial x^4} + A_{xz}^s \frac{\partial^2 w_s}{\partial x^2} - \Gamma_{31} \frac{\partial^2 H_z}{\partial x^2} \right] +
$$
$$
\left(1 - \mu^2 \nabla^2\right) \left[-I_0 \frac{\partial^2 (w_b + w_s)}{\partial t^2} - J_1 \frac{\partial^3 u}{\partial x \partial t^2} + J_2 \frac{\partial^4 w_b}{\partial x^2 \partial t^2} + K_2 \frac{\partial^4 w_s}{\partial x^2 \partial t^2} \right] = 0.
$$
(7.33)

Now, the above equations must be rewritten according to the definition of the H , presented in Section 7.1, Eq. (7.1), to obtain the governing equations in terms of displacement fields in a complete manner. Doing this substitution in the case of analyzing Euler–Bernoulli nanobeams with $u = w$, one can obtain the governing equations as follows:

$$\left(1 - \lambda^2 \nabla^2\right)\left[A_{xx}\frac{\partial^2 u}{\partial x^2} - B_{xx}\frac{\partial^3 w}{\partial x^3} - \Xi_{31}k_c c(t)\frac{\partial^2 w}{\partial t \partial x}\right] + \left(1 - \mu^2 \nabla^2\right)\left[-I_0\frac{\partial^2 u}{\partial t^2} + I_1\frac{\partial^3 w}{\partial x \partial t^2}\right] = 0,$$
(7.34)

$$\left(1 - \lambda^2 \nabla^2\right)\left[B_{xx}\frac{\partial^3 u}{\partial x^3} - D_{xx}\frac{\partial^4 w}{\partial x^4} - \Pi_{31}k_c c(t)\frac{\partial^3 w}{\partial t \partial x^2}\right] + \left(1 - \mu^2 \nabla^2\right)$$
$$\times \left[-I_0\frac{\partial^2 w}{\partial t^2} - I_1\frac{\partial^3 u}{\partial x \partial t^2} + I_2\frac{\partial^4 w}{\partial x^2 \partial t^2}\right] = 0.$$
(7.35)

Also, using Eq. (7.1) in association with $u = w_b + w_s$ in the case of probing refined sinusoidal smart nanobeams, Eqs. (7.31)–(7.33) can be rewritten as follows:

$$\left(1 - \lambda^2 \nabla^2\right)\left[A_{xx}\frac{\partial^2 u}{\partial x^2} - B_{xx}\frac{\partial^3 w_b}{\partial x^3} - B_{xx}^s\frac{\partial^3 w_s}{\partial x^3} - \Xi_{31}k_c c(t)\frac{\partial^2 \left(w_b + w_s\right)}{\partial t \partial x}\right]$$
$$+ \left(1 - \mu^2 \nabla^2\right)\left[-I_0\frac{\partial^2 u}{\partial t^2} + I_1\frac{\partial^3 w_b}{\partial x \partial t^2} + J_1\frac{\partial^3 w_s}{\partial x \partial t^2}\right] = 0,$$
(7.36)

$$\left(1 - \lambda^2 \nabla^2\right)\left[B_{xx}\frac{\partial^3 u}{\partial x^3} - D_{xx}\frac{\partial^4 w_b}{\partial x^4} - D_{xx}^s\frac{\partial^4 w_s}{\partial x^4} - \Pi_{31}k_c c(t)\frac{\partial^3 \left(w_b + w_s\right)}{\partial t \partial x^2}\right]$$
$$+ \left(1 - \mu^2 \nabla^2\right)\left[-I_0\frac{\partial^2 \left(w_b + w_s\right)}{\partial t^2} - I_1\frac{\partial^3 u}{\partial x \partial t^2} + I_2\frac{\partial^4 w_b}{\partial x^2 \partial t^2} + J_2\frac{\partial^4 w_s}{\partial x^2 \partial t^2}\right] = 0, \quad (7.37)$$

$$\left(1 - \lambda^2 \nabla^2\right)\left[B_{xx}^s\frac{\partial^3 u}{\partial x^3} - D_{xx}^s\frac{\partial^4 w_b}{\partial x^4} - H_{xx}^s\frac{\partial^4 w_s}{\partial x^4} + A_x^s\frac{\partial^2 w_s}{\partial x^2} - \Gamma_{31}k_c c(t)\frac{\partial^3 \left(w_b + w_s\right)}{\partial t \partial x^2}\right]$$
$$+ \left(1 - \mu^2 \nabla^2\right)\left[-I_0\frac{\partial^2 \left(w_b + w_s\right)}{\partial t^2} - J_1\frac{\partial^3 u}{\partial x \partial t^2} + J_2\frac{\partial^4 w_b}{\partial x^2 \partial t^2} + K_2\frac{\partial^4 w_s}{\partial x^2 \partial t^2}\right] = 0. \quad (7.38)$$

7.4.2 Governing Equations of Magnetostrictive Nanoplates

As in the previous section, herein, the equations of motion of both Kirchhoff–Love and refined sinusoidal plates will be incorporated into the constitutive equations of smart magnetostrictive nanomaterials based on the nonlocal stress–strain gradient to obtain the governing equations of smart nanoplates.

To derive the governing equations of smart magnetostrictive Kirchhoff–Love nanoplates, Eqs. (7.19) and (7.20) must be substituted in Eqs. (4.76)–(4.78). Once the above substitution is performed, the governing equations of smart magnetostrictive nanoplates on the basis of the Kirchhoff–Love plate model can be obtained in the following forms:

$$\left(1 - \lambda^2 \nabla^2\right)\left(A_{11}\frac{\partial^2 u}{\partial x^2} + (A_{12} + A_{66})\frac{\partial^2 v}{\partial x \partial y} + A_{66}\frac{\partial^2 u}{\partial y^2} - B_{11}\frac{\partial^3 w}{\partial x^3} - (B_{12} + 2B_{66})\frac{\partial^3 w}{\partial x \partial y^2}\right.$$
$$\left. - \left[\Xi_{31}\frac{\partial H}{\partial x} + \Xi_{34}\frac{\partial H}{\partial y}\right]\right) + \left(1 - \mu^2 \nabla^2\right)\left(-I_0\frac{\partial^2 u}{\partial t^2} + I_1\frac{\partial^3 w}{\partial x \partial t^2}\right) = 0, \quad (7.39)$$

$$\left(1 - \lambda^2 \nabla^2\right) \left(A_{22} \frac{\partial^2 v}{\partial y^2} + (A_{12} + A_{66}) \frac{\partial^2 u}{\partial x \partial y} + A_{66} \frac{\partial^2 v}{\partial x^2} - B_{22} \frac{\partial^3 w}{\partial y^3} - (B_{12} + 2B_{66}) \frac{\partial^3 w}{\partial x^2 \partial y} \right.$$

$$\left. - \left[\Xi_{34} \frac{\partial H}{\partial x} + \Xi_{32} \frac{\partial H}{\partial y} \right] \right) + \left(1 - \mu^2 \nabla^2\right) \left(-I_0 \frac{\partial^2 v}{\partial t^2} + I_1 \frac{\partial^3 w}{\partial y \partial t^2} \right) = 0, \quad (7.40)$$

$$\left(1 - \lambda^2 \nabla^2\right) \left(B_{11} \frac{\partial^3 u}{\partial x^3} + (B_{12} + 2B_{66}) \frac{\partial^3 u}{\partial x \partial y^2} + B_{22} \frac{\partial^3 v}{\partial y^3} + (B_{12} + 2B_{66}) \frac{\partial^3 v}{\partial x^2 \partial y} - D_{11} \frac{\partial^4 w}{\partial x^4} \right.$$

$$\left. - 2(D_{12} + 2D_{66}) \frac{\partial^4 w}{\partial x^2 \partial y^2} - D_{22} \frac{\partial^4 w}{\partial y^4} - \left[\Pi_{31} \frac{\partial^2 H}{\partial x^2} + 2\Pi_{34} \frac{\partial^2 H}{\partial x \partial y} + \Pi_{32} \frac{\partial^2 H}{\partial y^2} \right] \right)$$

$$+ \left(1 - \mu^2 \nabla^2\right) \left(-I_0 \frac{\partial w}{\partial t^2} - I_1 \left(\frac{\partial^3 u}{\partial x \partial t^2} + \frac{\partial^3 v}{\partial y \partial t^2} \right) + I_2 \left(\frac{\partial^4 w}{\partial x^2 \partial t^2} + \frac{\partial^4 w}{\partial y^2 \partial t^2} \right) \right) = 0. \tag{7.41}$$

Besides, the governing equations of refined sinusoidal smart magnetostrictive nanoplates are obtained by substituting the stress resultants in Eqs. (7.22)–(7.25) in Eqs. (4.85)–(4.88). Following this substitution, the governing equations can be expressed in the following forms:

$$\left(1 - \lambda^2 \nabla^2\right) \left(A_{11} \frac{\partial^2 u}{\partial x^2} + (A_{12} + A_{66}) \frac{\partial^2 v}{\partial x \partial y} + A_{66} \frac{\partial^2 u}{\partial y^2} - B_{11} \frac{\partial^3 w_b}{\partial x^3} - (B_{12} + 2B_{66}) \frac{\partial^3 w_b}{\partial x \partial y^2} \right.$$

$$\left. - B_{11}^s \frac{\partial^3 w_s}{\partial x^3} - (B_{12}^s + 2B_{66}^s) \frac{\partial^3 w_s}{\partial x \partial y^2} - \left[\Xi_{31} \frac{\partial H}{\partial x} + \Xi_{34} \frac{\partial H}{\partial y} \right] \right)$$

$$+ \left(1 - \mu^2 \nabla^2\right) \left(-I_0 \frac{\partial^2 u}{\partial t^2} + I_1 \frac{\partial^3 w_b}{\partial x \partial t^2} + J_1 \frac{\partial^3 w_s}{\partial x \partial t^2} \right) = 0, \tag{7.42}$$

$$\left(1 - \lambda^2 \nabla^2\right) \left(A_{22} \frac{\partial^2 v}{\partial y^2} + (A_{12} + A_{66}) \frac{\partial^2 u}{\partial x \partial y} + A_{66} \frac{\partial^2 v}{\partial x^2} - B_{22} \frac{\partial^3 w_b}{\partial y^3} - (B_{12} + 2B_{66}) \frac{\partial^3 w_b}{\partial x^2 \partial y} \right.$$

$$\left. - B_{22}^s \frac{\partial^3 w_s}{\partial y^3} - (B_{12}^s + 2B_{66}^s) \frac{\partial^3 w_s}{\partial x^2 \partial y} - \left[\Xi_{34} \frac{\partial H}{\partial x} + \Xi_{32} \frac{\partial H}{\partial y} \right] \right)$$

$$+ \left(1 - \mu^2 \nabla^2\right) \left(-I_0 \frac{\partial^2 v}{\partial t^2} + I_1 \frac{\partial^3 w_b}{\partial y \partial t^2} + J_1 \frac{\partial^3 w_s}{\partial y \partial t^2} \right) = 0, \tag{7.43}$$

$$\left(1 - \lambda^2 \nabla^2\right) \left(B_{11} \frac{\partial^3 u}{\partial x^3} + (B_{12} + 2B_{66}) \frac{\partial^3 u}{\partial x \partial y^2} + B_{22} \frac{\partial^3 v}{\partial y^3} + (B_{12} + 2B_{66}) \frac{\partial^3 v}{\partial x^2 \partial y} - D_{11} \frac{\partial^4 w_b}{\partial x^4} \right.$$

$$- 2(D_{12} + 2D_{66}) \frac{\partial^4 w_b}{\partial x^2 \partial y^2} - D_{22} \frac{\partial^4 w_b}{\partial y^4} - D_{11}^s \frac{\partial^4 w_s}{\partial x^4} - 2(D_{12}^s + 2D_{66}^s) \frac{\partial^4 w_s}{\partial x^2 \partial y^2}$$

$$- D_{22}^s \frac{\partial^4 w_s}{\partial y^4} \quad \left[\Pi_{31} \frac{\partial^2 H}{\partial x^2} + 2\Pi_{34} \frac{\partial^2 H}{\partial x \partial y} + \Pi_{32} \frac{\partial^2 H}{\partial y^2} \right] \right) + \left(1 - \mu^2 \nabla^2\right) \left(-I_0 \frac{\partial^2 (w_b + w_s)}{\partial t^2} \right.$$

$$\left. - I_1 \left(\frac{\partial^3 u}{\partial t^2 \partial x} + \frac{\partial^3 v}{\partial t^2 \partial y} \right) + I_2 \left(\frac{\partial^4 w_b}{\partial x^2 \partial t^2} + \frac{\partial^4 w_b}{\partial y^2 \partial t^2} \right) + J_2 \left(\frac{\partial^4 w_s}{\partial x^2 \partial t^2} + \frac{\partial^4 w_s}{\partial y^2 \partial t^2} \right) \right) = 0, \tag{7.44}$$

$$\left(1 - \lambda^2\nabla^2\right)\left(B_{11}^s\frac{\partial^3 u}{\partial x^3} + (B_{12}^s + 2B_{66}^s)\frac{\partial^3 u}{\partial x\partial y^2} + B_{22}^s\frac{\partial^3 v}{\partial y^3} + (B_{12}^s + 2B_{66}^s)\frac{\partial^3 v}{\partial x^2\partial y} - D_{11}^s\frac{\partial^4 w_b}{\partial x^4}\right.$$

$$- 2(D_{12}^s + 2D_{66}^s)\frac{\partial^4 w_b}{\partial x^2\partial y^2} - D_{22}^s\frac{\partial^4 w_b}{\partial y^4} - H_{11}^s\frac{\partial^4 w_s}{\partial x^4} - 2(H_{12}^s + 2H_{66}^s)\frac{\partial^4 w_s}{\partial x^2\partial y^2}$$

$$- H_{22}^s\frac{\partial^4 w_s}{\partial y^4}A_{44}^s\left(\frac{\partial^2(w_b + w_s)}{\partial x^2} + \frac{\partial^2(w_b + w_s)}{\partial y^2}\right) - \left[\Gamma_{31}\frac{\partial^2 H}{\partial x^2} + 2\Gamma_{34}\frac{\partial^2 H}{\partial x\partial y} + \Gamma_{32}\frac{\partial^2 H}{\partial y^2}\right]\right)$$

$$+ \left(1 - \mu^2\nabla^2\right)\left(-I_0\frac{\partial^2(w_b + w_s)}{\partial t^2} - J_1\left(\frac{\partial^3 u}{\partial t^2\partial x} + \frac{\partial^3 v}{\partial t^2\partial y}\right) + J_2\left(\frac{\partial^4 w_b}{\partial x^2\partial t^2} + \frac{\partial^4 w_b}{\partial y^2\partial t^2}\right)\right.$$

$$+ K_2\left(\frac{\partial^4 w_s}{\partial x^2\partial t^2} + \frac{\partial^4 w_s}{\partial y^2\partial t^2}\right)\right) = 0. \tag{7.45}$$

Now, the derived governing equations must be simplified using the definition of the magnetic field, mentioned in Eq. (7.1). Indeed, the governing equations expressed in Eqs. (7.39)–(7.45) must be transformed into other equations in terms of displacement components. Now, the governing equations of classical smart magnetostrictive nanoplates can be obtained in the following forms by modifying Eqs. (7.39)–(7.41) using $u = w$ as well as Eq. (7.1):

$$\left(1 - \lambda^2\nabla^2\right)\left(A_{11}\frac{\partial^2 u}{\partial x^2} + (A_{12} + A_{66})\frac{\partial^2 v}{\partial x\partial y} + A_{66}\frac{\partial^2 u}{\partial y^2} - B_{11}\frac{\partial^3 w}{\partial x^3} - (B_{12} + 2B_{66})\frac{\partial^3 w}{\partial x\partial y^2}\right.$$

$$\left. - k_c c(t)\left[\Xi_{31}\frac{\partial^2 w}{\partial t\partial x} + \Xi_{34}\frac{\partial^2 w}{\partial t\partial y}\right]\right) + \left(1 - \mu^2\nabla^2\right)\left(-I_0\frac{\partial^2 u}{\partial t^2} + I_1\frac{\partial^3 w}{\partial x\partial t^2}\right) = 0, \tag{7.46}$$

$$\left(1 - \lambda^2\nabla^2\right)\left(A_{22}\frac{\partial^2 v}{\partial y^2} + (A_{12} + A_{66})\frac{\partial^2 u}{\partial x\partial y} + A_{66}\frac{\partial^2 v}{\partial x^2} - B_{22}\frac{\partial^3 w}{\partial y^3} - (B_{12} + 2B_{66})\frac{\partial^3 w}{\partial x^2\partial y}\right.$$

$$\left. - k_c c(t)\left[\Xi_{34}\frac{\partial^2 w}{\partial t\partial x} + \Xi_{32}\frac{\partial^2 w}{\partial t\partial y}\right]\right) + \left(1 - \mu^2\nabla^2\right)\left(-I_0\frac{\partial^2 v}{\partial t^2} + I_1\frac{\partial^3 w}{\partial y\partial t^2}\right) = 0, \tag{7.47}$$

$$\left(1 - \lambda^2\nabla^2\right)\left(B_{11}\frac{\partial^3 u}{\partial x^3} + (B_{12} + 2B_{66})\frac{\partial^3 u}{\partial x\partial y^2} + B_{22}\frac{\partial^3 v}{\partial y^3} + (B_{12} + 2B_{66})\frac{\partial^3 v}{\partial x^2\partial y} - D_{11}\frac{\partial^4 w}{\partial x^4}\right.$$

$$\left. - 2(D_{12} + 2D_{66})\frac{\partial^4 w}{\partial x^2\partial y^2} - D_{22}\frac{\partial^4 w}{\partial y^4} - k_c c(t)\left[\Pi_{31}\frac{\partial^3 w}{\partial t\partial x^2} + 2\Pi_{34}\frac{\partial^3 w}{\partial t\partial x\partial y} + \Pi_{32}\frac{\partial^3 w}{\partial t\partial y^2}\right]\right)$$

$$+ \left(1 - \mu^2\nabla^2\right)\left(-I_0\frac{\partial w}{\partial t^2} - I_1\left(\frac{\partial^3 u}{\partial x\partial t^2} + \frac{\partial^3 v}{\partial y\partial t^2}\right) + I_2\left(\frac{\partial^4 w}{\partial x^2\partial t^2} + \frac{\partial^4 w}{\partial y^2\partial t^2}\right)\right) = 0. \tag{7.48}$$

On the other hand, the governing equations of refined sinusoidal magnetostrictive nanpoplates can be obtained by substituting H from Eq. (7.1) in Eqs. (7.42)–(7.45). Once this substitution is performed, the governing differential equations in terms of displacement field arrays can be presented as follows:

$$\left(1 - \lambda^2\nabla^2\right)\left(A_{11}\frac{\partial^2 u}{\partial x^2} + (A_{12} + A_{66})\frac{\partial^2 v}{\partial x\partial y} + A_{66}\frac{\partial^2 u}{\partial y^2} - B_{11}\frac{\partial^3 w_b}{\partial x^3} - (B_{12} + 2B_{66})\frac{\partial^3 w_b}{\partial x\partial y^2}\right.$$

$$\left. - B_{11}^s\frac{\partial^3 w_s}{\partial x^3} - (B_{12}^s + 2B_{66}^s)\frac{\partial^3 w_s}{\partial x\partial y^2} - k_c c(t)\left[\Xi_{31}\frac{\partial^2(w_b + w_s)}{\partial t\partial x} + \Xi_{34}\frac{\partial^2(w_b + w_s)}{\partial t\partial y}\right]\right)$$

$$+ \left(1 - \mu^2\nabla^2\right)\left(-I_0\frac{\partial^2 u}{\partial t^2} + I_1\frac{\partial^3 w_b}{\partial x\partial t^2} + J_1\frac{\partial^3 w_s}{\partial x\partial t^2}\right) = 0, \tag{7.49}$$

$$
\left(1 - \lambda^2 \nabla^2\right) \left(A_{22} \frac{\partial^2 v}{\partial y^2} + (A_{12} + A_{66}) \frac{\partial^2 u}{\partial x \partial y} + A_{66} \frac{\partial^2 v}{\partial x^2} - B_{22} \frac{\partial^3 w_b}{\partial y^3} - (B_{12} + 2B_{66}) \frac{\partial^3 w_b}{\partial x^2 \partial y} \right.
$$

$$
\left. - B_{22}^s \frac{\partial^3 w_s}{\partial y^3} - (B_{12}^s + 2B_{66}^s) \frac{\partial^3 w_s}{\partial x^2 \partial y} - k_c c(t) \left[\Xi_{34} \frac{\partial^2 (w_b + w_s)}{\partial t \partial x} + \Xi_{32} \frac{\partial^2 (w_b + w_s)}{\partial t \partial y} \right] \right)
$$

$$
+ \left(1 - \mu^2 \nabla^2\right) \left(-I_0 \frac{\partial^2 v}{\partial t^2} + I_1 \frac{\partial^3 w_b}{\partial y \partial t^2} + J_1 \frac{\partial^3 w_s}{\partial y \partial t^2} \right) = 0, \tag{7.50}
$$

$$
\left(1 - \lambda^2 \nabla^2\right) \left(B_{11} \frac{\partial^3 u}{\partial x^3} + (B_{12} + 2B_{66}) \frac{\partial^3 u}{\partial x \partial y^2} + B_{22} \frac{\partial^3 v}{\partial y^3} + (B_{12} + 2B_{66}) \frac{\partial^3 v}{\partial x^2 \partial y} - D_{11} \frac{\partial^4 w_b}{\partial x^4} \right.
$$

$$
- 2(D_{12} + 2D_{66}) \frac{\partial^4 w_b}{\partial x^2 \partial y^2} - D_{22} \frac{\partial^4 w_b}{\partial y^4} - D_{11}^s \frac{\partial^4 w_s}{\partial x^4} - 2(D_{12}^s + 2D_{66}^s) \frac{\partial^4 w_s}{\partial x^2 \partial y^2} - D_{22}^s \frac{\partial^4 w_s}{\partial y^4}
$$

$$
\left. - k_c c(t) \left[\Pi_{31} \frac{\partial^3 (w_b + w_s)}{\partial t \partial x^2} + 2\Pi_{34} \frac{\partial^3 (w_b + w_s)}{\partial t \partial x \partial y} + \Pi_{32} \frac{\partial^3 (w_b + w_s)}{\partial t \partial y^2} \right] \right)
$$

$$
+ \left(1 - \mu^2 \nabla^2\right) \left(-I_0 \frac{\partial^2 (w_b + w_s)}{\partial t^2} - I_1 \left(\frac{\partial^3 u}{\partial t^2 \partial x} + \frac{\partial^3 v}{\partial t^2 \partial y} \right) + I_2 \left(\frac{\partial^4 w_b}{\partial x^2 \partial t^2} + \frac{\partial^4 w_b}{\partial y^2 \partial t^2} \right) \right.
$$

$$
\left. + J_2 \left(\frac{\partial^4 w_s}{\partial x^2 \partial t^2} + \frac{\partial^4 w_s}{\partial y^2 \partial t^2} \right) \right) = 0, \tag{7.51}
$$

$$
\left(1 - \lambda^2 \nabla^2\right) \left(B_{11}^s \frac{\partial^3 u}{\partial x^3} + (B_{12}^s + 2B_{66}^s) \frac{\partial^3 u}{\partial x \partial y^2} + B_{22}^s \frac{\partial^3 v}{\partial y^3} + (B_{12}^s + 2B_{66}^s) \frac{\partial^3 v}{\partial x^2 \partial y} - D_{11}^s \frac{\partial^4 w_b}{\partial x^4} \right.
$$

$$
- 2(D_{12}^s + 2D_{66}^s) \frac{\partial^4 w_b}{\partial x^2 \partial y^2} - D_{22}^s \frac{\partial^4 w_b}{\partial y^4} - H_{11}^s \frac{\partial^4 w_s}{\partial x^4} - 2(H_{12}^s + 2H_{66}^s) \frac{\partial^4 w_s}{\partial x^2 \partial y^2}
$$

$$
- H_{22}^s \frac{\partial^4 w_s}{\partial y^4} A_{44}^s \left(\frac{\partial^2 (w_b + w_s)}{\partial x^2} + \frac{\partial^2 (w_b + w_s)}{\partial y^2} \right) - k_c c(t) \left[\Gamma_{31} \frac{\partial^3 (w_b + w_s)}{\partial t \partial x^2} \right.
$$

$$
\left. + 2\Gamma_{34} \frac{\partial^3 (w_b + w_s)}{\partial t \partial x \partial y} + \Gamma_{32} \frac{\partial^3 (w_b + w_s)}{\partial t \partial y^2} \right] \right) + \left(1 - \mu^2 \nabla^2\right) \left(-I_0 \frac{\partial^2 (w_b + w_s)}{\partial t^2} \right.
$$

$$
\left. - J_1 \left(\frac{\partial^3 u}{\partial t^2 \partial x} + \frac{\partial^3 v}{\partial t^2 \partial y} \right) + J_2 \left(\frac{\partial^4 w_b}{\partial x^2 \partial t^2} + \frac{\partial^4 w_b}{\partial y^2 \partial t^2} \right) + K_2 \left(\frac{\partial^4 w_s}{\partial x^2 \partial t^2} + \frac{\partial^4 w_s}{\partial y^2 \partial t^2} \right) \right) = 0. \tag{7.52}
$$

7.5 Solution Procedure

This section is dedicated to solve the governing equations achieved in the previous section in the framework of the well-known analytical wave solution method introduced in Chapter 1. It must be noted that, because of existence of first-order derivation of deflection components with respect to time, the analysis cannot be performed constructing conventional stiffness and mass matrices. Indeed, the damping phenomenon will be observed in this study because of this issue. Therefore, the damping matrix as well as stiffness and mass matrices must be constructed. Comparing the governing equations obtained with those of simple elastic nanosize beams and plates, one can observe that the mass and stiffness matrices are completely identical with those reported in Chapter 4. Thus, calculating the nonzero arrays of the damping matrix corresponds to solving the problem. In the mathematical framework, the final equation of such a problem is an eigenvalue problem as follows:

$$\left([K] + \omega[C] - \omega^2[M]\right)[\Delta] = 0 \tag{7.53}$$

in which C is the damping matrix and K and M are the known stiffness and mass matrices, respectively. It is worth mentioning that the dimension of the damping matrix is the same as that of either the stiffness or mass matrix. The components of stiffness and mass matrices are provided in Chapter 4 for both classical and shear deformable nanosize beams and plates, and they will not be expressed in this section anymore. In what follows, the components of damping matrix of nanobeams and nanoplates will be derived to complete the procedure of solving the externally affected wave propagation problem of nanobeams and nanoplates consisting of magnetostrictive materials.

7.5.1 Arrays of Damping Matrix of Nanobeams

On substituting the analytical solution in the set of governing equations of nanosizie beams, we can derive the nonzero components of the damping matrix. For an Euler–Bernoulli magnetostrictive nanobeam, the damping matrix will be a 2×2 matrix with nonzero C_{12} and C_{22} arrays. Therefore, the only nonzero component of the damping matrix for a classical smart magnetostrictive nanobeam can be written in the following form:

$$
\begin{aligned}
C_{12} &= \left(1 + \lambda^2\beta^2\right)\Xi_{31}k_c c(t)\beta, \\
C_{22} &= i\left(1 + \lambda^2\beta^2\right)\Pi_{31}k_c c(t)\beta^2.
\end{aligned}
\tag{7.54}
$$

Performing the same procedure for refined sinusoidal magnetostrictive nanobeams, one can find that the nonzero components of the damping matrix belong to bending and shear deflections. The damping matrix will be of 3×3 order for refined shear deformable nanobeams. Therefore, C_{12}, C_{13}, C_{22}, C_{23}, C_{32} and C_{33} arrays can be expressed in the following forms:

$$
\begin{aligned}
C_{12} = C_{13} &= \left(1 + \lambda^2\beta^2\right)\Xi_{31}k_c c(t)\beta, \\
C_{22} = C_{23} &= i\left(1 + \lambda^2\beta^2\right)\Pi_{31}k_c c(t)\beta^2, \\
C_{32} = C_{33} &= i\left(1 + \lambda^2\beta^2\right)\Gamma_{31}k_c c(t)\beta^2.
\end{aligned}
\tag{7.55}
$$

Now, the wave frequency and phase velocity of smart magnetostrictive sandwich nanobeams can be achieved using either classical or refined sinusoidal beam theories.

7.5.2 Arrays of Damping Matrix of Nanoplates

Following the same procedure for smart magnetostrictive nanoplates, it can be inferred that the damping matrix will be of 3×3 and 4×4 order in the cases using Kirchhoff–Love and shear deformable nanoplates, respectively. The nonzero components of the damping matrix of a Kirchhoff–Love nanoplate can be written in the following forms:

$$
\begin{aligned}
C_{13} &= k_c c(t)\left[1 + \lambda^2\left(\beta_1^2 + \beta_2^2\right)\right]\left(\Xi_{31}\beta_1 + \Xi_{34}\beta_2\right), \\
C_{23} &= k_c c(t)\left[1 + \lambda^2\left(\beta_1^2 + \beta_2^2\right)\right]\left(\Xi_{34}\beta_1 + \Xi_{32}\beta_2\right), \\
C_{33} &= i k_c c(t)\left[1 + \lambda^2\left(\beta_1^2 + \beta_2^2\right)\right]\left(\Pi_{31}\beta_1^2 + 2\Pi_{34}\beta_1\beta_2 + \Pi_{32}\beta_2^2\right).
\end{aligned}
\tag{7.56}
$$

In the case of investigating a refined sinusoidal magnetostrictive nanoplate, the nonzero arrays of the damping matrix can be formulated as follows:

$$C_{13} = C_{14} = k_c c(t) \left[1 + \lambda^2 \left(\beta_1^2 + \beta_2^2\right)\right] \left(\Xi_{31}\beta_1 + \Xi_{34}\beta_2\right),$$

$$C_{23} = C_{24} = k_c c(t) \left[1 + \lambda^2 \left(\beta_1^2 + \beta_2^2\right)\right] \left(\Xi_{34}\beta_1 + \Xi_{32}\beta_2\right),$$

$$C_{33} = C_{34} = i k_c c(t) \left[1 + \lambda^2 \left(\beta_1^2 + \beta_2^2\right)\right] \left(\Pi_{31}\beta_1^2 + 2\Pi_{34}\beta_1\beta_2 + \Pi_{32}\beta_2^2\right),$$

$$C_{43} = C_{44} = i k_c c(t) \left[1 + \lambda^2 \left(\beta_1^2 + \beta_2^2\right)\right] \left(\Gamma_{31}\beta_1^2 + 2\Gamma_{34}\beta_1\beta_2 + \Gamma_{32}\beta_2^2\right). \tag{7.57}$$

Using the above components, the damped, magnetically influenced wave dispersion responses of magnetostrictive nanosize plates can be obtained solving the eigenvalue problem introduced in Eq. (7.53).

7.6 Numerical Results and Discussion

In this section, numerical illustrations related to the smart magnetostrictive nanoscale beams and plates subjected to elastic waves will be presented to show the impact of each of the involved variants on the dispersion behaviors of such smart nanostructures. Herein, the expression $k_c c(t)$ is assumed to be a constant gain which controls the velocity feedback control system that governs the distribution of the magnetic field.

As the first illustration in this chapter, Figure 7.1 is presented to show the variation of wave frequency with wave number for different values of velocity control gain when the scale factor is assumed to be less than 1. It can be seen that the frequency of the dispersed waves will be decreased as the control gain increases. Indeed, this behavior originates from the damping effect of the material's magnetostriction on the dynamic behavior of the structure. Actually, wave frequency will increase as the wave number increases and reaches a maximum value followed by a sudden decline which leads to attenuation of the waves' natural frequencies to a stationary state. Overall, the curves will shift downward as the control gain of the velocity feedback control system increases.

The next result belongs to the investigation of the variation of the phase velocity of waves dispersed in smart magnetostrictive nanobeams with wave number for various values of control gain (see Figure 7.2). It can be easily found that the system's reaction to the variation of the phase velocity is nearly the same as its reaction to that of the wave frequency. In other words, again, the phase velocity follows a dummy path and experiences a continuous rise followed by a similar decline. However, it is clear that the phase velocity will be finally damped entirely, whereas the wave frequency reaches a fixed value. According to the diagram, the phase velocity will be greatly lessened as the control gain is increased. Similar to wave frequency, herein, the dependency of the dynamic response of the smart nanostructure on the control gain can be sensed better at mid-range wave numbers than very low or very high ones.

Furthermore, the same graphs will be discussed in the following paragraph to clarify how the velocity control gain can affect the dispersion curves of smart magnetostrictive nanoplates and the differences between the effect of the aforementioned term on the wave propagation responses of nanosize beams and plates. Figure 7.3 is presented in order to show the variation of the natural frequency of the waves dispersed in smart nanoplates with wave number. It is clear that the wave frequency will decrease as the control gain increases. This trend is observed in Figure 7.1, too; however, the difference is in the range of the wave

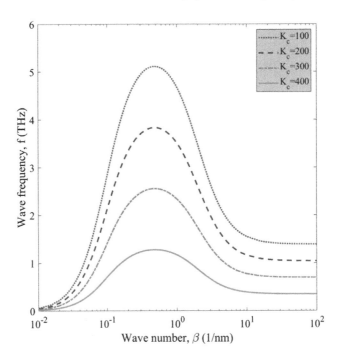

FIGURE 7.1
Variation of the frequency of the waves propagated inside smart magnetostrictive nanobeams with wave number for various amounts of control gain, K_c ($c < 1$).

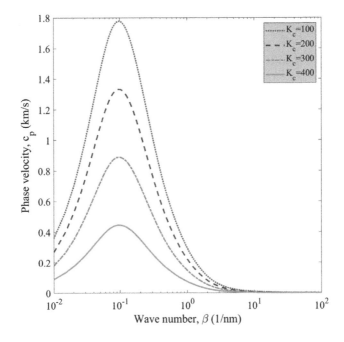

FIGURE 7.2
Variation of the phase speed of the waves propagated inside smart magnetostrictive nanobeams with wave number for various amounts of control gain, K_c ($c < 1$).

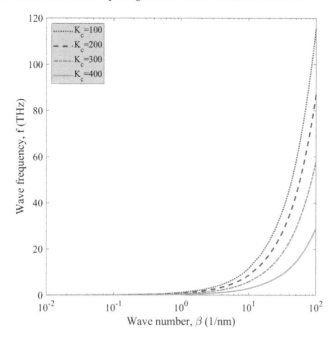

FIGURE 7.3
Variation of the frequency of the waves propagated inside smart magnetostrictive nanoplates with wave number for various amounts of control gain, K_c ($c < 1$).

numbers at which the dispersion curves will be influenced when the control gain is varied. In other words, in the case of monitoring the dispersion curves of smart nanobeams, the control gain is more powerful in mid-range wave numbers in decreasing the frequency of the propagated waves, whereas this influence can be seen at only very high wave numbers when the case of smart nanoplates is considered.

Eventually, the variation of the velocity of the waves propagated inside smart magnetostrictive nanoplates with wave number is depicted in Figure 7.4 for different amounts of control gain. It can be easily observed that the phase speed behaves in the same as the wave frequency and follows a gradual increasing path as the wave number increases. Actually, the effect of the control gain cannot be monitored at very low wave numbers, and this lessening impact can be seen better as the wave number exceeds $\beta = 5$ (1/nm). As stated in the interpretation of the previous illustration, the damping effect of the control gain will occur at high wave numbers. In other words, the phenomenon of damping in smart magnetostrictive nanoplates is relatively delayed in comparison with nanobeams constructed from the same smart material.

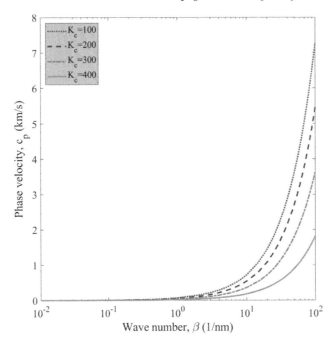

FIGURE 7.4
Variation of the phase speed of the waves propagated inside smart magnetostrictive nanoplates with wave number for various amounts of control gain, K_c ($c < 1$).

8

Wave Propagation Analysis of Magnetoelectroelastic Heterogeneous Nanostructures

In this chapter we will incorporate the MEE interactions while investigating the wave dispersion responses of smart functionally graded (FG) nanosize beams and plates within the framework of the nonlocal strain gradient magnetoelectroelasticity hypothesis. The homogenization method is presumed to be the same as that employed in Chapter6 for piezoelectric materials. Moreover, the equations of motion of nanostructures will be derived for both classical and higher-order beam and plate theories to show the effect of magnetoelectroelastic (MEE) interactions on the governing equations of both classical and higher-order nanostructures. Besides, a revised form of the nonlocal strain gradient elasticity will be discussed for smart MEE materials to obtain the constitutive equations of nanobeams and nanoplates. Afterward, a set of illustrations are studied in order to graphically explain the effects of electric voltage and magnetic potential on the wave dispersion characteristics of smart FG nanostructures. At the end, we will examine the wave dispersion curves of porous MEE nanostructures to figure out how porosities can affect the wave propagation responses of nanostructures.

8.1 Introduction

In the previous chapters, it was declared that the new trend of modern engineering designs across the world is moving toward utilization of smart multitasking materials to satisfy many industrial needs. One of the most efficient materials that can be employed in structures is an MEE material. As stated in Chapter 2, MEE materials are capable of converting any one form of excitation among electric, magnetic or elastic excitations to the any of the other forms. So, it is crucial to gain adequate knowledge about the mechanical responses of MEE structures. For example, Razavi and Shooshtari [145] studied the natural frequency behaviors of an MEE shell with dual curvature by means of the first-order shell theory. They have also probed the nonlinear vibrational responses of MEE shells by considering the effects of shear deflection up to first order [146]. They used a symmetric sandwich shell in their study. Razavi and Shooshtari [147] implemented a perturbation approach incorporated with the nonlinear Mindlin plate model in order to analyze the dynamic responses of MEE plates with simply supported edges. In another study, Xin and Hu [148] presented an elasticity solution for the vibrational responses of an MEE shell on the basis of the effective discrete singular convolution numerical method. The application of a higher-order plate hypothesis in the analysis of an MEE plate was studied by Shooshtari and Razavi [149]. Vinyas and Kattimani [150] studied the static behaviors of plates made from

MEE materials with respect to the external effects of hygrothermal environment employing a finite element method (FEM). Also, the finite element (FE) dynamic analysis of MEE plates in a hygrothermal environment was performed by Vinyas and Kattimani [151]. The stability problem of an MEE skew plate subjected to both monoaxial and biaxial excitations was investigated by Kiran and Kattimani [152]. Recently, Vinyas et al. [153] have conducted an FE-based study on the natural frequency characteristics of MEE plates. They presented a comparison of the numerical results of their FE modeling with those of their simulations using the COMSOL software to highlight the accuracy of their reported results.

It is very important to also be aware of the mechanical behaviors of smart MEE nanostructures due to their continued and increasing rate of use in nano-electro-mechanical-systems (NEMSs). For this reason, many researchers have tried to probe the mechanical responses of nanosize beams, plates and shells made from such smart materials. Li et al. [154] used the first-order beam model incorporated with the Eringen's nonlocal hypothesis for studying the static and dynamic responses of MEE nanobeams. Xu et al. [155] considered the effects of surface stresses on the general mechanical characteristics of MEE nanoscale beams. Ghadiri and Safarpour [156] analyzed the thermally affected vibration problem of an MEE nanoshell. A higher-order beam model was implemented by Ebrahimi and Barati [157] in order to survey the frequency behaviors of MEE-FG nanobeams at various values of temperature gradient. The effects of both magnetic and electric potentials on the stability limits of a curved FG nanobeam were considered by Ebrahimi and Barati [158]. Moreover, the effects of geometrical nonlinearities on the large-amplitude dynamic responses of MEE nanoplates are included in a nonlocal investigation conducted by Farajpour et al. [159]. The MEE analysis of waves traveling in an FG nanoplate was performed by Ebrahimi et al. [160]. In addition, Ebrahimi and Barati [161] used a refined higher-order plate theorem coupled with the nonlocal elasticity for studying the buckling properties of an MEE-FG nanoplate when the nanoplate is considered to be subjected to magnetoelectric potentials. The effects of a material's time dependency on the vibrational behaviors of viscoelastic FG nanobeams are covered by Ebrahimi and Barati [162]. They assumed the upper and lower constituents of the FGM to be made from MEE materials. The nanobeam was modeled via a refined shear deformation beam model. The critical buckling analysis of an MEE-FG nanobeam was performed by Ebrahimi and Barati [163] applying the nonlocal constitutive equations of Eringen for a shear deformable beam-type element. The wave dispersion responses of FG-MEE nanobeams were examined by Ebrahimi et al. [164] when the nanobeam is presumed to be subjected to axial excitation. Furthermore, Ebrahimi and Barati [165] employed a refined nonlocal plate theorem to probe the stability behaviors of MEE-FG nanoplates. The effects of thermal environment on the buckling characteristics of FG-MEE nanobeams were included in an investigation performed by Ebrahimi and Barati [166]. Ebrahimi and Barati [167] have also studied the frequency problem of an MEE-FG nanobeam by considering the effects of the existence of porosities in the smart medium. Ebrahimi and Dabbagh [168] studied the dispersion curves of waves propagating in an MEE-FG nanoplate on the basis of the nonlocal strain gradient hypothesis. These researchers have employed nonlocal strain gradient elasticity in another study dealing with the wave propagation properties of MEE-FG nanoplates when the nanoplate is assumed to rotate around an axis parallel to the plate's thickness direction [169]. In another endeavor, Ebrahimi and Barati [170] probed the thermally affected dynamic properties of MEE-FG nanobeams. The nonlocal strain gradient magnetoelectroelasticity has been recently utilized by Sahmani and Aghdam [171] while investigating the nonlinear instability responses of MEE nanocylinders in the framework of a perturbation method.

8.2 Analysis of MEE-FG Nanobeams

In this section, the wave propagation problem will be formulated in the framework of an energy-based method for smart MEE-FG nanobeams based upon both Euler–Bernoulli and shear deformation beam models. Actually, the problem will be explained for both the classical and higher-order beams to show how the MEE interactions can affect each of the mentioned beam models. Due to the fact that the nonlocal elasticity can be obtained from the general nonlocal strain gradient elasticity, the nonlocal strain gradient theory will be extended to the following formulations.

8.2.1 Euler–Bernoulli MEE Nanobeams

This section is dedicated to present the problem's formulation for an Euler–Bernoulli nanobeam according to the nonlocal strain gradient model for the constitutive equations of the MEE-FG nanobeam. The equations of motion will be obtained by using the Hamilton's principle for the nanobeam.

8.2.1.1 Motion Equations of MEE Euler–Bernoulli Beams

In this section, we will satisfy Maxwell's equation to relate the electromagnetic potentials to the undertaken problem via a set of coupled partial differential equations. For the purpose of satisfying Maxwell's equation in the quasi-static approximation, the applied external electric and magnetic potentials will be modeled as a function of combined linear and cosine variations. In Chapter 6, the electric potential was expressed in Eq. (6.1). Thus, this formula will not be presented again, and the magnetic potential of the beam can be expressed as

$$\bar{\Upsilon}(x, z, t) = -\cos(\xi z)\gamma(x, t) + \frac{2z}{h}\Omega \tag{8.1}$$

in which Ω is the magnetic potential. Now, the electric and magnetic potentials must be converted into electric and magnetic fields. This operation has been performed in Eqs. (6.2) and (6.3) for the electric field. Here, the magnetic field can be obtained by performing the same operations on the magnetic potential of the material. Based on the above statements, the magnetic field $(H_x, H\,)$ can be expressed as

$$H_x = -\frac{\partial \bar{\Upsilon}}{\partial x} = \cos(\xi z)\frac{\partial \gamma(x, t)}{\partial x}, \tag{8.2}$$

$$H\, = -\frac{\partial \bar{\Upsilon}}{\partial z} = -\xi \sin(\xi z)\gamma(x, t) - \frac{2\Omega}{h}. \tag{8.3}$$

Now, the equations of motion should be obtained for a beam made from an MEE material. The equations of motion must be obtained by applying the Hamilton's principle for an MEE nanobeam. Herein, the variation of strain energy can be expressed as

$$\delta \Pi_S = \int_V \left(\sigma_{xx}\delta\varepsilon_{xx} - D_x\delta E_x - D\,\delta E\, - B_x\delta H_x - B\,\delta H\, \right)dV \tag{8.4}$$

in which D and B are electric displacement and magnetic induction, respectively. The above relation can be simplified by inserting Eq. (4.12) instead of the normal strain of the beam and Eqs. (6.2), (6.3), (8.2) and (8.3) instead of electric and magnetic fields, respectively in the Eq. (8.4). Hence, the variation of strain energy of a classical beam is

$$\delta\Pi_S = \int_0^L \left(N\delta\varepsilon_{xx}^0 - M\delta\kappa_{xx}^0\right)dx + \int_0^L \int_A \left(-D_x\cos(\xi z)\frac{\partial\delta\phi}{\partial x} + D\ \xi\sin(\xi z)\delta\phi\right.$$
$$\left. -B_x\cos(\xi z)\frac{\partial\delta\gamma}{\partial x} + B\ \xi\sin(\xi z)\delta\gamma\right)dAdx \qquad (8.5)$$

in which the normal force and bending moment can be achieved by employing Eq. (4.25). Next, it is time to present an expression for the variation of work done by external loading for considering the effects of electromagnetic forces. The variation of work done by external normal in-plane forces applied to the beam can be written as

$$\delta\Pi_W = \int_0^L \left(N^E + N^H\right)\frac{\partial w}{\partial x}\frac{\partial\delta w}{\partial x}dx, \qquad (8.6)$$

where N^E and N^H stand for normal in-plane forces generated due to electric voltage and magnetic potential, respectively. The electromagnetic loads can be computed as

$$N^E = -\int_A \tilde{e}_{31}\frac{2V}{h}dA, \qquad (8.7)$$

$$N^H = -\int_A \tilde{q}_{31}\frac{2\Omega}{h}dA. \qquad (8.8)$$

The variation of kinetic energy of the beam can be obtained in the form presented in Eq. (4.26). Now, the Euler–Lagrange equations of a smart classical beam can be obtained by substituting Eqs. (8.5), (4.26) and (8.6) in Eq. (4.19) and using the nontrivial response of the obtained set of coupled differential equations in the following forms:

$$\frac{\partial N}{\partial x} = I_0\frac{\partial^2 u}{\partial t^2} - I_1\frac{\partial^3 w}{\partial x\partial t^2}, \qquad (8.9)$$

$$\frac{\partial^2 M}{\partial x^2} - \left(N^E + N^H\right)\frac{\partial^2 w}{\partial x^2} = I_0\frac{\partial^2 w}{\partial t^2} + I_1\frac{\partial^3 u}{\partial x\partial t^2} - I_2\frac{\partial^4 w}{\partial x^2\partial t^2}, \qquad (8.10)$$

$$\int_A \left(\cos(\xi z)\frac{\partial D_x}{\partial x} - \xi\sin(\xi z)D\right)dA = 0, \qquad (8.11)$$

$$\int_A \left(\cos(\xi z)\frac{\partial B_x}{\partial x} - \xi\sin(\xi z)B\right)dA = 0. \qquad (8.12)$$

8.2.1.2 Nonlocal Strain Gradient Magnetoelectroelasticity for Euler–Bernoulli Nanobeams

Here, the smart constitutive relationships of the nonlocal strain gradient magnetoelectroelasticity presented in Section 2.3.3 will be obtained for a smart MEE nanosize beam-type element. Using the fundamental relations expressed in Eqs. (2.20)–(2.22), the initial forms of the constitutive equations of an MEE Euler nanobeam can be formulated as

$$\left(1 - \mu^2\nabla^2\right)\sigma_{xx} = \left(1 - \lambda^2\nabla^2\right)\left[\tilde{c}_{11}\varepsilon_{xx} - \tilde{e}_{31}E - \tilde{q}_{31}H\right], \qquad (8.13)$$

$$\left(1 - \mu^2\nabla^2\right)D_x = \left(1 - \lambda^2\nabla^2\right)\left[\tilde{s}_{11}E_x + \tilde{d}_{11}H_x\right], \qquad (8.14)$$

$$\left(1 - \mu^2\nabla^2\right)D = \left(1 - \lambda^2\nabla^2\right)\left[\tilde{e}_{31}\varepsilon_{xx} + \tilde{s}_{33}E + \tilde{d}_{33}H\right], \quad (8.15)$$

$$\left(1 - \mu^2\nabla^2\right)B_x = \left(1 - \lambda^2\nabla^2\right)\left[\tilde{d}_{11}E_x + \tilde{\chi}_{11}H_x\right], \quad (8.16)$$

$$\left(1 - \mu^2\nabla^2\right)B = \left(1 - \lambda^2\nabla^2\right)\left[\tilde{q}_{31}\varepsilon_{xx} + \tilde{d}_{33}E + \tilde{\chi}_{33}H\right], \quad (8.17)$$

where $\tilde{c}_{ij}$, $\tilde{d}_{ij}$, $\tilde{e}_{ij}$, $\tilde{s}_{ij}$, $\tilde{q}_{ij}$ and $\tilde{\chi}_{ij}$ are reduced coefficients of an MEE-FG nanobeam when it is subjected to a plane stress state [164]:

$$\tilde{c}_{11} = c_{11} - \frac{c_{13}^2}{c_{33}}, \tilde{d}_{11} = d_{11}, \tilde{d}_{33} = d_{33} - \frac{e_{33}q_{33}}{c_{33}},$$

$$\tilde{e}_{31} = e_{31} - \frac{c_{13}e_{33}}{c_{33}}, \tilde{s}_{11} = s_{11}, \tilde{s}_{33} = s_{33} - \frac{e_{33}^2}{c_{33}},$$

$$\tilde{q}_{31} = q_{31} - \frac{c_{13}q_{33}}{c_{33}}, \tilde{\chi}_{11} = \chi_{11}, \tilde{\chi}_{33} = \chi_{33} - \frac{q_{33}^2}{c_{33}}. \quad (8.18)$$

On integrating the above equations over the cross-section area of the nanobeam, the following integral forms of the constitutive equations can be obtained:

$$\left(1 - \mu^2\nabla^2\right)N = \left(1 - \lambda^2\nabla^2\right)\left[A_{xx}\frac{\partial u}{\partial x} - B_{xx}\frac{\partial^2 w}{\partial x^2} + A_{31}^e\phi + A_{31}^m\gamma\right] - \left(N^E + N^H\right), \quad (8.19)$$

$$\left(1 - \mu^2\nabla^2\right)M = \left(1 - \lambda^2\nabla^2\right)\left[B_{xx}\frac{\partial u}{\partial x} - D_{xx}\frac{\partial^2 w}{\partial x^2} + E_{31}^e\phi + E_{31}^m\gamma\right] - \left(M^E + M^H\right), \quad (8.20)$$

$$\left(1 - \mu^2\nabla^2\right)\int_A D_x \cos(\xi z)dA = \left(1 - \lambda^2\nabla^2\right)\left[F_{11}^e\frac{\partial\phi}{\partial x} + F_{11}^m\frac{\partial\gamma}{\partial x}\right], \quad (8.21)$$

$$\left(1 - \mu^2\nabla^2\right)\int_A D \xi \sin(\xi z)dA = \left(1 - \lambda^2\nabla^2\right)\left[A_{31}^e\frac{\partial u}{\partial x} + E_{31}^e\frac{\partial^2 w}{\partial x^2} - F_{33}^e\phi + F_{33}^m\gamma\right], \quad (8.22)$$

$$\left(1 - \mu^2\nabla^2\right)\int_A B_x \cos(\xi z)dA = \left(1 - \lambda^2\nabla^2\right)\left[F_{11}^m\frac{\partial\phi}{\partial x} + \chi_{11}^m\frac{\partial\gamma}{\partial x}\right], \quad (8.23)$$

$$\left(1 - \mu^2\nabla^2\right)\int_A B \xi \sin(\xi z)dA = \left(1 - \lambda^2\nabla^2\right)\left[A_{31}^m\frac{\partial u}{\partial x} + E_{31}^m\frac{\partial^2 w}{\partial x^2} - F_{33}^m\phi + \chi_{33}^m\gamma\right]. \quad (8.24)$$

The cross-sectional rigidities and electrical integrals in Eqs. (8.19)–(8.24) were introduced in the Chapter 6. One can find these integrals in Eqs. (6.19)–(6.22). Here, the magnetic integrals can be calculated in the following forms:

$$[A_{31}^m, E_{31}^m] = \int_A [1, z]\tilde{q}_{31}\xi \sin(\xi z)dA, \quad (8.25)$$

$$A_{15}^m = \int_A \tilde{q}_{15} \cos(\xi z)dA, \quad (8.26)$$

$$[F_{11}^m, F_{33}^m] = \int_A [\tilde{d}_{11}\cos^2(\xi z), \tilde{d}_{33}\xi^2\sin^2(\xi z)]dA, \quad (8.27)$$

$$[\chi_{11}^m, \chi_{33}^m] = \int_A [\tilde{\chi}_{11}\cos^2(\xi z), \tilde{\chi}_{33}\xi^2\sin^2(\xi z)]dA. \quad (8.28)$$

Moreover, in Eq. (8.20), the electromagnetic moments can be expressed as

$$M^E = -\int_A \tilde{e}_{31}\frac{2V}{h}z\,dA, \tag{8.29}$$

$$M^H = -\int_A \tilde{q}_{31}\frac{2\Omega}{h}z\,dA. \tag{8.30}$$

8.2.1.3 Governing Equations of MEE Euler–Bernoulli Nanobeams

In this section, we are about to mix the nonlocal strain gradient magnetoelectroelasticity theory with the Euler–Lagrange equations of Euler–Bernoulli beams in order to obtain the final forms of the differential equations of the problem. In order to obtain the final equations, Eqs. (8.19)–(8.24) must be inserted in Eqs. (8.9)–(8.12). Finally, the governing equations can be written in the following forms:

$$\left(1 - \lambda^2\nabla^2\right)\left[A_{xx}\frac{\partial^2 u}{\partial x^2} - B_{xx}\frac{\partial^3 w}{\partial x^3} + A_{31}^e\frac{\partial\phi}{\partial x} + A_{31}^m\frac{\partial\gamma}{\partial x}\right] + \left(1 - \mu^2\nabla^2\right)$$
$$\times\left[-I_0\frac{\partial^2 u}{\partial t^2} + I_1\frac{\partial^3 w}{\partial x\partial t^2}\right] = 0, \tag{8.31}$$

$$\left(1 - \lambda^2\nabla^2\right)\left[B_{xx}\frac{\partial^3 u}{\partial x^3} - D_{xx}\frac{\partial^4 w}{\partial x^4} + E_{31}^e\frac{\partial^2\phi}{\partial x^2} + E_{31}^m\frac{\partial^2\gamma}{\partial x^2}\right] + \left(1 - \mu^2\nabla^2\right)$$
$$\times\left[-I_0\frac{\partial^2 w}{\partial t^2} - I_1\frac{\partial^3 u}{\partial x\partial t^2} + I_2\frac{\partial^4 w}{\partial x^2\partial t^2} - \left(N^E + N^H\right)\frac{\partial^2 w}{\partial x^2}\right] = 0, \tag{8.32}$$

$$\left(1 - \lambda^2\nabla^2\right)\left[A_{31}^e\frac{\partial u}{\partial x} - E_{31}^e\frac{\partial^2 w}{\partial x^2} + F_{11}^e\frac{\partial^2\phi}{\partial x^2} + F_{11}^m\frac{\partial^2\gamma}{\partial x^2} - F_{33}^e\phi - F_{33}^m\gamma\right] = 0, \tag{8.33}$$

$$\left(1 - \lambda^2\nabla^2\right)\left[A_{31}^m\frac{\partial u}{\partial x} - E_{31}^m\frac{\partial^2 w}{\partial x^2} + F_{11}^m\frac{\partial^2\phi}{\partial x^2} + \chi_{11}^m\frac{\partial^2\gamma}{\partial x^2} - F_{33}^m\phi - \chi_{33}^m\gamma\right] = 0. \tag{8.34}$$

8.2.1.4 Wave Solution of the MEE Euler–Bernoulli Nanobeams

In this section, the obtained equations will be solved with an analytical solution for obtaining the wave frequency and phase velocity values of MEE-FG nanobeams on the basis of the classical theory of the beams. As explained in Chapter 6, the solution function of the electric potential is considered to be the same as that of the displacement fields of the beam. Similarly, the magnetic potential can be assumed to have the following form:

$$\gamma = \hat{\Gamma}e^{i(\beta x - \omega t)}. \tag{8.35}$$

On substituting Eqs. (1.1), (1.2), (6.26) and (8.35) in Eqs. (8.31)–(8.34), an eigenvalue equation similar to Eq. (1.3) can be obtained. However, the Δ vector in this problem can be expressed as

$$\Delta = [U, W, \hat{\Phi}, \hat{\Gamma}]^T. \tag{8.36}$$

Due to the mentioned differences between this problem and problems of former chapters, the details of this solution are different from those of the previous analyses. Indeed, the stiffness matrix of the problem varies when analyzing an MEE-FG nanobeam comparison with an elastic nanobeam or a piezoelectric nanobeam. So, the solution functions must be substituted in the governing equations to obtain the matrix form of the problem. Therefore, the components of the stiffness matrix of an MEE-FG nanobeam can be written as

$$
\begin{aligned}
k_{11} &= -\left(1+\lambda^2\beta^2\right)A_{xx}\beta^2, \\
k_{12} &= i\left(1+\lambda^2\beta^2\right)B_{xx}\beta^3, \\
k_{13} &= i\left(1+\lambda^2\beta^2\right)\beta A_{31}^e, \\
k_{14} &= i\left(1+\lambda^2\beta^2\right)\beta A_{31}^m, \\
k_{22} &= -\left(1+\lambda^2\beta^2\right)D_{xx}\beta^4 + \left(1+\mu^2\beta^2\right)\beta^2\left(N^E + N^H\right), \\
k_{23} &= -\left(1+\lambda^2\beta^2\right)\beta^2 E_{31}^e, \\
k_{24} &= -\left(1+\lambda^2\beta^2\right)\beta^2 E_{31}^m, \\
k_{33} &= -\left(1+\lambda^2\beta^2\right)\left[F_{33}^e + F_{11}^e\beta^2\right], \\
k_{34} &= -\left(1+\lambda^2\beta^2\right)\left[F_{33}^m + F_{11}^m\beta^2\right], \\
k_{44} &= -\left(1+\lambda^2\beta^2\right)\left[\chi_{33}^m + \chi_{11}^m\beta^2\right].
\end{aligned}
\tag{8.37}
$$

Also, the nonzero arrays of mass matrix can be assumed to be the same as those in Eq. (4.52).

8.2.2 Refined Sinusoidal MEE Nanobeams

Our objective in this section is to extend the Hamilton's principle to an MEE-FG nanobeam within the framework of a refined sinusoidal beam hypothesis. The effects of shear deflection will be covered in this new formulation.

8.2.2.1 Equations of Motion of MEE Refined Sinusoidal Beams

In this section, the equations of motion of beam-type elements will be derived with respect to the effects of shear deflection. We are now familiar with the electromagnetic potentials of an MEE beam. Thus, in this section, the derivation procedure starts from calculating the variations of the system's energies. First of all, the variation of strain energy can be expressed as

$$
\delta\Pi_S = \int_V \left(\sigma_{xx}\delta\varepsilon_{xx} + \sigma_x\ \delta\varepsilon_x\ - D_x\delta E_x - D\ \delta E\ - B_x\delta H_x - B\ \delta H\ \right)dV.
\tag{8.38}
$$

Using the nonzero strains of the refined theory of beams (see Eq. (4.17)) and the relations between the electromagnetic fields and potentials (see Eqs. (6.2) and (6.3) for electric relationships and Eqs. (8.2) and (8.2) for magnetic relationships), Eq. (8.38) can be rewritten in the following form:

$$
\begin{aligned}
\delta\Pi_S &= \int_0^L \left(N\delta\varepsilon_{xx}^0 - M^b\delta\kappa_{xx}^b - M^s\delta\kappa_{xx}^s + Q_x\ \delta\varepsilon_x^0\ \right)dx \\
&+ \int_0^L \int_A \left(-D_x\cos(\xi z)\frac{\partial\delta\phi}{\partial x} + D\ \xi\sin(\xi z)\delta\phi\right)dA dx \\
&+ \int_0^L \int_A \left(-B_x\cos(\xi z)\frac{\partial\delta\gamma}{\partial x} + B\ \xi\sin(\xi z)\delta\gamma\right)dA dx.
\end{aligned}
\tag{8.39}
$$

In the above equation, the stress resultants N, M^b, M^s and Q_x can be calculated using their primary definition in Eq. (4.32). In addition to the variation of strain energy, the variation of work done by external forces can be written in the following form:

$$\delta\Pi_W = \int_0^L \left(N^E + N^H\right)\frac{\partial(w_b + w_s)}{\partial x}\frac{\partial\delta(w_b + w_s)}{\partial x}dx \tag{8.40}$$

in which N^E and N^H are the forces generated from electric and magnetic fields, respectively. Also, the variation of kinetic energy is in the same form as presented in Eq. (4.33). Incorporating these new variations in the definition of the Hamilton's principle, the Euler–Lagrange equations of the higher-order beam can be obtained. These coupled equations can be obtained by inserting Eqs. (8.39), (8.40) and (4.33) in Eq. (4.19) in the following forms:

$$\frac{\partial N}{\partial x} = I_0\frac{\partial^2 u}{\partial t^2} - I_1\frac{\partial^3 w_b}{\partial x\partial t^2} - J_1\frac{\partial^3 w_s}{\partial x\partial t^2}, \tag{8.41}$$

$$\frac{\partial^2 M^b}{\partial x^2} - \left(N^E + N^H\right)\frac{\partial^2(w_b + w_s)}{\partial x^2} = I_0\frac{\partial^2(w_b + w_s)}{\partial t^2} + I_1\frac{\partial^3 u}{\partial x\partial t^2}$$
$$- I_2\frac{\partial^4 w_b}{\partial x^2\partial t^2} - J_2\frac{\partial^4 w_s}{\partial x^2\partial t^2}, \tag{8.42}$$

$$\frac{\partial^2 M^s}{\partial x^2} + \frac{\partial Q_x}{\partial x} - \left(N^E + N^H\right)\frac{\partial^2(w_b + w_s)}{\partial x^2} = I_0\frac{\partial^2(w_b + w_s)}{\partial t^2}$$
$$+ J_1\frac{\partial^3 u}{\partial x\partial t^2} - J_2\frac{\partial^4 w_b}{\partial x^2\partial t^2} - K_2\frac{\partial^4 w_s}{\partial x^2\partial t^2}, \tag{8.43}$$

$$\int_A \left(\cos(\xi z)\frac{\partial D_x}{\partial x} - \xi\sin(\xi z)D\right)dA = 0, \tag{8.44}$$

$$\int_A \left(\cos(\xi z)\frac{\partial B_x}{\partial x} - \xi\sin(\xi z)B\right)dA = 0. \tag{8.45}$$

8.2.2.2 Nonlocal Strain Gradient Magnetoelectroelasticity for Refined Sinusoidal Nanobeams

Herein, the effects of the small size of the nanobeam on the constitutive equations of the beam will be applied. As in the previous section on Euler–Bernoulli nanobeams (see Section 8.2.1), Eqs. (8.20)–(8.22) will be expanded in this section for the higher-order sinusoidal nanobeams. Thus, the MEE equations of the nanobeam can be written as

$$\left(1 - \mu^2\nabla^2\right)\sigma_{xx} = \left(1 - \lambda^2\nabla^2\right)\left[\tilde{c}_{11}\varepsilon_{xx} - \tilde{e}_{31}E - \tilde{q}_{31}H\right], \tag{8.46}$$

$$\left(1 - \mu^2\nabla^2\right)\sigma_x = \left(1 - \lambda^2\nabla^2\right)\left[\tilde{c}_{55}\varepsilon_x - \tilde{e}_{15}E_x - \tilde{q}_{15}H_x\right], \tag{8.47}$$

$$\left(1 - \mu^2\nabla^2\right)D_x = \left(1 - \lambda^2\nabla^2\right)\left[\tilde{e}_{15}\varepsilon_x + \tilde{s}_{11}E_x + \tilde{d}_{11}H_x\right], \tag{8.48}$$

$$\left(1 - \mu^2\nabla^2\right)D = \left(1 - \lambda^2\nabla^2\right)\left[\tilde{e}_{31}\varepsilon_{xx} + \tilde{s}_{33}E + \tilde{d}_{33}H\right], \tag{8.49}$$

$$\left(1 - \mu^2\nabla^2\right)B_x = \left(1 - \lambda^2\nabla^2\right)\left[\tilde{q}_{15}\varepsilon_x + \tilde{d}_{11}E_x + \tilde{\chi}_{11}H_x\right], \tag{8.50}$$

$$\left(1 - \mu^2\nabla^2\right)B = \left(1 - \lambda^2\nabla^2\right)\left[\tilde{q}_{31}\varepsilon_{xx} + \tilde{d}_{33}E + \tilde{\chi}_{33}H\right]. \tag{8.51}$$

Now, the final relations required can be obtained by integrating the aforementioned equations over the cross-sectional area of the nanobeam. The following relations can be obtained on integration:

$$
\left(1 - \mu^2 \nabla^2\right) N = \left(1 - \lambda^2 \nabla^2\right) \left[A_{xx} \frac{\partial u}{\partial x} - B_{xx} \frac{\partial^2 w_b}{\partial x^2} - B_{xx}^s \frac{\partial^2 w_s}{\partial x^2} + A_{31}^e \phi + A_{31}^m \gamma \right]
$$
$$
- \left(N^E + N^H\right), \tag{8.52}
$$

$$
\left(1 - \mu^2 \nabla^2\right) M^b = \left(1 - \lambda^2 \nabla^2\right) \left[B_{xx} \frac{\partial u}{\partial x} - D_{xx} \frac{\partial^2 w_b}{\partial x^2} - D_{xx}^s \frac{\partial^2 w_s}{\partial x^2} + E_{31}^e \phi + E_{31}^m \gamma \right]
$$
$$
- \left(M_b^E + M_b^H\right), \tag{8.53}
$$

$$
\left(1 - \mu^2 \nabla^2\right) M^s = \left(1 - \lambda^2 \nabla^2\right) \left[B_{xx}^s \frac{\partial u}{\partial x} - D_{xx}^s \frac{\partial^2 w_b}{\partial x^2} - H_{xx}^s \frac{\partial^2 w_s}{\partial x^2} + F_{31}^e \phi + F_{31}^m \gamma \right]
$$
$$
- \left(M_s^E + M_s^H\right), \tag{8.54}
$$

$$
\left(1 - \mu^2 \nabla^2\right) Q_x = \left(1 - \lambda^2 \nabla^2\right) \left[A_x^s \frac{\partial w_s}{\partial x} - A_{15}^e \frac{\partial \phi}{\partial x} - A_{15}^m \frac{\partial \gamma}{\partial x} \right], \tag{8.55}
$$

$$
\left(1 - \mu^2 \nabla^2\right) \int_A D_x \cos(\xi z) dA = \left(1 - \lambda^2 \nabla^2\right) \left[E_{15}^e \frac{\partial w_s}{\partial x} + F_{11}^e \frac{\partial \phi}{\partial x} + F_{11}^m \frac{\partial \gamma}{\partial x} \right], \tag{8.56}
$$

$$
\left(1 - \mu^2 \nabla^2\right) \int_A D\,\xi \sin(\xi z) dA = \left(1 - \lambda^2 \nabla^2\right) \left[A_{31}^e \frac{\partial u}{\partial x} + E_{31}^e \frac{\partial^2 w_b}{\partial x^2} + F_{31}^e \frac{\partial^2 w_s}{\partial x^2} - F_{33}^e \phi - F_{33}^m \gamma \right],
$$
$$
\tag{8.57}
$$

$$
\left(1 - \mu^2 \nabla^2\right) \int_A B_x \cos(\xi z) dA = \left(1 - \lambda^2 \nabla^2\right) \left[E_{15}^m \frac{\partial w_s}{\partial x} + F_{11}^m \frac{\partial \phi}{\partial x} + \chi_{11}^m \frac{\partial \gamma}{\partial x} \right], \tag{8.58}
$$

$$
\left(1 - \mu^2 \nabla^2\right) \int_A B\,\xi \sin(\xi z) dA = \left(1 - \lambda^2 \nabla^2\right) \left[A_{31}^m \frac{\partial u}{\partial x} + E_{31}^m \frac{\partial^2 w_b}{\partial x^2} + F_{31}^m \frac{\partial^2 w_s}{\partial x^2} - F_{33}^m \phi - \chi_{33}^m \gamma \right].
$$
$$
\tag{8.59}
$$

In Eqs. (8.52)–(8.59), there are some integrals that are related to the elastic, electric and magnetic material properties of the refined nanobeam. Also, some of them belong to the shear approximation shape function of the sinusoidal beam model. Some of these integrals related to the electromechanical properties were defined in Chapter 6. Herein, the other integrals related to the magnetic properties of the smart FGM are defined as

$$
[A_{31}^m, E_{31}^m, F_{31}^m] = \int_A [1, z, f(z)] \tilde{q}_{31} \xi \sin(\xi z) dA, \tag{8.60}
$$

$$
[A_{15}^m, E_{15}^m] = \int_A [1, g(z)] \tilde{q}_{15} \cos(\xi z) dA, \tag{8.61}
$$

$$
[F_{11}^m, F_{33}^m] = \int_A [\tilde{d}_{11} \cos^2(\xi z), \tilde{d}_{33} \xi^2 \sin^2(\xi z)] dA, \tag{8.62}
$$

$$
[\chi_{11}^m, \chi_{33}^m] = \int_A [\tilde{\chi}_{11} \cos^2(\xi z), \tilde{\chi}_{33} \xi^2 \sin^2(\xi z)] dA. \tag{8.63}
$$

In addition to the above equations, the electromagnetic resultants used in Eqs. (8.52)–(8.54) can be defined as follows:

$$[N^E, M_b^E, M_s^E] = -\int_A [1, z, f(z)] \tilde{e}_{31} \frac{2V}{h} dA, \tag{8.64}$$

$$[N^H, M_b^H, M_s^H] = -\int_A [1, z, f(z)] \tilde{q}_{31} \frac{2\Omega}{h} dA. \tag{8.65}$$

8.2.2.3 Governing Equations of MEE Refined Sinusoidal Nanobeams

Here, the constitutive equations obtained in the previous section will be used in order to obtain the nonlocal governing equations of the refined shear deformable smart nanobeam. By substituting Eqs. (8.52)–(8.59) in Eqs. (8.41)–(8.45), the final governing equations of the refined sinusoidal MEE-FG nanobeams can be obtained as follows:

$$\left(1 - \lambda^2 \nabla^2\right) \left[A_{xx} \frac{\partial^2 u}{\partial x^2} - B_{xx} \frac{\partial^3 w_b}{\partial x^3} - B_{xx}^s \frac{\partial^3 w_s}{\partial x^3} + A_{31}^e \frac{\partial \phi}{\partial x} + A_{31}^m \frac{\partial \gamma}{\partial x} \right]$$
$$+ \left(1 - \mu^2 \nabla^2\right) \left[-I_0 \frac{\partial^2 u}{\partial t^2} + I_1 \frac{\partial^3 w_b}{\partial x \partial t^2} + J_1 \frac{\partial^3 w_s}{\partial x \partial t^2} \right] = 0, \tag{8.66}$$

$$\left(1 - \lambda^2 \nabla^2\right) \left[B_{xx} \frac{\partial^3 u}{\partial x^3} - D_{xx} \frac{\partial^4 w_b}{\partial x^4} - D_{xx}^s \frac{\partial^4 w_s}{\partial x^4} + E_{31}^e \frac{\partial^2 \phi}{\partial x^2} + E_{31}^m \frac{\partial^2 \gamma}{\partial x^2} \right] + \left(1 - \mu^2 \nabla^2\right)$$
$$\left[-I_0 \frac{\partial^2 (w_b + w_s)}{\partial t^2} - I_1 \frac{\partial^3 u}{\partial x \partial t^2} + I_2 \frac{\partial^4 w_b}{\partial x^2 \partial t^2} + J_2 \frac{\partial^4 w_s}{\partial x^2 \partial t^2} - \left(N^E + N^H\right) \frac{\partial^2 (w_b + w_s)}{\partial x^2} \right] = 0, \tag{8.67}$$

$$\left(1 - \lambda^2 \nabla^2\right) \left[B_{xx}^s \frac{\partial^3 u}{\partial x^3} - D_{xx}^s \frac{\partial^4 w_b}{\partial x^4} - H_{xx}^s \frac{\partial^4 w_s}{\partial x^4} + A_x^s \frac{\partial^2 w_s}{\partial x^2} + \left(F_{31}^e - A_{15}^e\right) \frac{\partial^2 \phi}{\partial x^2} \right.$$
$$+ \left(F_{31}^m - A_{15}^m\right) \frac{\partial^2 \gamma}{\partial x^2} \right] + \left(1 - \mu^2 \nabla^2\right) \left[-I_0 \frac{\partial^2 (w_b + w_s)}{\partial t^2} - J_1 \frac{\partial^3 u}{\partial x \partial t^2} + J_2 \frac{\partial^4 w_b}{\partial x^2 \partial t^2} \right.$$
$$+ K_2 \frac{\partial^4 w_s}{\partial x^2 \partial t^2} - \left(N^E + N^H\right) \frac{\partial^2 (w_b + w_s)}{\partial x^2} \right] = 0, \tag{8.68}$$

$$\left(1 - \lambda^2 \nabla^2\right) \left[A_{31}^e \frac{\partial u}{\partial x} + E_{31}^e \frac{\partial^2 w_b}{\partial x^2} + \left(F_{31}^e - E_{15}^e\right) \frac{\partial^2 w_s}{\partial x^2} - F_{11}^e \frac{\partial^2 \phi}{\partial x^2} \right.$$
$$\left. - F_{33}^e \phi - F_{11}^m \frac{\partial^2 \gamma}{\partial x^2} - F_{33}^m \gamma \right] = 0, \tag{8.69}$$

$$\left(1 - \lambda^2 \nabla^2\right) \left[A_{31}^m \frac{\partial u}{\partial x} + E_{31}^m \frac{\partial^2 w_b}{\partial x^2} + \left(F_{31}^m - E_{15}^m\right) \frac{\partial^2 w_s}{\partial x^2} - F_{11}^m \frac{\partial^2 \phi}{\partial x^2} \right.$$
$$\left. - F_{33}^m \phi - \chi_{11}^m \frac{\partial^2 \gamma}{\partial x^2} - \chi_{33}^m \gamma \right] = 0. \tag{8.70}$$

8.2.2.4 Wave Solution of MEE Refined Sinusoidal Nanobeams

Now, Eqs. (8.66)–(8.70) will be solved via the analytical solution presented in the Chapter 1. For this reason, Eqs. (1.5)–(1.7), (6.26) and (8.35) must be substituted in Eqs. (8.66)–(8.70). Once the aforementioned substitution is performed, the formulated problem will get transformed into an eigenvalue problem with the following amplitude vector:

$$\Delta = [U, W_b, W_s, \hat{\Phi}, \hat{\Gamma}]^T. \tag{8.71}$$

The arrays of the stiffness matrix of this problem are as follows:

$$k_{11} = -(1 + \lambda^2\beta^2)A_{xx}\beta^2,$$
$$k_{12} = i(1 + \lambda^2\beta^2)B_{xx}\beta^3,$$
$$k_{13} = i(1 + \lambda^2\beta^2)B_{xx}^s\beta^3,$$
$$k_{14} = i(1 + \lambda^2\beta^2)\beta A_{31}^e,$$
$$k_{15} = i(1 + \lambda^2\beta^2)\beta A_{31}^m,$$
$$k_{22} = -(1 + \lambda^2\beta^2)D_{xx}\beta^4 + (1 + \mu^2\beta^2)\beta^2(N^E + N^H),$$
$$k_{23} = -(1 + \lambda^2\beta^2)D_{xx}^s\beta^4 + (1 + \mu^2\beta^2)\beta^2(N^E + N^H),$$
$$k_{24} = -(1 + \lambda^2\beta^2)\beta^2 E_{31}^e,$$
$$k_{25} = -(1 + \lambda^2\beta^2)\beta^2 E_{31}^m,$$
$$k_{33} = -(1 + \lambda^2\beta^2)\left[H_{xx}^s\beta^4 + A_x^s\ \beta^2\right] + (1 + \mu^2\beta^2)\beta^2(N^E + N^H),$$
$$k_{34} = -(1 + \lambda^2\beta^2)(F_{31}^e - E_{15}^e)\beta^2,$$
$$k_{35} = -(1 + \lambda^2\beta^2)(F_{31}^m - E_{15}^m)\beta^2,$$
$$k_{44} = -(1 + \lambda^2\beta^2)(F_{33}^e - F_{11}^e\beta^2),$$
$$k_{45} = -(1 + \lambda^2\beta^2)(F_{33}^m - F_{11}^m\beta^2),$$
$$k_{55} = -(1 + \lambda^2\beta^2)(\chi_{33}^m - \chi_{11}^m\beta^2). \tag{8.72}$$

It is worth mentioning that the nonzero arrays of the mass matrix can be found in Eq. (4.54).

8.2.3 Numerical Results about MEE Nanobeams

In this section, the wave propagation characteristics of MEE-FG nanobeams will be studied with numerical illustrations. The smart nanobeam is presumed to be made from $BaTiO_3$ and $CoFe_2O_3$. The material properties can be found in reference [164]. Due to the limited effects of electromagnetic potentials on the wave dispersion responses of MEE-FG nanobeams at small wave numbers, the phase velocity of the nanobeams is selected for the investigations, and the wave frequency is not employed in this section. Actually, the wave frequency curves of the nanostructures coincide at small wave numbers at various amounts of any desired variant.

Figure 8.1 is presented to probe the wave dispersion curves of smart MEE nanobeams at various amounts of magnetic potential at a specific applied electric voltage. It can be found that the behavior of the nanobeam is the same as in previous chapters when the scale factor is varied. Actually, an increase in this parameter means enhancement of the system's stiffness, and this reality results in a jump in the phase velocity curve of the nanobeam. On the other hand, it is obvious that the phase velocity of the nanobeam can be intensified by increasing the magnetic potential in a finite range of wave numbers, particularly at wave numbers less than 0.05 (1/nm).

In the next illustration (Figure 8.2), the variation of phase velocity with the magnetic potential is plotted at various gradient indices with respect to both positive and negative applied electric voltages. It is clear that the phase speed of the dispersed waves decreases when using high gradient indices. Actually, the stiffness of the nanobeam reduces as the gradient index increases. Besides, the phase velocity increases when a greater value is assigned to the magnetic potential of the smart MEE-FG nanobeam. Indeed, it can be

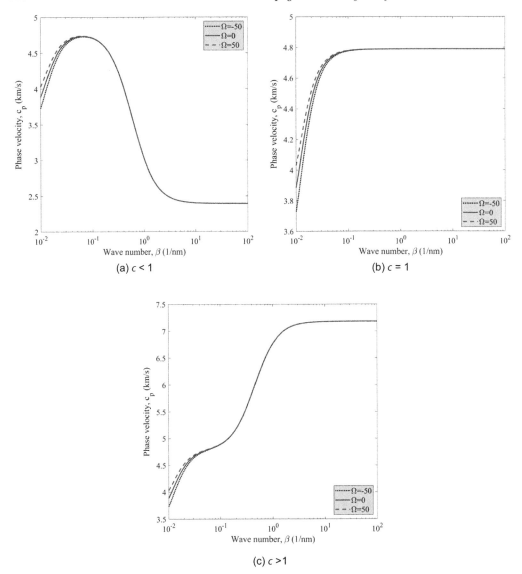

FIGURE 8.1
Variation of phase velocity of MEE-FG nanobeam with wave number at various amounts
of magnetic potential $(p = 2, V = 0)$.

realized that the external force generated from changing the magnetic potential enhances the
stiffness of the nanobeam. Thus, it is natural to see that the dynamic response of the
nanobeam increases at a higher magnetic potential. Also, one should pay attention to
the fact that the system shows greater phase speeds when a negative voltage is connected
to the nanostructure compared with the case of connecting the nanobeam to a positive
electric voltage. In fact, the heat generated from raising the voltage results in a decrease in
the stiffness of the nanobeam.

Moreover, the effect of the nonlocal parameter on the phase speed of the dispersed waves
in a MEE nanobeam can be seen in Figure 8.3. From this diagram, it is evident that the
phase velocity of the nanobeam can be decreased by increasing either the nonlocal parameter

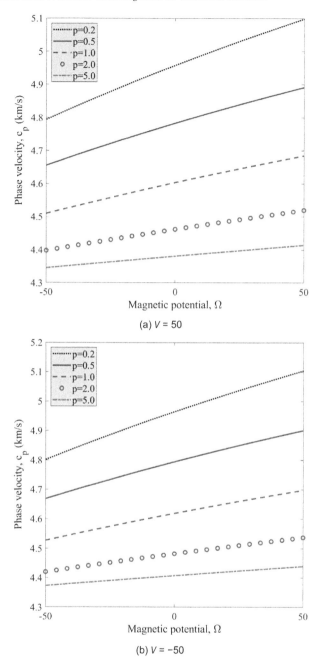

FIGURE 8.2
Variation of phase velocity of MEE-FG nanobeam with magnetic potential at various gradient indices ($p = 2$, $\beta = 0.02$ (1/nm), $c < 1$).

or the applied electric voltage. As stated in previous explanations, both these variants can lessen the stiffness of the nanobeam when they are increased. It is worth mentioning that the phase velocity of the nanobeam can be intensified at any desired nonlocal coefficient when the magnetic potential of the nanobeam is increased. One should note the fact that the changes in this diagram are observable because of the chosen wave number which is in the domain $\beta < 0.05$ (1/nm).

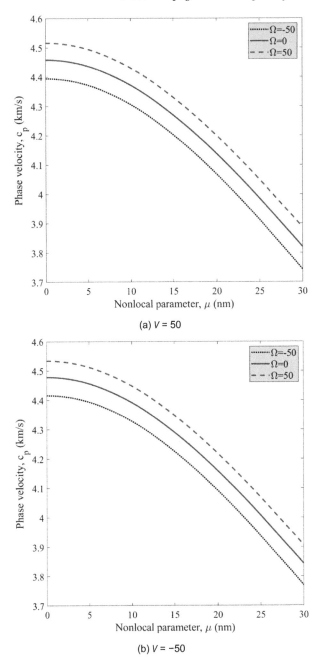

FIGURE 8.3
Variation of phase velocity of MEE-FG nanobeam with nonlocal parameter for various amounts of magnetic potential ($p = 2$, $\beta = 0.02$ (1/nm), $\lambda = 1$ nm).

The effect of length-scale parameter on the phase speed curves of smart MEE nanobeams can be observed in Figure 8.4. As stated in the Chapter 2, the length-scale parameter describes a completely different trend compared with the nonlocal parameter. Henceforward, it is natural to see that the phase velocity increases on increasing the length-scale parameter, but this speed decreases when the nonlocal parameter increases. Again it can be seen that the mechanical response of the nanobeam can be increased by increasing the value of the

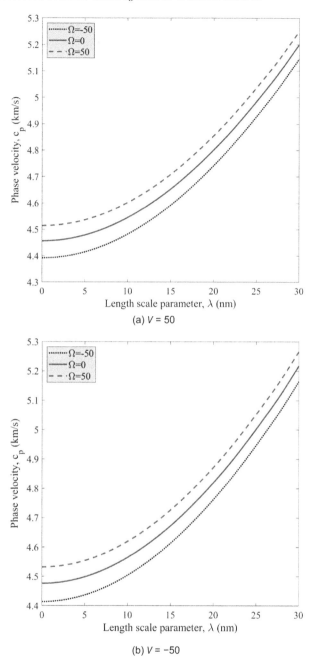

FIGURE 8.4

Variation of phase velocity of MEE-FG nanobeam with length-scale parameter at various amounts of magnetic potential ($p = 2$, $\beta = 0.02$ (1/nm), $\mu = 1$ nm).

magnetic potential. In addition to the aforementioned issues, it is clear that applying positive electric voltages can result in waves with reduced speeds in comparison with situations in which a negative voltage is chosen.

The effect of using a porous smart MEE-FGM in the nanobeam on the phase velocity behaviors of the nanostructure is shown in Figure 8.5. In this diagram, the variation of the phase velocity of the nanobeam with the applied electric voltage is plotted at various

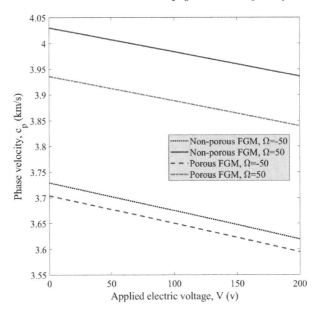

FIGURE 8.5
Variation of phase velocity of MEE-FG nanobeam with applied electric voltage at various amounts of magnetic potential for both porous and nonporous materials ($p = 2$, $\beta = 0.01$ (1/nm), $c < 1$).

magnetic potentials. It is observable that the speed of the scattered waves decreases as the applied voltage increases. In addition, it is clear that at any amount of the magnetic potential, the nonporous nanobeam possesses a higher phase speed compared with the porous one. The reason for this phenomenon is that in porous materials, the existence of defects results in a decrease in the stiffness of the material. Therefore, the phase velocity of the nanobeam reduces due to the direct relation between dynamic response of a continuous system and its stiffness.

Finally, Figure 8.6 shows the effect of the existence of probable porosities in the smart MEE-FGM on the wave propagation curves of nanobeams. This diagram consists of two subplots for both positive and negative applied voltages. It is clear that the phase velocity of the smart nanobeam has a higher magnitude whenever a lower voltage is selected. This trend is observable more at small wave numbers. Besides, it can be observed that at any magnetic potential, nonporous smart nanobeams can support higher phase speeds compared with porous ones. In other words, in porous smart nanostructures, the speed of the scattered waves can decrease due to defects that can be found in the microstructure of porous materials. Also, it must be noted that the effect of porosity on the dispersion curves of nanobeams is permanent. Indeed, one can see the effect of the existence of porosities in the medium on the dispersion curves of the nanobeam at all wave numbers considered.

8.3   Analysis of MEE-FG Nanoplates

Now, the wave propagation analysis will be presented in the following sections for an MEE-FG nanoplate. The equations of motion will be obtained for the nanoplate employing the dynamic form of the principle of virtual work. A new expression will be presented for the

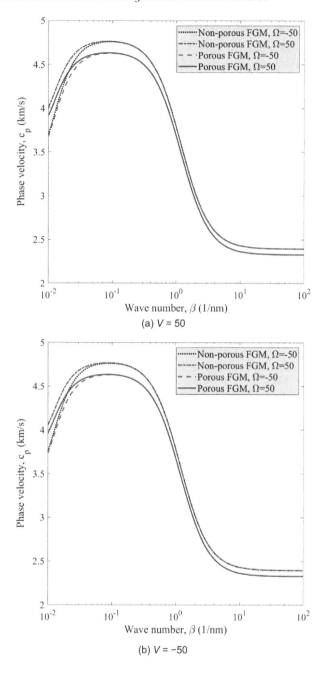

(a) $V = 50$

(b) $V = -50$

FIGURE 8.6
Variation of phase velocity of MEE-FG nanobeam with wave number at various amounts of magnetic potential ($p = 2$, $c < 1$).

magnetic potential of the nanoplate to obtain the magnetic fields of the nanostructure. As in Section 8.2, the effects of tiny dimensions of the structure will be covered by extending the nonlocal stress–strain gradient theory. It is worth mentioning that, at first, the formulations will be presented for Kirchhoff-Love plates, and afterward, the formulations will be obtained for the refined shear deformable plates.

8.3.1   Kirchhoff–Love MEE Nanoplates

Here, we will employ the classical theory of the plates incorporated with the Hamilton's principle for deriving the Euler–Lagrange equations of the MEE-FG plate. Next, the effects of size dependency will be applied on the basis of the nonlocal strain gradient elasticity theorem.

8.3.1.1   Equations of Motion of MEE Kirchhoff–Love Plates

Similar to the previous section about nanobeams, here we must start with the introduction of electromagnetic potentials for a MEE plate. As discussed in Chapter 6, the electric potential of a plate differs from that of a beam. In fact, the effect of y-direction must be included in the analysis of a plate-type structure. Here, we will not present the equations of electric potential again. These equations (Eqs. (6.57)–(6.60)) can be found in Chapter 6. The magnetic potential of an MEE plate can be written as

$$\bar{\Upsilon}(x,y,z,t) = -\cos(\xi z)\gamma(x,y,t) + \frac{2z}{h}\Omega \tag{8.73}$$

in which Ω is the magnetic potential applied to the structure in terms of Wb/m. By differentiating the magnetic potential with respect to the components of the coordinate system, the magnetic fields $(H_x, H_y, H\ )$ can be expressed as

$$H_x = -\frac{\partial\bar{\Upsilon}}{\partial x} = \cos(\xi z)\frac{\partial\gamma(x,y,t)}{\partial x}, \tag{8.74}$$

$$H_y = -\frac{\partial\bar{\Upsilon}}{\partial y} = \cos(\xi z)\frac{\partial\gamma(x,y,t)}{\partial x}, \tag{8.75}$$

$$H\ = -\frac{\partial\bar{\Upsilon}}{\partial z} = -\xi\sin(\xi z)\gamma(x,y,t) - \frac{2\Omega}{h}. \tag{8.76}$$

Now, the expression for the variation of strain energy can be obtained for an MEE-FG plate by means of the classical kinematic theory of the plates. The variation of the strain energy of an MEE-FG plate is

$$\delta\Pi_S = \int\limits_V \big(\sigma_{xx}\delta\varepsilon_{xx} + \sigma_{yy}\delta\varepsilon_{yy} + \sigma_{xy}\delta\varepsilon_{xy} - D_x\delta E_x - D_y\delta E_y$$

$$- D\ \delta E\ - B_x\delta H_x - B_y\delta H_y - B\ \delta H\ )dV. \tag{8.77}$$

The above equation can be rewritten by substituting the strains of the classical theory and the relations of electromagnetic fields in it. To do this simplification, Eqs. (4.59)–(4.61), (6.58)–(6.60) and (8.74)–(8.76) must be inserted in Eq. (8.77). Thus, the variation of strain energy can be expressed as

$$\delta\Pi_S = \int\limits_A \Bigg[\big(N_{xx}\delta\varepsilon_{xx}^0 - M_{xx}\delta\kappa_{xx}^0 + N_{yy}\delta\varepsilon_{yy}^0 - M_{yy}\delta\kappa_{yy}^0 + N_{xy}\delta\varepsilon_{xy}^0 - M_{xy}\delta\kappa_{xy}^0\big)$$

$$+ \int\limits_{-\frac{h}{2}}^{\frac{h}{2}} \Big(-D_x\cos(\xi z)\frac{\partial\delta\phi}{\partial x} - D_y\cos(\xi z)\frac{\partial\delta\phi}{\partial y} + D\ \xi\sin(\xi z)\delta\phi \Big)dz$$

$$+ \int\limits_{-\frac{h}{2}}^{\frac{h}{2}} \Big(-B_x\cos(\xi z)\frac{\partial\delta\gamma}{\partial x} - B_y\cos(\xi z)\frac{\partial\delta\gamma}{\partial y} + B\ \xi\sin(\xi z)\delta\gamma \Big)dz \Bigg]dA. \tag{8.78}$$

N_{ij}, M_{ij} $(i, j = x, y)$ in the above equation have been defined in Eq. (4.73). Moreover, the effects of external electromagnetic forces on the Lagrangian of the plate will be included in determining the variation of work done by external forces. Therefore, the variation of work done by external forces can be expressed as

$$\delta \Pi_W = \int_A \left[N^E + N^H \right] \left(\frac{\partial w}{\partial x} \frac{\partial \delta w}{\partial x} + \frac{\partial w}{\partial y} \frac{\partial \delta w}{\partial y} \right) dA \tag{8.79}$$

in which N^E and N^H are the applied electric and magnetic forces that are produced because of the difference between voltage and magnetic potential. Now, the Euler–Lagrange equations of the plate can be obtained. It must be mentioned that the variation of kinetic energy of the plate can be written as presented in Eq. (4.74). Once Eqs. (8.78), (8.79) and (4.74) are inserted in Eq. (4.19), the Euler–Lagrange equations of the plate can be written as follows:

$$\frac{\partial N_{xx}}{\partial x} + \frac{\partial N_{xy}}{\partial y} = I_0 \frac{\partial^2 u}{\partial t^2} - I_1 \frac{\partial^3 w}{\partial x \partial t^2}, \tag{8.80}$$

$$\frac{\partial N_{xy}}{\partial x} + \frac{\partial N_{yy}}{\partial y} = I_0 \frac{\partial^2 v}{\partial t^2} - I_1 \frac{\partial^3 w}{\partial y \partial t^2}, \tag{8.81}$$

$$\frac{\partial^2 M_{xx}}{\partial x^2} + 2 \frac{\partial^2 M_{xy}}{\partial x \partial y} + \frac{\partial^2 M_{yy}}{\partial y^2} - \left(N^E + N^H \right) \left(\frac{\partial^2 w}{\partial x^2} + \frac{\partial^2 w}{\partial y^2} \right)$$
$$= I_0 \frac{\partial^2 w}{\partial t^2} + I_1 \left(\frac{\partial^3 u}{\partial x \partial t^2} + \frac{\partial^3 v}{\partial y \partial t^2} \right) - I_2 \left(\frac{\partial^4 w}{\partial x^2 \partial t^2} + \frac{\partial^4 w}{\partial y^2 \partial t^2} \right), \tag{8.82}$$

$$\int_A \left(\cos(\xi z) \frac{\partial D_x}{\partial x} + \cos(\xi z) \frac{\partial D_y}{\partial y} - \xi \sin(\xi z) D \right) dA = 0, \tag{8.83}$$

$$\int_A \left(\cos(\xi z) \frac{\partial B_x}{\partial x} + \cos(\xi z) \frac{\partial B_y}{\partial y} - \xi \sin(\xi z) B \right) dA = 0. \tag{8.84}$$

8.3.1.2 Nonlocal Strain Gradient Magnetoelectroelasticity for Kirchhoff–Love Nanoplates

Herein, the small scale effects of nanoplate will be applied by extending the nonlocal strain gradient magnetoelectroelasticity to the nanoplates modeled according to classical plate model. By extending the constitutive relations presented in Eqs. (2.20)–(2.22) to a classical plate, one can obtain the following constitutive relations:

$$(1 - \mu^2 \nabla^2) \sigma_{xx} = (1 - \lambda^2 \nabla^2) \left[\tilde{c}_{11} \varepsilon_{xx} + \tilde{c}_{12} \varepsilon_{yy} - \tilde{e}_{31} E - \tilde{q}_{31} H \right], \tag{8.85}$$

$$(1 - \mu^2 \nabla^2) \sigma_{yy} = (1 - \lambda^2 \nabla^2) \left[\tilde{c}_{12} \varepsilon_{xx} + \tilde{c}_{11} \varepsilon_{yy} - \tilde{e}_{31} E - \tilde{q}_{31} H \right], \tag{8.86}$$

$$(1 - \mu^2 \nabla^2) \sigma_{xy} = (1 - \lambda^2 \nabla^2) \tilde{c}_{66} \varepsilon_{xy}, \tag{8.87}$$

$$(1 - \mu^2 \nabla^2) D_x = (1 - \lambda^2 \nabla^2) \left[\tilde{s}_{11} E_x + \tilde{d}_{11} H_x \right], \tag{8.88}$$

$$(1 - \mu^2 \nabla^2) D_y = (1 - \lambda^2 \nabla^2) \left[\tilde{s}_{11} E_y + \tilde{d}_{11} H_y \right], \tag{8.89}$$

$$(1 - \mu^2 \nabla^2) D = (1 - \lambda^2 \nabla^2) \left[\tilde{e}_{31} \varepsilon_{xx} + \tilde{e}_{31} \varepsilon_{yy} + \tilde{s}_{33} E + \tilde{d}_{33} H \right], \tag{8.90}$$

$$(1 - \mu^2 \nabla^2) B_x = (1 - \lambda^2 \nabla^2) \left[\tilde{d}_{11} E_x + \tilde{\chi}_{11} H_x \right], \tag{8.91}$$

$$\left(1 - \mu^2\nabla^2\right)B_y = \left(1 - \lambda^2\nabla^2\right)\left[\tilde{d}_{11}E_y + \tilde{\chi}_{11}H_y\right], \tag{8.92}$$

$$\left(1 - \mu^2\nabla^2\right)B = \left(1 - \lambda^2\nabla^2\right)\left[\tilde{e}_{31}\varepsilon_{xx} + \tilde{e}_{31}\varepsilon_{yy} + \tilde{d}_{33}E + \tilde{\chi}_{33}H\right]. \tag{8.93}$$

Integrating Eqs. (8.85)–(8.93) over the thickness of the nanoplate, the following equations can be obtained:

$$
\left(1 - \mu^2\nabla^2\right)\begin{bmatrix}N_{xx}\\N_{yy}\\N_{xy}\end{bmatrix} = \left(1 - \lambda^2\nabla^2\right)\left(\begin{bmatrix}A_{11}A_{12} & 0\\A_{12} & A_{22} & 0\\0 & 0 & A_{66}\end{bmatrix}\begin{bmatrix}\frac{\partial u}{\partial x}\\\frac{\partial v}{\partial y}\\\frac{\partial u}{\partial y} + \frac{\partial v}{\partial x}\end{bmatrix}\right.
$$
$$
\left.+\begin{bmatrix}B_{11} & B_{12} & 0\\B_{12} & B_{22} & 0\\0 & 0 & B_{66}\end{bmatrix}\begin{bmatrix}-\frac{\partial^2 w}{\partial x^2}\\-\frac{\partial^2 w}{\partial y^2}\\-2\frac{\partial^2 w}{\partial x\partial y}\end{bmatrix} + \begin{bmatrix}A_{31}^e\\A_{31}^e\\0\end{bmatrix}\phi + \begin{bmatrix}A_{31}^m\\A_{31}^m\\0\end{bmatrix}\gamma - \begin{bmatrix}N^E + N^H\\N^E + N^H\\0\end{bmatrix}\right), \tag{8.94}
$$

$$
\left(1 - \mu^2\nabla^2\right)\begin{bmatrix}M_{xx}\\M_{yy}\\M_{xy}\end{bmatrix} = \left(1 - \lambda^2\nabla^2\right)\left(\begin{bmatrix}B_{11} & B_{12} & 0\\B_{12} & B_{22} & 0\\0 & 0 & B_{66}\end{bmatrix}\begin{bmatrix}\frac{\partial u}{\partial x}\\\frac{\partial v}{\partial y}\\\frac{\partial u}{\partial y} + \frac{\partial v}{\partial x}\end{bmatrix}\right.
$$
$$
\left.\times\begin{bmatrix}D_{11} & D_{12} & 0\\D_{12} & D_{22} & 0\\0 & 0 & D_{66}\end{bmatrix}\begin{bmatrix}-\frac{\partial^2 w}{\partial x^2}\\-\frac{\partial^2 w}{\partial y^2}\\-2\frac{\partial^2 w}{\partial x\partial y}\end{bmatrix} + \begin{bmatrix}E_{31}^e\\E_{31}^e\\0\end{bmatrix}\phi + \begin{bmatrix}E_{31}^m\\E_{31}^m\\0\end{bmatrix}\gamma - \begin{bmatrix}M^E + M^H\\M^E + M^H\\0\end{bmatrix}\right), \tag{8.95}
$$

$$
\left(1 - \mu^2\nabla^2\right)\int_A \begin{bmatrix}D_x\\D_y\end{bmatrix}\cos(\xi z)dz = \left(1 - \lambda^2\nabla^2\right)\left(F_{11}^e\begin{bmatrix}\frac{\partial \phi}{\partial x}\\\frac{\partial \phi}{\partial y}\end{bmatrix} + F_{11}^m\begin{bmatrix}\frac{\partial \gamma}{\partial x}\\\frac{\partial \gamma}{\partial y}\end{bmatrix}\right), \tag{8.96}
$$

$$
\left(1 - \mu^2\nabla^2\right)\int_A D\,\xi\sin(\xi z)dz = \left(1 - \lambda^2\nabla^2\right)\left(A_{31}^e\left(\frac{\partial u}{\partial x} + \frac{\partial v}{\partial y}\right)\right.
$$
$$
\left.- E_{31}^e\left(\frac{\partial^2 w}{\partial x^2} + \frac{\partial^2 w}{\partial y^2}\right) - F_{33}^e\phi - F_{33}^m\gamma\right), \tag{8.97}
$$

$$
\left(1 - \mu^2\nabla^2\right)\int_A \begin{bmatrix}B_x\\B_y\end{bmatrix}\cos(\xi z)dz = \left(1 - \lambda^2\nabla^2\right)\left(F_{11}^m\begin{bmatrix}\frac{\partial \phi}{\partial x}\\\frac{\partial \phi}{\partial y}\end{bmatrix} + \chi_{11}^m\begin{bmatrix}\frac{\partial \gamma}{\partial x}\\\frac{\partial \gamma}{\partial y}\end{bmatrix}\right), \tag{8.98}
$$

$$
\left(1 - \mu^2\nabla^2\right)\int_A B\,\xi\sin(\xi z)dz = \left(1 - \lambda^2\nabla^2\right)\left(A_{31}^m\left(\frac{\partial u}{\partial x} + \frac{\partial v}{\partial y}\right)\right.
$$
$$
\left.- E_{31}^m\left(\frac{\partial^2 w}{\partial x^2} + \frac{\partial^2 w}{\partial y^2}\right) - F_{33}^m\phi - \chi_{33}^m\gamma\right). \tag{8.99}
$$

In the above equations, there are some electric, magnetic and elastic terms that can be calculated via some integrations over the thickness of the nanoplate. Some of these integrals are presented in Eqs. (6.79)–(6.82). The rest of them can be expressed as:

$$[A_{31}^m, E_{31}^m] = \int_{-\frac{h}{2}}^{\frac{h}{2}} [1, z]\tilde{q}_{31}\xi\sin(\xi z)dz, \tag{8.100}$$

$$A_{15}^m = \int\limits_{-\frac{h}{2}}^{\frac{h}{2}} \tilde{q}_{15} \cos(\xi z) dz,$$ (8.101)

$$[F_{11}^m, F_{33}^m] = \int\limits_{-\frac{h}{2}}^{\frac{h}{2}} \left[\tilde{d}_{11} \cos^2(\xi z), \tilde{d}_{33} \xi^2 \sin^2(\xi z) \right] dz,$$ (8.102)

$$[\chi_{11}^m, \chi_{33}^m] = \int\limits_{-\frac{h}{2}}^{\frac{h}{2}} \left[\tilde{\chi}_{11} \cos^2(\xi z), \tilde{\chi}_{33} \xi^2 \sin^2(\xi z) \right] dz.$$ (8.103)

Also, in Eqs. (8.94) and (8.95), the electromagnetic forces and moments can be written as follows:

$$[N^E, M^E] = - \int\limits_{-\frac{h}{2}}^{\frac{h}{2}} \tilde{e}_{31} \frac{2V}{h} [1, z] dz,$$ (8.104)

$$[N^H, M^H] = - \int\limits_{-\frac{h}{2}}^{\frac{h}{2}} \tilde{q}_{31} \frac{2\Omega}{h} [1, z] dz.$$ (8.105)

8.3.1.3 Governing Equations of MEE Kirchhoff–Love Nanoplates

In this section, the constitutive equations of the nanoplates will be substituted in the Euler–Lagrange equations of plates to obtain the final governing equations of the MEE-FG nanoplates. Thus, Eqs. (8.94)–(8.99) must be inserted in Eqs. (8.80)–(8.84). Finally, the following equations will be obtained for the MEE-FG nanoplates:

$$\left(1 - \lambda^2 \nabla^2\right) \left(A_{11} \frac{\partial^2 u}{\partial x^2} + (A_{12} + A_{66}) \frac{\partial^2 v}{\partial x \partial y} + A_{66} \frac{\partial^2 u}{\partial y^2} - B_{11} \frac{\partial^3 w}{\partial x^3} - (B_{12} + 2B_{66}) \frac{\partial^3 w}{\partial x \partial y^2} \right.$$
$$\left. + A_{31}^e \frac{\partial \phi}{\partial x} + A_{31}^m \frac{\partial \gamma}{\partial x} \right) + \left(1 - \mu^2 \nabla^2\right) \left(-I_0 \ddot{u} + I_1 \frac{\partial \ddot{w}}{\partial x} \right) = 0,$$ (8.106)

$$\left(1 - \lambda^2 \nabla^2\right) \left(A_{22} \frac{\partial^2 v}{\partial y^2} + (A_{12} + A_{66}) \frac{\partial^2 u}{\partial x \partial y} + A_{66} \frac{\partial^2 v}{\partial x^2} - B_{22} \frac{\partial^3 w}{\partial y^3} - (B_{12} + 2B_{66}) \frac{\partial^3 w}{\partial x^2 \partial y} \right.$$
$$\left. + A_{31}^e \frac{\partial \phi}{\partial y} + A_{31}^m \frac{\partial \gamma}{\partial y} \right) + \left(1 - \mu^2 \nabla^2\right) \left(-I_0 \ddot{v} + I_1 \frac{\partial \ddot{w}}{\partial y} \right) = 0,$$ (8.107)

$$\left(1 - \lambda^2 \nabla^2\right) \left(B_{11} \frac{\partial^3 u}{\partial x^3} + (B_{12} + 2B_{66}) \frac{\partial^3 u}{\partial x \partial y^2} + B_{22} \frac{\partial^3 v}{\partial y^3} + (B_{12} + 2B_{66}) \frac{\partial^3 v}{\partial x^2 \partial y} \right.$$
$$- D_{11} \frac{\partial^4 w}{\partial x^4} - 2(D_{12} + 2D_{66}) \frac{\partial^4 w}{\partial x^2 \partial y^2} - D_{22} \frac{\partial^4 w}{\partial y^4} + E_{31}^e \left(\frac{\partial^2 \phi}{\partial x^2} + \frac{\partial^2 \phi}{\partial y^2} \right)$$
$$+ E_{31}^m \left(\frac{\partial^2 \gamma}{\partial x^2} + \frac{\partial^2 \gamma}{\partial y^2} \right) \right) + \left(1 - \mu^2 \nabla^2\right) \left(-I_0 \ddot{w} - I_1 \left(\frac{\partial \ddot{u}}{\partial x} + \frac{\partial \ddot{v}}{\partial y} \right) + I_2 \left(\frac{\partial^2 \ddot{w}}{\partial x^2} + \frac{\partial^2 \ddot{w}}{\partial y^2} \right) \right.$$
$$\left. - \left[N^E + N^H \right] \left(\frac{\partial^2 w}{\partial x^2} + \frac{\partial^2 w}{\partial y^2} \right) \right) = 0,$$ (8.108)

$$\left(1 - \lambda^2 \nabla^2\right) \left(A_{31}^e \left(\frac{\partial u}{\partial x} + \frac{\partial v}{\partial y}\right) - E_{31}^e \left(\frac{\partial^2 w}{\partial x^2} + \frac{\partial^2 w}{\partial y^2}\right) - F_{11}^e \left(\frac{\partial^2 \phi}{\partial x^2} + \frac{\partial^2 \phi}{\partial y^2}\right)\right.$$
$$\left. - F_{11}^m \left(\frac{\partial^2 \gamma}{\partial x^2} + \frac{\partial^2 \gamma}{\partial y^2}\right) - F_{33}^e \phi - F_{33}^m \gamma\right) = 0, \tag{8.109}$$

$$\left(1 - \lambda^2 \nabla^2\right) \left(A_{31}^m \left(\frac{\partial u}{\partial x} + \frac{\partial v}{\partial y}\right) - E_{31}^m \left(\frac{\partial^2 w}{\partial x^2} + \frac{\partial^2 w}{\partial y^2}\right) - F_{11}^m \left(\frac{\partial^2 \phi}{\partial x^2} + \frac{\partial^2 \phi}{\partial y^2}\right)\right.$$
$$\left. - \chi_{11}^m \left(\frac{\partial^2 \gamma}{\partial x^2} + \frac{\partial^2 \gamma}{\partial y^2}\right) - F_{33}^m \phi - \chi_{33}^m \gamma\right) = 0. \tag{8.110}$$

8.3.1.4 Wave Solution of MEE Kirchhoff–Love Nanoplates

Here, the solution of the governing equations of an MEE-FG nanoplate will be presented. The wave solution for the electric potential of a nanoplate was introduced in Eq. (6.89). The magnetic potential of the MEE nanoplate can be formulated as follows:

$$\gamma = \hat{\Gamma} e^{i(\beta_1 x + \beta_2 y - \omega t)}. \tag{8.111}$$

Now, the final equation of the problem, as presented in Eq. (1.3), can be obtained by substituting Eqs. (1.9)–(1.11), (6.89) and (8.111) in Eqs. (8.106)–(8.110). The amplitude vector of an MEE-FG Kirchhoff nanoplate can be written in the following form:

$$\Delta = [U, V, W, \hat{\Phi}, \hat{\Gamma}]^T. \tag{8.112}$$

Besides, due to the smart features of the nanoplate, the stiffness and mass matrices of this nanostructure differ from those of elastic or piezoelectric ones. Here, the arrays of the stiffness matrix of the problem can be expressed as follows:

$$k_{11} = -\left(1 + \lambda^2(\beta_1^2 + \beta_2^2)\right) \left[A_{11}\beta_1^2 + A_{66}\beta_1\beta_2\right],$$
$$k_{12} = -\left(1 + \lambda^2(\beta_1^2 + \beta_2^2)\right) \left[A_{12} + A_{66}\right]\beta_1\beta_2,$$
$$k_{13} = i\left(1 + \lambda^2(\beta_1^2 + \beta_2^2)\right) \left[B_{11}\beta_1^3 + (B_{12} + 2B_{66})\beta_1\beta_2^2\right],$$
$$k_{14} = \left(1 + \lambda^2(\beta_1^2 + \beta_2^2)\right) i\beta_1 A_{31}^e,$$
$$k_{15} = \left(1 + \lambda^2(\beta_1^2 + \beta_2^2)\right) i\beta_1 A_{31}^m,$$
$$k_{22} = -\left(1 + \lambda^2(\beta_1^2 + \beta_2^2)\right) \left[A_{66}\beta_1^2 + A_{22}\beta_2^2\right],$$
$$k_{23} = i\left(1 + \lambda^2(\beta_1^2 + \beta_2^2)\right) \left[B_{22}\beta_2^3 + (B_{12} + 2B_{66})\beta_1^2\beta_2\right],$$
$$k_{24} = \left(1 + \lambda^2(\beta_1^2 + \beta_2^2)\right) i\beta_2 A_{31}^e,$$
$$k_{25} = \left(1 + \lambda^2(\beta_1^2 + \beta_2^2)\right) i\beta_2 A_{31}^m,$$
$$k_{33} = -\left(1 + \lambda^2(\beta_1^2 + \beta_2^2)\right) \left[D_{11}\beta_1^4 + 2(D_{12} + 2D_{66})\beta_1^2\beta_2^2 + D_{22}\beta_2^4\right]$$
$$\qquad + \left(1 + \mu^2(\beta_1^2 + \beta_2^2)\right) \left[N^E + N^H\right](\beta_1^2 + \beta_2^2),$$
$$k_{34} = -\left(1 + \lambda^2(\beta_1^2 + \beta_2^2)\right) E_{31}^e (\beta_1^2 + \beta_2^2),$$
$$k_{35} = -\left(1 + \lambda^2(\beta_1^2 + \beta_2^2)\right) E_{31}^m (\beta_1^2 + \beta_2^2),$$
$$k_{44} = \left(1 + \lambda^2(\beta_1^2 + \beta_2^2)\right) \left[F_{11}^e (\beta_1^2 + \beta_2^2) - F_{33}^e\right],$$
$$k_{45} = \left(1 + \lambda^2(\beta_1^2 + \beta_2^2)\right) \left[F_{11}^m (\beta_1^2 + \beta_2^2) - F_{33}^m\right],$$
$$k_{55} = \left(1 + \lambda^2(\beta_1^2 + \beta_2^2)\right) \left[\chi_{11}^m (\beta_1^2 + \beta_2^2) - \chi_{33}^m\right]. \tag{8.113}$$

It must be mentioned that the size of the mass matrix is the same as that of the stiffness one; however, the nonzero components of this matrix are exactly those given in Eq. (4.108). The rest of the components of the mass matrix are all zero.

8.3.2 Refined Sinusoidal MEE Nanoplate

The purpose of this section is to derive the governing equations of an MEE-FG nanoplate on the basis of a higher-order plate hypothesis free from the additional shear modification coefficient. As in the previous sections presented for considering the effects of shear deformation in nanostructures, herein the sinusoidal shear approximation function will be utilized again. The derivation procedure starts from the calculation of the variations required for obtaining the Euler–Lagrange equations of a plate, and then, the size dependency of the nanoplate will be added to the obtained equations based on the nonlocal strain gradient model.

8.3.2.1 Equations of Motion of MEE Refined Sinusoidal Plates

In Section 8.3.1, the equation needed for the magnetic potential of an MEE plate was derived in a way to satisfy Maxwell's equation. Hence, the results of that section will be used here, and those relations will not be discussed again. Using a refined plate model, the variation of strain energy of a plate can be written in the following form:

$$
\delta\Pi_S = \int_V (\sigma_{xx}\delta\varepsilon_{xx} + \sigma_{yy}\delta\varepsilon_{yy} + \sigma_{xy}\delta\varepsilon_{xy} + \sigma_x \ \delta\varepsilon_x \ + \sigma_y \ \delta\varepsilon_y \ - D_x\delta E_x - D_y\delta E_y
$$

$$
- D \ \delta E \ - B_x\delta H_x - B_y\delta H_y - B \ \delta H \ )\, dV. \tag{8.114}
$$

Once the nonzero strains of the refined plate (Eqs. (4.66)–(4.69)) and the electromagnetic fields (Eqs. (6.58)–(6.60) and (8.74)–(8.76)) are substituted in Eq. (8.114), the following equation can be written for the variation of strain energy:

$$
\delta\Pi_S = \int_A \left[N_{xx}\delta\varepsilon^0_{xx} - M^b_{xx}\delta\kappa^b_{xx} - M^s_{xx}\delta\kappa^s_{xx} + N_{yy}\delta\varepsilon^0_{yy} - M^b_{yy}\delta\kappa^b_{yy} - M^s_{yy}\delta\kappa^s_{yy} \right.
$$

$$
+ N_{xy}\delta\varepsilon^0_{xy} - M^b_{xy}\delta\kappa^b_{xy} - M^s_{xy}\delta\kappa^s_{xy} + Q_x \ \varepsilon^0_x \ + Q_y \ \varepsilon^0_y \ + \int_{-\frac{h}{2}}^{\frac{h}{2}} \left(-D_x \cos(\xi z)\frac{\partial\delta\phi}{\partial x} \right.
$$

$$
\left. - D_y \cos(\xi z)\frac{\partial\delta\phi}{\partial y} + D \ \xi\sin(\xi z)\delta\phi \right) dz + \int_{-\frac{h}{2}}^{\frac{h}{2}} \left(-B_x \cos(\xi z)\frac{\partial\delta\gamma}{\partial x} - B_y \cos(\xi z)\frac{\partial\delta\gamma}{\partial y} \right.
$$

$$
\left. + B \ \xi\sin(\xi z)\delta\gamma \right) dz \Bigg]\, dA. \tag{8.115}
$$

N_i, M^b_i, M^s_i, Q_j ($i = xx, yy, xy$ and $j = xz, yz$) in the above equation have been defined in Eqs. (4.82) and (4.83). Also, the variation of work done by external forces can be expressed in the following form:

$$
\delta\Pi_W = \int_A [N^E + N^H] \left(\frac{\partial(w_b + w_s)}{\partial x}\frac{\partial\delta(w_b + w_s)}{\partial x} + \frac{\partial(w_b + w_s)}{\partial y}\frac{\partial\delta(w_b + w_s)}{\partial y} \right) dA. \tag{8.116}
$$

One should pay attention to the fact that the variation of the kinetic energy of the plate is exactly the same as that expressed in Eq. (4.84). Now, the Euler–Lagrange equations of the plate can be obtained when Eqs. (8.115), (8.116) and (4.84) are inserted in Eq. (4.19) as follows:

$$
\frac{\partial N_{xx}}{\partial x} + \frac{\partial N_{xy}}{\partial y} = I_0\frac{\partial^2 u}{\partial t^2} - I_1\frac{\partial^3 w_b}{\partial x\partial t^2} - J_1\frac{\partial^3 w_s}{\partial x\partial t^2}, \tag{8.117}
$$

$$\frac{\partial N_{xy}}{\partial x} + \frac{\partial N_{yy}}{\partial y} = I_0 \frac{\partial^2 v}{\partial t^2} - I_1 \frac{\partial^3 w_b}{\partial y \partial t^2} - J_1 \frac{\partial^3 w_s}{\partial y \partial t^2}, \tag{8.118}$$

$$\frac{\partial^2 M_{xx}^b}{\partial x^2} + 2\frac{\partial^2 M_{xy}^b}{\partial x \partial y} + \frac{\partial^2 M_{yy}^b}{\partial y^2} - \left[N^E + N^H\right]\left(\frac{\partial^2 (w_b + w_s)}{\partial x^2} + \frac{\partial^2 (w_b + w_s)}{\partial y^2}\right)$$
$$= I_0 \frac{\partial^2 (w_b + w_s)}{\partial t^2} + I_1\left(\frac{\partial^3 u}{\partial x \partial t^2} + \frac{\partial^3 v}{\partial y \partial t^2}\right) - I_2\left(\frac{\partial^4 w_b}{\partial x^2 \partial t^2} + \frac{\partial^4 w_b}{\partial y^2 \partial t^2}\right)$$
$$- J_2\left(\frac{\partial^4 w_s}{\partial x^2 \partial t^2} + \frac{\partial^4 w_s}{\partial y^2 \partial t^2}\right), \tag{8.119}$$

$$\frac{\partial^2 M_{xx}^s}{\partial x^2} + 2\frac{\partial^2 M_{xy}^s}{\partial x \partial y} + \frac{\partial^2 M_{yy}^s}{\partial y^2} + \frac{\partial Q_x}{\partial x} + \frac{\partial Q_y}{\partial y} - \left[N^E + N^H\right]$$
$$\times \left(\frac{\partial^2 (w_b + w_s)}{\partial x^2} + \frac{\partial^2 (w_b + w_s)}{\partial y^2}\right) = I_0 \frac{\partial^2 (w_b + w_s)}{\partial t^2} + J_1\left(\frac{\partial^3 u}{\partial x \partial t^2} + \frac{\partial^3 v}{\partial y \partial t^2}\right)$$
$$- J_2\left(\frac{\partial^4 w_b}{\partial x^2 \partial t^2} + \frac{\partial^4 w_b}{\partial y^2 \partial t^2}\right) - K_2\left(\frac{\partial^4 w_s}{\partial x^2 \partial t^2} + \frac{\partial^4 w_s}{\partial y^2 \partial t^2}\right), \tag{8.120}$$

$$\int_A \left(-\cos(\xi z)\frac{\partial D_x}{\partial x} - \cos(\xi z)\frac{\partial D_y}{\partial y} + \xi \sin(\xi z)D\right) dA = 0, \tag{8.121}$$

$$\int_A \left(-\cos(\xi z)\frac{\partial B_x}{\partial x} - \cos(\xi z)\frac{\partial B_y}{\partial y} + \xi \sin(\xi z)B\right) dA = 0. \tag{8.122}$$

8.3.2.2 Nonlocal Strain Gradient Magnetoelectroelasticity for Refined Sinusoidal Nanoplates

In this section, the nonlocal constitutive equations will be derived for an MEE-FG refined higher-order nanoplate. Again, Eqs. (2.20)–(2.22) will be employed for deriving the constitutive equations of the nanoplate. Extending the above-referenced equations to a refined shear deformable nanoplate, one can obtain the following equations:

$$\left(1 - \mu^2 \nabla^2\right)\sigma_{xx} = \left(1 - \lambda^2 \nabla^2\right)\left[\tilde{c}_{11}\varepsilon_{xx} + \tilde{c}_{12}\varepsilon_{yy} - \tilde{e}_{31}E - \tilde{q}_{31}H\right], \tag{8.123}$$

$$\left(1 - \mu^2 \nabla^2\right)\sigma_{yy} = \left(1 - \lambda^2 \nabla^2\right)\left[\tilde{c}_{12}\varepsilon_{xx} + \tilde{c}_{11}\varepsilon_{yy} - \tilde{e}_{31}E - \tilde{q}_{31}H\right], \tag{8.124}$$

$$\left(1 - \mu^2 \nabla^2\right)\sigma_{xy} = \left(1 - \lambda^2 \nabla^2\right)\tilde{c}_{66}\varepsilon_{xy}, \tag{8.125}$$

$$\left(1 - \mu^2 \nabla^2\right)\sigma_x = \left(1 - \lambda^2 \nabla^2\right)\left[\tilde{c}_{55}\varepsilon_x - \tilde{e}_{15}E_x - \tilde{q}_{15}H_x\right], \tag{8.126}$$

$$\left(1 - \mu^2 \nabla^2\right)\sigma_y = \left(1 - \lambda^2 \nabla^2\right)\left[\tilde{c}_{55}\varepsilon_y - \tilde{e}_{15}E_y - \tilde{q}_{15}H_y\right], \tag{8.127}$$

$$\left(1 - \mu^2 \nabla^2\right)D_x = \left(1 - \lambda^2 \nabla^2\right)\left[\tilde{e}_{15}\varepsilon_x + \tilde{s}_{11}E_x + \tilde{d}_{11}H_x\right], \tag{8.128}$$

$$\left(1 - \mu^2 \nabla^2\right)D_y = \left(1 - \lambda^2 \nabla^2\right)\left[\tilde{e}_{15}\varepsilon_y + \tilde{s}_{11}E_y + \tilde{d}_{11}H_y\right], \tag{8.129}$$

$$\left(1 - \mu^2 \nabla^2\right)D = \left(1 - \lambda^2 \nabla^2\right)\left[\tilde{e}_{31}\varepsilon_{xx} + \tilde{e}_{31}\varepsilon_{yy} + \tilde{s}_{33}E + \tilde{d}_{33}H\right], \tag{8.130}$$

$$\left(1 - \mu^2 \nabla^2\right)B_x = \left(1 - \lambda^2 \nabla^2\right)\left[\tilde{q}_{15}\varepsilon_x + \tilde{d}_{11}E_x + \tilde{\chi}_{11}H_x\right], \tag{8.131}$$

$$\left(1 - \mu^2 \nabla^2\right)B_y = \left(1 - \lambda^2 \nabla^2\right)\left[\tilde{q}_{15}\varepsilon_y + \tilde{d}_{11}E_y + \tilde{\chi}_{11}H_y\right], \tag{8.132}$$

$$(1 - \mu^2 \nabla^2)B = (1 - \lambda^2 \nabla^2)\left[\tilde{q}_{31}\varepsilon_{xx} + \tilde{q}_{31}\varepsilon_{yy} + \tilde{d}_{33}E + \tilde{\chi}_{33}H\right]. \tag{8.133}$$

Once the above equations are integrated over the thickness of the nanoplate, we can obtain the following equations:

$$(1 - \mu^2 \nabla^2)\begin{bmatrix} N_{xx} \\ N_{yy} \\ N_{xy} \end{bmatrix} = (1 - \lambda^2 \nabla^2)\left(\begin{bmatrix} A_{11} & A_{12} & 0 \\ A_{12} & A_{22} & 0 \\ 0 & 0 & A_{66} \end{bmatrix}\begin{bmatrix} \frac{\partial u}{\partial x} \\ \frac{\partial v}{\partial y} \\ \frac{\partial u}{\partial y} + \frac{\partial v}{\partial x} \end{bmatrix}\right.$$

$$+ \begin{bmatrix} B_{11} & B_{12} & 0 \\ B_{12} & B_{22} & 0 \\ 0 & 0 & B_{66} \end{bmatrix}\begin{bmatrix} -\frac{\partial^2 w_b}{\partial x^2} \\ -\frac{\partial^2 w_b}{\partial y^2} \\ -2\frac{\partial^2 w_b}{\partial x \partial y} \end{bmatrix} + \begin{bmatrix} B_{11}^s & B_{12}^s & 0 \\ B_{12}^s & B_{22}^s & 0 \\ 0 & 0 & B_{66}^s \end{bmatrix}\begin{bmatrix} -\frac{\partial^2 w_s}{\partial x^2} \\ -\frac{\partial^2 w_s}{\partial y^2} \\ -2\frac{\partial^2 w_s}{\partial x \partial y} \end{bmatrix}$$

$$\left.+ \begin{bmatrix} A_{31}^e \\ A_{31}^e \\ 0 \end{bmatrix}\phi + \begin{bmatrix} A_{31}^m \\ A_{31}^m \\ 0 \end{bmatrix}\gamma - \begin{bmatrix} N^E + N^H \\ N^E + N^H \\ 0 \end{bmatrix}\right), \tag{8.134}$$

$$(1 - \mu^2 \nabla^2)\begin{bmatrix} M_{xx}^b \\ M_{yy}^b \\ M_{xy}^b \end{bmatrix} = (1 - \lambda^2 \nabla^2)\left(\begin{bmatrix} B_{11} & B_{12} & 0 \\ B_{12} & B_{22} & 0 \\ 0 & 0 & B_{66} \end{bmatrix}\begin{bmatrix} \frac{\partial u}{\partial x} \\ \frac{\partial v}{\partial y} \\ \frac{\partial u}{\partial y} + \frac{\partial v}{\partial x} \end{bmatrix}\right.$$

$$+ \begin{bmatrix} D_{11} & D_{12} & 0 \\ D_{12} & D_{22} & 0 \\ 0 & 0 & D_{66} \end{bmatrix}\begin{bmatrix} -\frac{\partial^2 w_b}{\partial x^2} \\ -\frac{\partial^2 w_b}{\partial y^2} \\ -2\frac{\partial^2 w_b}{\partial x \partial y} \end{bmatrix} + \begin{bmatrix} D_{11}^s & D_{12}^s & 0 \\ D_{12}^s & D_{22}^s & 0 \\ 0 & 0 & D_{66}^s \end{bmatrix}\begin{bmatrix} -\frac{\partial^2 w_s}{\partial x^2} \\ -\frac{\partial^2 w_s}{\partial y^2} \\ -2\frac{\partial^2 w_s}{\partial x \partial y} \end{bmatrix}$$

$$\left.+ \begin{bmatrix} E_{31}^e \\ E_{31}^e \\ 0 \end{bmatrix}\phi + \begin{bmatrix} E_{31}^m \\ E_{31}^m \\ 0 \end{bmatrix}\gamma - \begin{bmatrix} M_b^E + M_b^H \\ M_b^E + M_b^H \\ 0 \end{bmatrix}\right), \tag{8.135}$$

$$(1 - \mu^2 \nabla^2)\begin{bmatrix} M_{xx}^s \\ M_{yy}^s \\ M_{xy}^s \end{bmatrix} = (1 - \lambda^2 \nabla^2)\left(\begin{bmatrix} B_{11}^s & B_{12}^s & 0 \\ B_{12}^s & B_{22}^s & 0 \\ 0 & 0 & B_{66}^s \end{bmatrix}\begin{bmatrix} \frac{\partial u}{\partial x} \\ \frac{\partial v}{\partial y} \\ \frac{\partial u}{\partial y} + \frac{\partial v}{\partial x} \end{bmatrix}\right.$$

$$+ \begin{bmatrix} D_{11}^s & D_{12}^s & 0 \\ D_{12}^s & D_{22}^s & 0 \\ 0 & 0 & D_{66}^s \end{bmatrix}\begin{bmatrix} -\frac{\partial^2 w_b}{\partial x^2} \\ -\frac{\partial^2 w_b}{\partial y^2} \\ -2\frac{\partial^2 w_b}{\partial x \partial y} \end{bmatrix} + \begin{bmatrix} H_{11}^s & H_{12}^s & 0 \\ H_{12}^s & H_{22}^s & 0 \\ 0 & 0 & H_{66}^s \end{bmatrix}\begin{bmatrix} -\frac{\partial^2 w_s}{\partial x^2} \\ -\frac{\partial^2 w_s}{\partial y^2} \\ -2\frac{\partial^2 w_s}{\partial x \partial y} \end{bmatrix}$$

$$\left.+ \begin{bmatrix} F_{31}^e \\ F_{31}^e \\ 0 \end{bmatrix}\phi + \begin{bmatrix} F_{31}^m \\ F_{31}^m \\ 0 \end{bmatrix}\gamma - \begin{bmatrix} M_s^E + M_s^H \\ M_s^E + M_s^H \\ 0 \end{bmatrix}\right), \tag{8.136}$$

$$(1 - \mu^2 \nabla^2)\begin{bmatrix} Q_x \\ Q_y \end{bmatrix} = (1 - \lambda^2 \nabla^2)\left(\begin{bmatrix} A_{44}^s & 0 \\ 0 & A_{55}^s \end{bmatrix}\begin{bmatrix} \frac{\partial w_s}{\partial x} \\ \frac{\partial w_s}{\partial y} \end{bmatrix} - A_{15}^e \begin{bmatrix} \frac{\partial \phi}{\partial x} \\ \frac{\partial \phi}{\partial y} \end{bmatrix} - A_{15}^m \begin{bmatrix} \frac{\partial \gamma}{\partial x} \\ \frac{\partial \gamma}{\partial y} \end{bmatrix}\right), \tag{8.137}$$

$$(1 - \mu^2 \nabla^2)\int_A \begin{bmatrix} D_x \\ D_y \end{bmatrix}cos(\xi z)dz = (1 - \lambda^2 \nabla^2)\left(E_{15}^e \begin{bmatrix} \frac{\partial w_s}{\partial x} \\ \frac{\partial w_s}{\partial y} \end{bmatrix} + F_{11}^e \begin{bmatrix} \frac{\partial \phi}{\partial x} \\ \frac{\partial \phi}{\partial y} \end{bmatrix} + F_{11}^m \begin{bmatrix} \frac{\partial \gamma}{\partial x} \\ \frac{\partial \gamma}{\partial y} \end{bmatrix}\right),$$
$$\tag{8.138}$$

$$\left(1 - \mu^2 \nabla^2\right) \int\limits_A D \, \xi \sin(\xi z) dz = \left(1 - \lambda^2 \nabla^2\right) \left(A_{31}^e \left(\frac{\partial u}{\partial x} + \frac{\partial v}{\partial y}\right) - E_{31}^e \left(\frac{\partial^2 w_b}{\partial x^2} + \frac{\partial^2 w_b}{\partial y^2}\right) \right.$$

$$\left. - F_{31}^e \left(\frac{\partial^2 w_s}{\partial x^2} + \frac{\partial^2 w_s}{\partial y^2}\right) - F_{33}^e \phi - F_{33}^m \gamma \right), \qquad (8.139)$$

$$\left(1 - \mu^2 \nabla^2\right) \int\limits_A \begin{bmatrix} B_x \\ B_y \end{bmatrix} \cos(\xi z) dz = \left(1 - \lambda^2 \nabla^2\right) \left(E_{15}^m \begin{bmatrix} \frac{\partial w_s}{\partial x} \\ \frac{\partial w_s}{\partial y} \end{bmatrix} + F_{11}^m \begin{bmatrix} \frac{\partial \phi}{\partial x} \\ \frac{\partial \phi}{\partial y} \end{bmatrix} + \chi_{11}^m \begin{bmatrix} \frac{\partial \gamma}{\partial x} \\ \frac{\partial \gamma}{\partial y} \end{bmatrix} \right),$$
$$(8.140)$$

$$\left(1 - \mu^2 \nabla^2\right) \int\limits_A B \, \xi \sin(\xi z) dz = \left(1 - \lambda^2 \nabla^2\right) \left(A_{31}^m \left(\frac{\partial u}{\partial x} + \frac{\partial v}{\partial y}\right) - E_{31}^m \left(\frac{\partial^2 w_b}{\partial x^2} + \frac{\partial^2 w_b}{\partial y^2}\right) \right.$$

$$\left. - F_{31}^m \left(\frac{\partial^2 w_s}{\partial x^2} + \frac{\partial^2 w_s}{\partial y^2}\right) - F_{33}^m \phi - \chi_{33}^m \gamma \right). \qquad (8.141)$$

There are some variants in Eqs. (8.134)–(8.141) that must be introduced. These variants are some integrals obtained by integration over the thickness of the nanoplate. In Chapter 6, some of these integrals were presented (see Eqs. (6.79)–(6.82) and (6.114)–(6.117). The integrals that have not been introduced yet can be expressed as follows:

$$[A_{31}^m, E_{31}^m, F_{31}^m] = \int\limits_{-\frac{h}{2}}^{\frac{h}{2}} [1, z, f(z)] \tilde{q}_{31} \xi \sin(\xi z) dz, \qquad (8.142)$$

$$[A_{15}^m, E_{15}^m] = \int\limits_{-\frac{h}{2}}^{\frac{h}{2}} [1, g(z)] \tilde{q}_{15} \cos(\xi z) dz, \qquad (8.143)$$

$$[F_{11}^m, F_{33}^m] = \int\limits_{-\frac{h}{2}}^{\frac{h}{2}} [\tilde{d}_{11} \cos^2(\xi z), \tilde{d}_{33} \xi^2 \sin^2(\xi z)] dz, \qquad (8.144)$$

$$[\chi_{11}^m, \chi_{33}^m] = \int\limits_{-\frac{h}{2}}^{\frac{h}{2}} [\tilde{\chi}_{11} \cos^2(\xi z), \tilde{\chi}_{33} \xi^2 \sin^2(\xi z)] dz. \qquad (8.145)$$

Also, the magnetic resultants can be expressed in the following form:

$$[N^H, M_b^H, M_s^H] = - \int\limits_{-\frac{h}{2}}^{\frac{h}{2}} [1, z, f(z)] \tilde{q}_{31} \frac{2\Omega}{h} dz. \qquad (8.146)$$

8.3.2.3 Governing Equations of MEE Refined Sinusoidal Nanoplates

In this section, the governing equations of an MEE-FG shear deformable nanoplate will be derived by inserting Eqs. (8.134)–(8.141) in Eqs. (8.117)–(8.122). Finally, the governing equations of the refined nanoplate can be written as

$$\left(1 - \lambda^2 \nabla^2\right) \left(A_{11} \frac{\partial^2 u}{\partial x^2} + (A_{12} + A_{66}) \frac{\partial^2 v}{\partial x \partial y} + A_{66} \frac{\partial^2 u}{\partial y^2} - B_{11} \frac{\partial^3 w_b}{\partial x^3} \right.$$

$$\left. - (B_{12} + 2B_{66}) \frac{\partial^3 w_b}{\partial x \partial y^2} - B_{11}^s \frac{\partial^3 w_s}{\partial x^3} - (B_{12}^s + 2B_{66}^s) \frac{\partial^3 w_s}{\partial x \partial y^2} + A_{31}^e \frac{\partial \phi}{\partial x} + A_{31}^m \frac{\partial \gamma}{\partial x} \right)$$

$$+ \left(1 - \mu^2 \nabla^2\right) \left(-I_0 \frac{\partial^2 u}{\partial t^2} + I_1 \frac{\partial^3 w_b}{\partial x \partial t^2} + J_1 \frac{\partial^3 w_s}{\partial x \partial t^2} \right) = 0, \qquad (8.147)$$

$$\left(1 - \lambda^2 \nabla^2\right) \left(A_{22} \frac{\partial^2 v}{\partial y^2} + (A_{12} + A_{66}) \frac{\partial^2 u}{\partial x \partial y} + A_{66} \frac{\partial^2 v}{\partial x^2} - B_{22} \frac{\partial^3 w_b}{\partial y^3} \right.$$

$$\left. - (B_{12} + 2B_{66}) \frac{\partial^3 w_b}{\partial x^2 \partial y} - B_{22}^s \frac{\partial^3 w_s}{\partial y^3} - (B_{12}^s + 2B_{66}^s) \frac{\partial^3 w_s}{\partial x^2 \partial y} + A_{31}^e \frac{\partial \phi}{\partial y} + A_{31}^m \frac{\partial \gamma}{\partial y} \right)$$

$$+ \left(1 - \mu^2 \nabla^2\right) \left(-I_0 \frac{\partial^2 v}{\partial t^2} + I_1 \frac{\partial^3 w_b}{\partial y \partial t^2} + J_1 \frac{\partial^3 w_s}{\partial y \partial t^2} \right) = 0, \qquad (8.148)$$

$$\left(1 - \lambda^2 \nabla^2\right) \left(B_{11} \frac{\partial^3 u}{\partial x^3} + (B_{12} + 2B_{66}) \frac{\partial^3 u}{\partial x \partial y^2} + B_{22} \frac{\partial^3 v}{\partial y^3} + (B_{12} + 2B_{66}) \frac{\partial^3 v}{\partial x^2 \partial y} - D_{11} \frac{\partial^4 w_b}{\partial x^4} \right.$$

$$- 2(D_{12} + 2D_{66}) \frac{\partial^4 w_b}{\partial x^2 \partial y^2} - D_{22} \frac{\partial^4 w_b}{\partial y^4} - D_{11}^s \frac{\partial^4 w_s}{\partial x^4} - 2(D_{12}^s + 2D_{66}^s) \frac{\partial^4 w_s}{\partial x^2 \partial y^2} - D_{22}^s \frac{\partial^4 w_s}{\partial y^4}$$

$$+ E_{31}^e \left(\frac{\partial^2 \phi}{\partial x^2} + \frac{\partial^2 \phi}{\partial y^2} \right) + E_{31}^m \left(\frac{\partial^2 \gamma}{\partial x^2} + \frac{\partial^2 \gamma}{\partial y^2} \right) \right) + \left(1 - \mu^2 \nabla^2\right) \left(-I_0 \frac{\partial^2 (w_b + w_s)}{\partial t^2} \right.$$

$$- I_1 \left(\frac{\partial^3 u}{\partial x \partial t^2} + \frac{\partial^3 v}{\partial y \partial t^2} \right) + I_2 \left(\frac{\partial^4 w_b}{\partial x^2 \partial t^2} + \frac{\partial^4 w_b}{\partial y^2 \partial t^2} \right) + J_2 \left(\frac{\partial^4 w_s}{\partial x^2 \partial t^2} + \frac{\partial^4 w_s}{\partial y^2 \partial t^2} \right)$$

$$\left. - \left[N^E + N^H \right] \left(\frac{\partial^2 (w_b + w_s)}{\partial x^2} + \frac{\partial^2 (w_b + w_s)}{\partial y^2} \right) \right) = 0, \qquad (8.149)$$

$$\left(1 - \lambda^2 \nabla^2\right) \left(B_{11}^s \frac{\partial^3 u}{\partial x^3} + (B_{12}^s + 2B_{66}^s) \frac{\partial^3 u}{\partial x \partial y^2} + B_{22}^s \frac{\partial^3 v}{\partial y^3} + (B_{12}^s + 2B_{66}^s) \frac{\partial^3 v}{\partial x^2 \partial y} - D_{11}^s \frac{\partial^4 w_b}{\partial x^4} \right.$$

$$- 2(D_{12}^s + 2D_{66}^s) \frac{\partial^4 w_b}{\partial x^2 \partial y^2} - D_{22}^s \frac{\partial^4 w_b}{\partial y^4} - H_{11}^s \frac{\partial^4 w_s}{\partial x^4} - 2(H_{12}^s + 2H_{66}^s) \frac{\partial^4 w_s}{\partial x^2 \partial y^2} - H_{22}^s \frac{\partial^4 w_s}{\partial y^4}$$

$$+ A_{44}^s \left(\frac{\partial^2 (w_b + w_s)}{\partial x^2} + \frac{\partial^2 (w_b + w_s)}{\partial y^2} \right) + (F_{31}^e - A_{15}^e) \left(\frac{\partial^2 \phi}{\partial x^2} + \frac{\partial^2 \phi}{\partial y^2} \right)$$

$$+ (F_{31}^m - A_{15}^m) \left(\frac{\partial^2 \gamma}{\partial x^2} + \frac{\partial^2 \gamma}{\partial y^2} \right) \right) + \left(1 - \mu^2 \nabla^2\right) \left(-I_0 \frac{\partial^2 (w_b + w_s)}{\partial t^2} - J_1 \left(\frac{\partial^3 u}{\partial x \partial t^2} + \frac{\partial^3 v}{\partial y \partial t^2} \right) \right.$$

$$+ J_2 \left(\frac{\partial^4 w_b}{\partial x^2 \partial t^2} + \frac{\partial^4 w_b}{\partial y^2 \partial t^2} \right) + K_2 \left(\frac{\partial^4 w_s}{\partial x^2 \partial t^2} + \frac{\partial^4 w_s}{\partial y^2 \partial t^2} \right) - \left[N^E + N^H \right]$$

$$\times \left. \left(\frac{\partial^2 (w_b + w_s)}{\partial x^2} + \frac{\partial^2 (w_b + w_s)}{\partial y^2} \right) \right) = 0, \qquad (8.150)$$

$$\left(1 - \lambda^2 \nabla^2\right) \left(A_{31}^e \left(\frac{\partial u}{\partial x} + \frac{\partial v}{\partial y} \right) - E_{31}^e \left(\frac{\partial^2 w_b}{\partial x^2} + \frac{\partial^2 w_b}{\partial y^2} \right) - (F_{31}^e + E_{15}^e) \left(\frac{\partial^2 w_s}{\partial x^2} + \frac{\partial^2 w_s}{\partial y^2} \right) \right.$$

$$\left. - F_{11}^e \left(\frac{\partial^2 \phi}{\partial x^2} + \frac{\partial^2 \phi}{\partial y^2} \right) - F_{33}^e \phi - F_{11}^m \left(\frac{\partial^2 \gamma}{\partial x^2} + \frac{\partial^2 \gamma}{\partial y^2} \right) - F_{33}^m \gamma \right) = 0, \qquad (8.151)$$

$$\left(1 - \lambda^2 \nabla^2\right) \left(A_{31}^m \left(\frac{\partial u}{\partial x} + \frac{\partial v}{\partial y}\right) - E_{31}^m \left(\frac{\partial^2 w_b}{\partial x^2} + \frac{\partial^2 w_b}{\partial y^2}\right) - \left(F_{31}^m + E_{15}^m\right) \left(\frac{\partial^2 w_s}{\partial x^2} + \frac{\partial^2 w_s}{\partial y^2}\right)\right.$$

$$\left. - F_{11}^m \left(\frac{\partial^2 \phi}{\partial x^2} + \frac{\partial^2 \phi}{\partial y^2}\right) - F_{33}^m \phi - \chi_{11}^m \left(\frac{\partial^2 \gamma}{\partial x^2} + \frac{\partial^2 \gamma}{\partial y^2}\right) - \chi_{33}^m \gamma\right) = 0. \tag{8.152}$$

8.3.2.4 Wave Solution of MEE Refined Sinusoidal Nanoplates

Now, the governing equations obtained in Eqs. (8.147)–(8.152) will be solved. In order to solve the governing equations, Eqs. (1.12)–(1.15), (6.89) and (8.111) must be substituted in Eqs. (8.147)–(8.152). On doing this substitution, the final equation will be an equation similar to Eq. (1.3). However, in this case, the dimensions of the equation will be 6×6 instead of 2×2. Also, the amplitude vector will be

$$\Delta = \left[U, V, W_b, W_s, \hat{\Phi}, \hat{\Gamma}\right]^T. \tag{8.153}$$

The components of the stiffness matrix for a refined sinusoidal smart FG nanoplate can be written as follows:

$$k_{11} = -\left(1 + \lambda^2\left(\beta_1^2 + \beta_2^2\right)\right)\left[A_{11}\beta_1^2 + A_{66}\beta_1\beta_2\right],$$

$$k_{12} = -\left(1 + \lambda^2\left(\beta_1^2 + \beta_2^2\right)\right)\left[A_{12} + A_{66}\right]\beta_1\beta_2,$$

$$k_{13} = i\left(1 + \lambda^2\left(\beta_1^2 + \beta_2^2\right)\right)\left[B_{11}\beta_1^3 + \left(B_{12} + 2B_{66}\right)\beta_1\beta_2^2\right],$$

$$k_{14} = i\left(1 + \lambda^2\left(\beta_1^2 + \beta_2^2\right)\right)\left[B_{11}^s\beta_1^3 + \left(B_{12}^s + 2B_{66}^s\right)\beta_1\beta_2^2\right],$$

$$k_{15} = \left(1 + \lambda^2\left(\beta_1^2 + \beta_2^2\right)\right)i\beta_1 A_{31}^e, k_{16} = \left(1 + \lambda^2\left(\beta_1^2 + \beta_2^2\right)\right)i\beta_1 A_{31}^m,$$

$$k_{22} = -\left(1 + \lambda^2\left(\beta_1^2 + \beta_2^2\right)\right)\left[A_{66}\beta_1^2 + A_{22}\beta_2^2\right],$$

$$k_{23} = i\left(1 + \lambda^2\left(\beta_1^2 + \beta_2^2\right)\right)\left[B_{22}\beta_2^3 + \left(B_{12} + 2B_{66}\right)\beta_1^2\beta_2\right],$$

$$k_{24} = i\left(1 + \lambda^2\left(\beta_1^2 + \beta_2^2\right)\right)\left[B_{22}^s\beta_2^3 + \left(B_{12}^s + 2B_{66}^s\right)\beta_1^2\beta_2\right],$$

$$k_{25} = \left(1 + \lambda^2\left(\beta_1^2 + \beta_2^2\right)\right)i\beta_2 A_{31}^e, k_{26} = \left(1 + \lambda^2\left(\beta_1^2 + \beta_2^2\right)\right)i\beta_2 A_{31}^m,$$

$$k_{33} = -\left(1 + \lambda^2\left(\beta_1^2 + \beta_2^2\right)\right)\left[D_{11}\beta_1^4 + 2\left(D_{12} + 2D_{66}\right)\beta_1^2\beta_2^2 + D_{22}\beta_2^4\right]$$
$$\qquad + \left(1 + \mu^2\left(\beta_1^2 + \beta_2^2\right)\right)\left[N^E + N^H\right]\left(\beta_1^2 + \beta_2^2\right),$$

$$k_{34} = -\left(1 + \lambda^2\left(\beta_1^2 + \beta_2^2\right)\right)\left[D_{11}^s\beta_1^4 + 2\left(D_{12}^s + 2D_{66}^s\right)\beta_1^2\beta_2^2 + D_{22}^s\beta_2^4\right]$$
$$\qquad + \left(1 + \mu^2\left(\beta_1^2 + \beta_2^2\right)\right)\left[N^E + N^H\right]\left(\beta_1^2 + \beta_2^2\right),$$

$$k_{35} = -\left(1 + \lambda^2\left(\beta_1^2 + \beta_2^2\right)\right)E_{31}^e\left(\beta_1^2 + \beta_2^2\right),$$

$$k_{36} = -\left(1 + \lambda^2\left(\beta_1^2 + \beta_2^2\right)\right)E_{31}^m\left(\beta_1^2 + \beta_2^2\right),$$

$$k_{44} = -\left(1 + \lambda^2\left(\beta_1^2 + \beta_2^2\right)\right)\left[H_{11}^s\beta_1^4 + 2\left(H_{12}^s + 2H_{66}^s\right)\beta_1^2\beta_2^2 + H_{22}^s\beta_2^4\right.$$
$$\qquad \left. + A_{44}^s\left(\beta_1^2 + \beta_2^2\right)\right] + \left(1 + \mu^2\left(\beta_1^2 + \beta_2^2\right)\right)\left[N^E + N^H\right]\left(\beta_1^2 + \beta_2^2\right),$$

$$k_{45} = -\left(1 + \lambda^2\left(\beta_1^2 + \beta_2^2\right)\right)\left(F_{31}^e - A_{15}^e\right)\left(\beta_1^2 + \beta_2^2\right),$$

$$k_{46} = -\left(1 + \lambda^2\left(\beta_1^2 + \beta_2^2\right)\right)\left(F_{31}^m - A_{15}^m\right)\left(\beta_1^2 + \beta_2^2\right),$$

$$k_{55} = \left(1 + \lambda^2\left(\beta_1^2 + \beta_2^2\right)\right)\left[F_{11}^e\left(\beta_1^2 + \beta_2^2\right) - F_{33}^e\right],$$

$$k_{56} = \left(1 + \lambda^2\left(\beta_1^2 + \beta_2^2\right)\right)\left[F_{11}^m\left(\beta_1^2 + \beta_2^2\right) - F_{33}^m\right],$$

$$k_{66} = \left(1 + \lambda^2\left(\beta_1^2 + \beta_2^2\right)\right)\left[\chi_{11}^m\left(\beta_1^2 + \beta_2^2\right) - \chi_{33}^m\right]. \tag{8.154}$$

The nonzero arrays of the mass matrix are the same as the components presented previously in Eq. (4.110).

8.3.3 Numerical Results for MEE Nanoplates

Herein, we will illustrate the effects of various variants on the wave propagation behaviors of smart MEE-FG nanoplates. As in Section 8.2.3, which was presented for the numerical analysis of MEE nanobeams, here the phase velocity of the dispersed waves will be studied. Also, various scale factors will be employed to show the effect of the relative value of length-scale and nonlocal parameters on the phase speed behaviors of smart nanoplates. Besides, the effect of using smart porous materials will be covered after the presented numerical graphs.

Figure 8.7 is plotted in order to emphasize the crucial role of the magnetic potential and scale factor on the propagation responses of smart MEE composite nanoplates. It is clear that the phase velocity of the nanoplate follows a continually increasing path as the scale factor increases. Thus, the higher the assigned length-scale parameter is, the greater will be the speed of the dispersed waves. Meanwhile, it is obvious that, in all of the subplots, there is a prevalent effect related to changes in the amount of the magnetic potential of the nanoplate. In fact, the phase velocity of the nanoplate can be increased

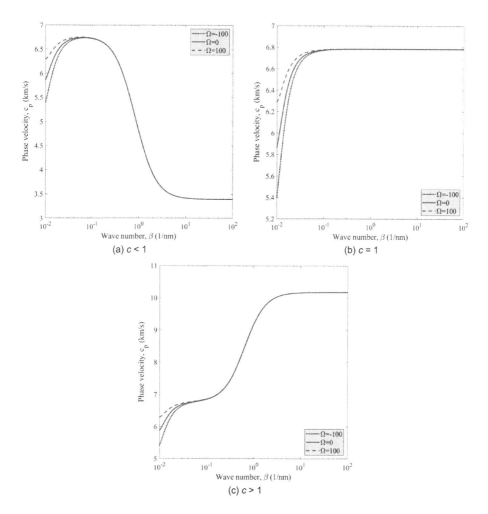

FIGURE 8.7
Variation of phase velocity of MEE-FG nanoplate with wave number at various amounts of magnetic potential ($p = 2$, $V = 0$).

at small wave numbers when the applied magnetic potential is increased. Thus, the use of higher magnetic potentials at wave numbers less than 0.05 (1/nm) corresponds to observing a higher phase speed.

Furthermore, the variation of the phase speed of the propagated waves with magnetic potential at different gradient indices can be observed in Figure 8.8. This diagram is plotted for both positive and negative applied voltages, and it can be figured out that the mechanical response can be intensified when a lower voltage is applied to the nanoplate. However, it can be seen another time that the phase speed of the propagated waves increases when applying

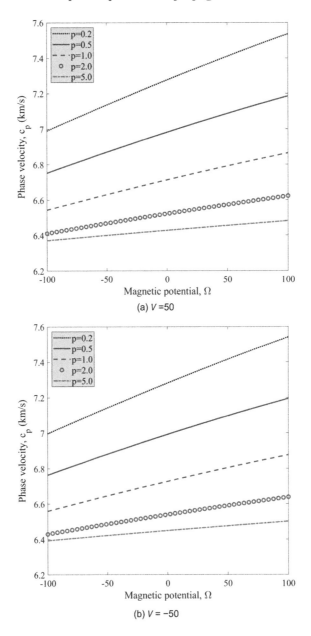

(a) $V = 50$

(b) $V = -50$

FIGURE 8.8

Variation of phase velocity of an MEE-FG nanoplate with magnetic potential at various gradient indices ($p = 2$, $\beta = 0.02$ (1/nm), $c < 1$).

a higher magnetic potential. Therefore, smart behaviors of the nanoplate, generated from variations in the electric voltage or magnetic potential, will be different. In other words, a similar rise in the value of each of them can produce two completely different results. Also, it is observable that phase velocity decreases at high gradient indices. The reason is that, in this condition, the composite moves toward becoming less stiff, and this fact can reduce the dynamic responses of a continuous system.

On the other hand, Figure 8.9 shows the effect of the nonlocal parameter on the phase speed of the dispersed waves. It can be realized that the phase velocity decreases gradually as

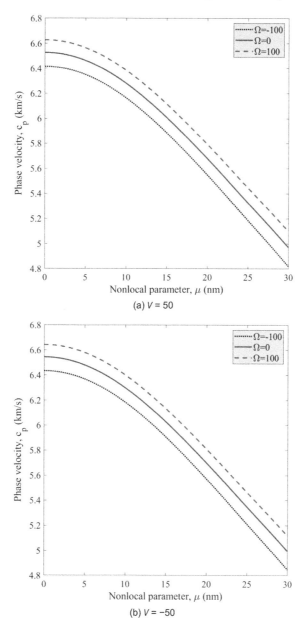

FIGURE 8.9

Variation of phase velocity of an MEE-FG nanoplate with nonlocal parameter at various amounts of magnetic potential ($p = 2$, $\beta = 0.02$ (1/nm), $\lambda = 1$ nm).

the nonlocal term increases. This decreasing effect, which is known as the stiffness-softening effect, was also estimated earlier. It should be considered that phase velocity values of the nanoplate will decrease when a higher electric voltage is connected to the nanoplate. This decreasing effect can be confronted with an amplifying influence originating from increasing the magnetic potential of the nanoplate. It is worth mentioning that, in this case, using a high magnetic potential is much more effective than selecting a lower voltage.

Next, the stiffness-hardening influence of the length-scale parameter on the phase speed of the scattered waves in an MEE-FG nanoplate is illustrated in Figure 8.10. In this diagram,

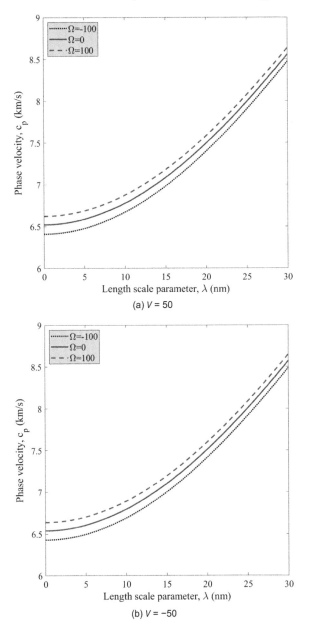

(a) $V = 50$

(b) $V = -50$

FIGURE 8.10

Variation of phase velocity of MEE-FG nanoplate with length-scale parameter at various amounts of magnetic potential ($p = 2$, $\beta = 0.02$ (1/nm), $\mu = 1$ nm).

the variation of the phase velocity with length-scale parameter is plotted at various magnetic potentials with respect to both positive and negative applied electric voltages. As in the case of previously discussed figures, in this figure, a higher mechanical response is again used when employing a lower applied voltage. However, once again it can be seen that the force generated by applying a magnetic potential to the plate affects in a way that corresponds with making the structure stiffer, and due to this fact, the nanoplate has a greater phase velocity when a high magnetic potential is used. Furthermore, one should be aware of the effect of length-scale term, which is an intensifying effect. Actually, the nanoplate shows a greater speed when increasing the length-scale parameter. This increasing effect are due to the stresses which are identical in value but opposite in direction. Thus, these stresses produce a couple that makes the nanostructure stiffer.

Now, it is time to study reciprocal electromagnetic interactions in a smart MEE nanoplate with respect to the effects of porosities on the mechanical response of the nanoplate (see Figure 8.11). It can be found that phase velocity of the smart nanoplate reduces whenever the continuum is connected to a higher electric voltage. The reason for this fact was discussed previously in Chapter 6. Also, as in previous case studies, the phase velocity of the dispersed waves can be increased by applying higher magnetic potentials. In addition to these smart behaviors, it must be mentioned that the nanoplate reveals lower phase speeds at any desired electromagnetic inputs when a porous smart material is selected to fabricate the nanoplate instead of a nonporous one. This trend is due to the fact that, in porous materials, the existence of porosities has a destructive impact on the stiffness of the material. This phenomenon can finally result in a decrease in the dynamic responses of any element that is fabricated with the porous material compared with a nonporous one.

Finally, the variation of the phase velocity of the MEE nanoplates with wave number is depicted in Figure 8.12 at different magnetic potentials for both porous and nonporous

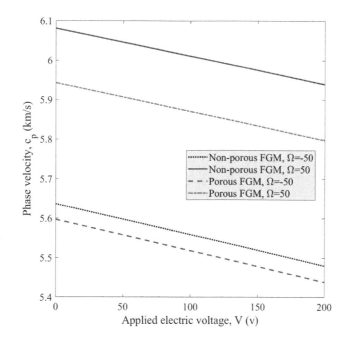

FIGURE 8.11

Variation of phase velocity of MEE-FG nanoplate with applied electric voltage at various amounts of magnetic potential for both porous and nonporous materials ($p = 2$, $\beta = 0.01$ $(1/nm)$, $c < 1$)

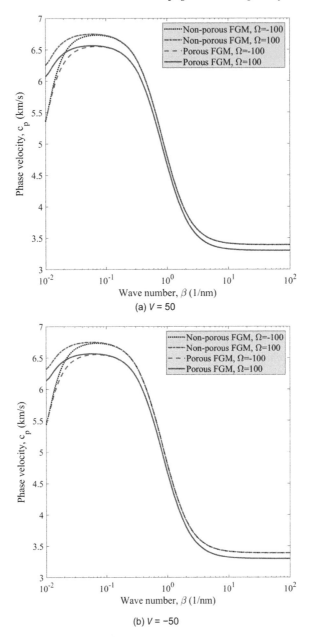

FIGURE 8.12
Variation of phase velocity of MEE-FG nanoplate with wave number at various amounts of magnetic potential ($p = 2$, $c < 1$).

materials. The results of the previous diagram can be observed again in this diagram. According to this figure, the greatest response belongs to the case of selecting a nonporous smart FGM which is connected to the maximum positive magnetic potential. It is clear that, the porous nanoplates can provide lower phase velocities at all the studied wave numbers. Also, it is worth mentioning that the nanoplate which is subjected to the lower electric voltage reveals greater wave speeds in a limited range of wave numbers, particularly when $\beta < 0.05$ (1/nm).

9

Effect of Various Resting Media on Wave Dispersion Characteristics of Smart Nanostructures

In this chapter, the wave propagation problems of nanobeams and nanoplates are solved for various resting conditions of nanostructures. In fact, the previously discussed formulations for nanostructures will be updated for cases when the nanostructure is rested on a medium. To provide complete results, different types of elastic foundations will be employed as well as the well-known viscoelastic model, the visco-Pasternak medium. The wave propagation responses of the nanostructures for various types of resting media are compared. The presented results show that embedding the nanostructure on an elastic substrate can improve the stiffness of the system and finally strengthen the dynamic response of the nanostructure, whereas the mechanical response is reduced whenever the nanostructure is placed on a visco-Pasternak foundation. It is important to be aware of the elastic behaviors of the nanostructures while they are rested on various types of elastic and viscoelastic mediums. The mechanical elements will not be implemented in different applications without resting them on a medium. Hence, a mechanical engineer should know about the mechanical reaction of the structures once they are rested on a substrate.

9.1 Winkler Foundation

Once a structure like a beam or a plate is rested on a Winkler medium, a group of discrete linear springs is used under the structure. In this type of foundation, there is no coupling between the springs. Indeed, each of the springs work independent from the variations generated in the other springs. This modeling can present a simple simulation about the resting condition of a structure on a foundation, however, it cannot completely model the foundation because springs are assumed to be independent from each other in this model. In what follows, the changes in the formulation of both nanobeams and nanoplates rested on this type of substrate will be presented and the solution will be completed for embedded nanobeams and nanoplates.

9.1.1 Analysis of Nanobeams Embedded on Winkler Foundation

The effect of resting the nanobeam on a Winkler substrate can be applied on the Lagrangian of the nanobeam in the framework of the variation of work done by external forces. Before presenting the variation of the external work, the work done by the Winkler foundation on a beam must be introduced. First, the work done by a desired force on a structure can be expressed as:

$$V = \frac{1}{2} \int_V \mathbf{F}.\mathbf{u}dV \qquad (9.1)$$

139

where $\mathbf{F}$ is the external force vector applied on the structure, and $\mathbf{u}$ is the displacement vector of the structure. For a beam rested on a Winkler foundation, the force and displacement vectors are:

$$\mathbf{F} = (0, 0, k_w u \ ) \tag{9.2}$$

$$\mathbf{u} = (u_x, 0, u \ ) \tag{9.3}$$

where k_w is the linear spring of the Winkler substrate and u is the component of the displacement vector in the $z-$ direction. Once on substituting Eqs. (9.2) and (9.3) in Eq. (9.1), the work done by external forces can be presented in the following form:

$$V = \frac{1}{2} \int_V k_w u^2 dV \tag{9.4}$$

Now, the variation of the above equation can be easily calculated for both Euler–Bernoulli and refined sinusoidal beams by applying the variation operator on Eq. (9.4). For an Euler–Bernoulli beam with displacement field explained in Eq. (4.10), the variation of work done by the Winkler medium can be expressed as:

$$\delta V = \int_V k_w w \delta w dV \tag{9.5}$$

Employing Eq. (4.15) for the displacement field of a refined sinusoidal beam and using the fundamental relation presented in Eq. (9.4), the variation of work done by the Winkler substrate on a refined higher-order beam is as follows:

$$\delta V = \int_V k_w (w_b + w_s) \delta (w_b + w_s) dV \tag{9.6}$$

Now, the influences of the Winkler medium on the motion equations and the governing equations of both classical and shear deformable nanobeams must be included. First, the effect of an elastic medium on the Euler–Lagrange equations of the beams must be applied. For studying an Euler–Bernoulli beam, the Euler–Lagrange equations of a beam embedded on a Winkler foundation can be achieved by substituting Eqs. (4.24), (4.26) and (9.5) in Eq. (4.19) as follows:

$$\frac{\partial N}{\partial x} = I_0 \frac{\partial^2 u}{\partial t^2} - I_1 \frac{\partial^3 w}{\partial x \partial t^2} \tag{9.7}$$

$$\frac{\partial^2 M}{\partial x^2} - k_w w = I_0 \frac{\partial^2 w}{\partial t^2} + I_1 \frac{\partial^3 u}{\partial x \partial t^2} - I_2 \frac{\partial^4 w}{\partial x^2 \partial t^2} \tag{9.8}$$

Also, the Euler–Lagrange equations of refined beams can be achieved by substituting Eqs. (4.31), (4.33) and (9.6) in Eq. (4.19). These coupled equations can be expressed as:

$$\frac{\partial N}{\partial x} = I_0 \frac{\partial^2 u}{\partial t^2} - I_1 \frac{\partial^3 w_b}{\partial x \partial t^2} - J_1 \frac{\partial^3 w_s}{\partial x \partial t^2} \tag{9.9}$$

$$\frac{\partial^2 M^b}{\partial x^2} - k_w (w_b + w_s) = I_0 \frac{\partial^2 (w_b + w_s)}{\partial t^2} + I_1 \frac{\partial^3 u}{\partial x \partial t^2} - I_2 \frac{\partial^4 w_b}{\partial x^2 \partial t^2} - J_2 \frac{\partial^4 w_s}{\partial x^2 \partial t^2} \tag{9.10}$$

$$\frac{\partial^2 M^s}{\partial x^2} + \frac{\partial Q_x}{\partial x} - k_w (w_b + w_s) = I_0 \frac{\partial^2 (w_b + w_s)}{\partial t^2} + J_1 \frac{\partial^3 u}{\partial x \partial t^2} - J_2 \frac{\partial^4 w_b}{\partial x^2 \partial t^2} - K_2 \frac{\partial^4 w_s}{\partial x^2 \partial t^2} \tag{9.11}$$

Now, the governing equations of nanobeams embedded on a Winkler substrate can be expressed using the above equations. For an Euler–Bernoulli nanobeam, the governing equations can be obtained once Eqs. (4.38) and (4.39) are substituted in Eqs. (9.7) and (9.8). Thus, the governing equations of the Euler–Bernoulli FG nanobeam can be expressed as:

$$\left(1 - \lambda^2 \nabla^2\right) \left[A_{xx} \frac{\partial^2 u}{\partial x^2} - B_{xx} \frac{\partial^3 w}{\partial x^3}\right] + \left(1 - \mu^2 \nabla^2\right) \left[-I_0 \frac{\partial^2 u}{\partial t^2} + I_1 \frac{\partial^3 w}{\partial x \partial t^2}\right] = 0 \qquad (9.12)$$

$$\left(1 - \lambda^2 \nabla^2\right) \left[B_{xx} \frac{\partial^3 u}{\partial x^3} - D_{xx} \frac{\partial^4 w}{\partial x^4}\right] + \left(1 - \mu^2 \nabla^2\right) \left[k_w w - I_0 \frac{\partial^2 w}{\partial t^2} - I_1 \frac{\partial^3 u}{\partial x \partial t^2} + I_2 \frac{\partial^4 w}{\partial x^2 \partial t^2}\right] = 0 \qquad (9.13)$$

Moreover, the governing equations of refined higher-order nanobeams can be achieved by substituting Eqs. (4.40)–(4.43) in Eqs. (9.9)–(9.11) in the following form:

$$\left(1 - \lambda^2 \nabla^2\right) \left[A_{xx} \frac{\partial^2 u}{\partial x^2} - B_{xx} \frac{\partial^3 w_b}{\partial x^3} - B_{xx}^s \frac{\partial^3 w_s}{\partial x^3}\right] + \left(1 - \mu^2 \nabla^2\right)$$
$$\times \left[-I_0 \frac{\partial^2 u}{\partial t^2} + I_1 \frac{\partial^3 w_b}{\partial x \partial t^2} + J_1 \frac{\partial^3 w_s}{\partial x \partial t^2}\right] = 0 \qquad (9.14)$$

$$\left(1 - \lambda^2 \nabla^2\right) \left[B_{xx} \frac{\partial^3 u}{\partial x^3} - D_{xx} \frac{\partial^4 w_b}{\partial x^4} - D_{xx}^s \frac{\partial^4 w_s}{\partial x^4}\right] + \left(1 - \mu^2 \nabla^2\right)$$
$$\times \left[k_w (w_b + w_s) - I_0 \frac{\partial^2 (w_b + w_s)}{\partial t^2} - I_1 \frac{\partial^3 u}{\partial x \partial t^2} + I_2 \frac{\partial^4 w_b}{\partial x^2 \partial t^2} + J_2 \frac{\partial^4 w_s}{\partial x^2 \partial t^2}\right] = 0 \quad (9.15)$$

$$\left(1 - \lambda^2 \nabla^2\right) \left[B_{xx}^s \frac{\partial^3 u}{\partial x^3} - D_{xx}^s \frac{\partial^4 w_b}{\partial x^4} - H_{xx}^s \frac{\partial^4 w_s}{\partial x^4} + A_x^s \frac{\partial^2 w_s}{\partial x^2}\right] + \left(1 - \mu^2 \nabla^2\right)$$
$$\times \left[k_w (w_b + w_s) - I_0 \frac{\partial^2 (w_b + w_s)}{\partial t^2} - J_1 \frac{\partial^3 u}{\partial x \partial t^2} + J_2 \frac{\partial^4 w_b}{\partial x^2 \partial t^2} + K_2 \frac{\partial^4 w_s}{\partial x^2 \partial t^2}\right] = 0 \quad (9.16)$$

After deriving the governing equations, it must be noted that the solution of this type of nanobeams are the same as that procured for FG nanobeams in Chapter 4. The only difference is the change that must be applied in the expression of the component(s) of the stiffness matrix that is (are) affected by the Winkler coefficient. In the Euler–Bernoulli nanobeams, the k_{22} component of the stiffness matrix must be updated. In fact, the wave propagation problem of an FG nanobeam rested on an elastic Winkler substrate can be simply solved using Eq. (1.3) with stiffness and mass matrices expressed in Eqs. (4.51) and (4.52), respectively. The only change that must be applied in Eq. (4.51) is:

$$k_{22} = - \left(1 - \lambda^2 \nabla^2\right) D_{xx} \beta^4 - \left(1 - \mu^2 \nabla^2\right) k_w \qquad (9.17)$$

Furthermore, Eq. (1.8) can be again used for solving the wave propagation problem of a refined FG nanobeam embedded on a Winkler substrate with stiffness and mass matrices and arrays defined in Eqs. (4.53) and (4.54), respectively. Because in the refined beams, arrays with index 2 and 3 are related to the bending and shear deflections of the beam, respectively, the following components of the stiffness matrix must be modified and substituted instead of those reported in Eq. (4.53):

$$k_{22} = - \left(1 - \lambda^2 \nabla^2\right) D_{xx} \beta^4 - \left(1 - \mu^2 \nabla^2\right) k_w,$$
$$k_{23} = - \left(1 - \lambda^2 \nabla^2\right) D_{xx}^s \beta^4 - \left(1 - \mu^2 \nabla^2\right) k_w,$$
$$k_{33} = - \left(1 - \lambda^2 \nabla^2\right) \left[H_{xx}^s \beta^4 + A_x^s \beta^2\right] - \left(1 - \mu^2 \nabla^2\right) k_w \qquad (9.18)$$

9.1.2 Analysis of Nanoplates Embedded on Winkler Foundation

Here, the effect of resting the nanoplate on the Winkler medium will be studied similar to the nanobeams discussed above. Both Kirchhoff–Love and refined higher-order plates will be used for this. The work done by external forces was presented in Section 9.1.1 (see Eq. (9.1)). For a plate embedded on the Winkler medium, the displacement vector can be expressed as:

$$\mathbf{u} = (u_x, u_y, u\ ) \tag{9.19}$$

It is worth mentioning that the force vector is the same as that introduced in Eq. (9.2). Due to the zero value of the x and y components of the force vector, the work done by the Winkler foundation on a plate-type structure will be the same as that presented in Eq. (9.4). The variation of work done by the foundation for a Kirchhoff-Love plate is similar as that presented in Eq. (9.5). Moreover, the variation can be calculated for a refined sinusoidal plate by using Eq. (9.6). It is worth mentioning that Eq. (9.5) for a classical plate can be obtained by implementing Eq. (4.58) instead of the u of the plate and applying the variation operator on Eq. (9.4). Using a similar procedure, Eq. (4.65) must be selected to be inserted as u in Eq. (9.4) for a refined sinusoidal plate.

The Euler–Lagrange equations of the plates rested on a simple Winkler medium can be achieved by employing the variation of work performed by the foundation in the fundamental equation introduced in Chapter 4 for calculating the systems' governing equations using a variational-based method. For a Kirchhoff–Love plate, Eqs. (4.72), (4.74) and (9.5) must be substituted in Eq. (4.19). The final Euler–Lagrange equations can be written in the following form:

$$\frac{\partial N_{xx}}{\partial x} + \frac{\partial N_{xy}}{\partial y} = I_0 \frac{\partial^2 u}{\partial t^2} - I_1 \frac{\partial^3 w}{\partial x \partial t^2} \tag{9.20}$$

$$\frac{\partial N_{xy}}{\partial x} + \frac{\partial N_{yy}}{\partial y} = I_0 \frac{\partial^2 v}{\partial t^2} - I_1 \frac{\partial^3 w}{\partial y \partial t^2} \tag{9.21}$$

$$\frac{\partial^2 M_{xx}}{\partial x^2} + 2\frac{\partial^2 M_{xy}}{\partial x \partial y} + \frac{\partial^2 M_{yy}}{\partial y^2} - k_w w = I_0 \frac{\partial^2 w}{\partial t^2}$$
$$+ I_1 \left(\frac{\partial^3 u}{\partial x \partial t^2} + \frac{\partial^3 v}{\partial y \partial t^2} \right) - I_2 \left(\frac{\partial^4 w}{\partial x^2 \partial t^2} + \frac{\partial^4 w}{\partial y^2 \partial t^2} \right) \tag{9.22}$$

Moreover, the Euler–Lagrange equations of refined plates rested on a Winkler substrate can be obtained by inserting Eqs. (4.80), (4.83) and (9.6) in Eq. (4.19). The final equations can be expressed as:

$$\frac{\partial N_{xx}}{\partial x} + \frac{\partial N_{xy}}{\partial y} = I_0 \frac{\partial^2 u}{\partial t^2} - I_1 \frac{\partial^3 w_b}{\partial x \partial t^2} - J_1 \frac{\partial^3 w_s}{\partial x \partial t^2} \tag{9.23}$$

$$\frac{\partial N_{xy}}{\partial x} + \frac{\partial N_{yy}}{\partial y} = I_0 \frac{\partial^2 v}{\partial t^2} - I_1 \frac{\partial^3 w_b}{\partial y \partial t^2} - J_1 \frac{\partial^3 w_s}{\partial y \partial t^2} \tag{9.24}$$

$$\frac{\partial^2 M_{xx}^b}{\partial x^2} + 2\frac{\partial^2 M_{xy}^b}{\partial x \partial y} + \frac{\partial^2 M_{yy}^b}{\partial y^2} - k_w (w_b + w_s) = I_0 \frac{\partial^2 (w_b + w_s)}{\partial t^2}$$
$$+ I_1 \left(\frac{\partial^3 u}{\partial x \partial t^2} + \frac{\partial^3 v}{\partial y \partial t^2} \right) - I_2 \left(\frac{\partial^4 w_b}{\partial x^2 \partial t^2} + \frac{\partial^4 w_b}{\partial y^2 \partial t^2} \right) - J_2 \left(\frac{\partial^4 w_s}{\partial x^2 \partial t^2} + \frac{\partial^4 w_s}{\partial y^2 \partial t^2} \right) \tag{9.25}$$

$$\frac{\partial^2 M_{xx}^s}{\partial x^2} + 2\frac{\partial^2 M_{xy}^s}{\partial x \partial y} + \frac{\partial^2 M_{yy}^s}{\partial y^2} + \frac{\partial Q_x}{\partial x} + \frac{\partial Q_y}{\partial y} - k_w (w_b + w_s) = I_0 \frac{\partial^2 (w_b + w_s)}{\partial t^2}$$
$$+ J_1 \left(\frac{\partial^3 u}{\partial x \partial t^2} + \frac{\partial^3 v}{\partial y \partial t^2} \right) - J_2 \left(\frac{\partial^4 w_b}{\partial x^2 \partial t^2} + \frac{\partial^4 w_b}{\partial y^2 \partial t^2} \right) - K_2 \left(\frac{\partial^4 w_s}{\partial x^2 \partial t^2} + \frac{\partial^4 w_s}{\partial y^2 \partial t^2} \right) \tag{9.26}$$

Afterward, the effects of the small scale of the nanoplate must be added to the obtained Euler–Lagrange equations of the plate to achieve the partial differential equations which can control the problem. These equations can be obtained for a Kirchhoff–Love nanoplate by substituting Eqs. (4.89) and (4.90) in Eqs. (9.20)–(9.22). The final equations can be expressed as:

$$
\left(1 - \lambda^2 \nabla^2\right) \left[A_{11} \frac{\partial^2 u}{\partial x^2} + \left(A_{12} + A_{66}\right) \frac{\partial^2 v}{\partial x \partial y} + A_{66} \frac{\partial^2 u}{\partial y^2} - B_{11} \frac{\partial^3 w}{\partial x^3} - \left(B_{12} + 2B_{66}\right) \frac{\partial^3 w}{\partial x \partial y^2} \right]
$$

$$
+ \left(1 - \mu^2 \nabla^2\right) \left[-I_0 \frac{\partial^2 u}{\partial t^2} + I_1 \frac{\partial^3 w}{\partial x \partial t^2} \right] = 0 \tag{9.27}
$$

$$
\left(1 - \lambda^2 \nabla^2\right) \left[A_{22} \frac{\partial^2 v}{\partial y^2} + \left(A_{12} + A_{66}\right) \frac{\partial^2 u}{\partial x \partial y} + A_{66} \frac{\partial^2 v}{\partial x^2} - B_{22} \frac{\partial^3 w}{\partial y^3} - \left(B_{12} + 2B_{66}\right) \frac{\partial^3 w}{\partial x^2 \partial y} \right]
$$

$$
+ \left(1 - \mu^2 \nabla^2\right) \left[-I_0 \frac{\partial^2 v}{\partial t^2} + I_1 \frac{\partial^3 w}{\partial y \partial t^2} \right] = 0 \tag{9.28}
$$

$$
\left(1 - \lambda^2 \nabla^2\right) \left[B_{11} \frac{\partial^3 u}{\partial x^3} + \left(B_{12} + 2B_{66}\right) \frac{\partial^3 u}{\partial x \partial y^2} + B_{22} \frac{\partial^3 v}{\partial y^3} + \left(B_{12} + 2B_{66}\right) \frac{\partial^3 v}{\partial x^2 \partial y} - D_{11} \frac{\partial^4 w}{\partial x^4} \right.
$$

$$
\left. - 2\left(D_{12} + 2D_{66}\right) \frac{\partial^4 w}{\partial x^2 \partial y^2} - D_{22} \frac{\partial^4 w}{\partial y^4} \right] + \left(1 - \mu^2 \nabla^2\right) \left[k_w w - I_0 \frac{\partial^2 w}{\partial t^2} \right.
$$

$$
\left. - I_1 \left(\frac{\partial^3 u}{\partial x \partial t^2} + \frac{\partial^3 v}{\partial y \partial t^2} \right) + I_2 \left(\frac{\partial^4 w}{\partial x^2 \partial t^2} + \frac{\partial^4 w}{\partial y^2 \partial t^2} \right) \right] = 0 \tag{9.29}
$$

Also, the nonlocal governing equations of refined nanoplates can be achieved by substituting Eqs. (4.93)–(4.96) in Eqs. (9.23)–(9.26). Finally, the governing equations are

$$
\left(1 - \lambda^2 \nabla^2\right) \left[A_{11} \frac{\partial^2 u}{\partial x^2} + \left(A_{12} + A_{66}\right) \frac{\partial^2 v}{\partial x \partial y} + A_{66} \frac{\partial^2 u}{\partial y^2} - B_{11} \frac{\partial^3 w_b}{\partial x^3} - \left(B_{12} + 2B_{66}\right) \frac{\partial^3 w_b}{\partial x \partial y^2} \right.
$$

$$
\left. - B_{11}^s \frac{\partial^3 w_s}{\partial x^3} - \left(B_{12}^s + 2B_{66}^s\right) \frac{\partial^3 w_s}{\partial x \partial y^2} \right] + \left(1 - \mu^2 \nabla^2\right) \left[-I_0 \frac{\partial^2 u}{\partial t^2} + I_1 \frac{\partial^3 w_b}{\partial x \partial t^2} + J_1 \frac{\partial^3 w_s}{\partial x \partial t^2} \right] = 0 \tag{9.30}
$$

$$
\left(1 - \lambda^2 \nabla^2\right) \left[A_{22} \frac{\partial^2 v}{\partial y^2} + \left(A_{12} + A_{66}\right) \frac{\partial^2 u}{\partial x \partial y} + A_{66} \frac{\partial^2 v}{\partial x^2} - B_{22} \frac{\partial^3 w_b}{\partial y^3} - \left(B_{12} + 2B_{66}\right) \frac{\partial^3 w_b}{\partial x^2 \partial y} \right.
$$

$$
\left. - B_{22}^s \frac{\partial^3 w_s}{\partial y^3} - \left(B_{12}^s + 2B_{66}^s\right) \frac{\partial^3 w_s}{\partial x^2 \partial y} \right] + \left(1 - \mu^2 \nabla^2\right) \left[-I_0 \frac{\partial^2 v}{\partial t^2} + I_1 \frac{\partial^3 w_b}{\partial y \partial t^2} + J_1 \frac{\partial^3 w_s}{\partial y \partial t^2} \right] = 0 \tag{9.31}
$$

$$
\left(1 - \lambda^2 \nabla^2\right) \left[B_{11} \frac{\partial^3 u}{\partial x^3} + \left(B_{12} + 2B_{66}\right) \frac{\partial^3 u}{\partial x \partial y^2} + B_{22} \frac{\partial^3 v}{\partial y^3} + \left(B_{12} + 2B_{66}\right) \frac{\partial^3 v}{\partial x^2 \partial y} - D_{11} \frac{\partial^4 w_b}{\partial x^4} \right.
$$

$$
\left. - 2\left(D_{12} + 2D_{66}\right) \frac{\partial^4 w_b}{\partial x^2 \partial y^2} - D_{22} \frac{\partial^4 w_b}{\partial y^4} - D_{11}^s \frac{\partial^4 w_s}{\partial x^4} - 2\left(D_{12}^s + 2D_{66}^s\right) \frac{\partial^4 w_s}{\partial x^2 \partial y^2} - D_{22}^s \frac{\partial^4 w_s}{\partial y^4} \right)
$$

$$
+ \left(1 - \mu^2 \nabla^2\right) \left[k_w \left(w_b + w_s\right) - I_0 \frac{\partial^2 \left(w_b + w_s\right)}{\partial t^2} - I_1 \left(\frac{\partial^3 u}{\partial x \partial t^2} + \frac{\partial^3 v}{\partial y \partial t^2} \right) \right.
$$

$$
\left. + I_2 \left(\frac{\partial^4 w_b}{\partial x^2 \partial t^2} + \frac{\partial^4 w_b}{\partial y^2 \partial t^2} \right) + J_2 \left(\frac{\partial^4 w_s}{\partial x^2 \partial t^2} + \frac{\partial^4 w_s}{\partial y^2 \partial t^2} \right) \right] = 0 \tag{9.32}
$$

$$\left(1 - \lambda^2 \nabla^2\right) \left[B_{11}^s \frac{\partial^3 u}{\partial x^3} + \left(B_{12}^s + 2B_{66}^s\right) \frac{\partial^3 u}{\partial x \partial y^2} + B_{22}^s \frac{\partial^3 v}{\partial y^3} + \left(B_{12}^s + 2B_{66}^s\right) \frac{\partial^3 v}{\partial x^2 \partial y} - D_{11}^s \frac{\partial^4 w_b}{\partial x^4} \right.$$

$$- 2\left(D_{12}^s + 2D_{66}^s\right) \frac{\partial^4 w_b}{\partial x^2 \partial y^2} - D_{22}^s \frac{\partial^4 w_b}{\partial y^4} - H_{11}^s \frac{\partial^4 w_s}{\partial x^4} - 2\left(H_{12}^s + 2H_{66}^s\right) \frac{\partial^4 w_s}{\partial x^2 \partial y^2} - H_{22}^s \frac{\partial^4 w_s}{\partial y^4}$$

$$\left. + A_{44}^s \left(\frac{\partial^2 (w_b + w_s)}{\partial x^2} + \frac{\partial^2 (w_b + w_s)}{\partial y^2} \right) \right] + \left(1 - \mu^2 \nabla^2\right) \left[k_w \left(w_b + w_s\right) - I_0 \frac{\partial^2 (w_b + w_s)}{\partial t^2} \right.$$

$$\left. - J_1 \left(\frac{\partial^3 u}{\partial x \partial t^2} + \frac{\partial^3 v}{\partial y \partial t^2} \right) + J_2 \left(\frac{\partial^4 w_b}{\partial x^2 \partial t^2} + \frac{\partial^4 w_b}{\partial y^2 \partial t^2} \right) + K_2 \left(\frac{\partial^4 w_s}{\partial x^2 \partial t^2} + \frac{\partial^4 w_s}{\partial y^2 \partial t^2} \right) \right] = 0$$

$$(9.33)$$

Now, the governing equations must be solved to obtain the wave propagation responses of FG nanoplates rested on a Winkler medium. Similar to the previous section on nanobeams (see Section 9.1.1), the solution of the embedded nanoplates is similar with that of simple nanoplates that are not rested on an elastic medium. The difference between these two types of nanoplates belongs to those components of stiffness matrix involved with the Winkler coefficient. Thus, for a Kirchhoff–Love nanoplate, the solution is the same as Eq. (1.8) with stiffness and mass matrices and their arrays are defined in Eqs. (4.107) and (4.108), respectively. The difference between embedded and not embedded nanoplates can be seen in k_{33}. The updated expression of this array must be used as below and substituted instead of the conventional one available in Eq. (4.107):

$$k_{33} = -\left(1 + \lambda^2\left(\beta_1^2 + \beta_2^2\right)\right)\left[D_{11}\beta_1^4 + 2\left(D_{12} + 2D_{66}\right)\beta_1^2\beta_2^2 + D_{22}\beta_2^4\right] - \left(1 + \mu^2\left(\beta_1^2 + \beta_2^2\right)\right)k_w$$

$$(9.34)$$

Furthermore, the solution of embedded refined nanoplates is similar as that of the simple nanoplate, which is not rested on an elastic medium. Similar to the Kirchhoff–Love plates, here the differences are in the some components of the stiffness matrix involved if the nanoplate is rested on a foundation. Because of the employment of a refined plate model, the arrays with 3 and 4 indices together must be modified. Indeed, the following arrays must be selected from below and inserted in Eq. (4.109):

$$k_{33} = -\left(1 + \lambda^2\left(\beta_1^2 + \beta_2^2\right)\right)\left[D_{11}\beta_1^4 + 2\left(D_{12} + 2D_{66}\right)\beta_1^2\beta_2^2 + D_{22}\beta_2^4\right] - \left(1 + \mu^2\left(\beta_1^2 + \beta_2^2\right)\right)k_w,$$

$$k_{34} = -\left(1 + \lambda^2\left(\beta_1^2 + \beta_2^2\right)\right)\left[D_{11}^s\beta_1^4 + 2\left(D_{12}^s + 2D_{66}^s\right)\beta_1^2\beta_2^2 + D_{22}^s\beta_2^4\right] - \left(1 + \mu^2\left(\beta_1^2 + \beta_2^2\right)\right)k_w,$$

$$k_{44} = -\left(1 + \lambda^2\left(\beta_1^2 + \beta_2^2\right)\right)\left[H_{11}^s\beta_1^4 + 2\left(H_{12}^s + 2H_{66}^s\right)\beta_1^2\beta_2^2 + H_{22}^s\beta_2^4 + A_{44}^s\left(\beta_1^2 + \beta_2^2\right)\right]$$

$$- \left(1 + \mu^2\left(\beta_1^2 + \beta_2^2\right)\right)k_w$$

$$(9.35)$$

The components of the mass matrix remain the same as those reported in Eq. (4.110).

9.2   Winkler–Pasternak Foundation

In Section 9.1, the mathematical and physical representation of embedding a beam or a plate on a Winkler-type elastic substrate was discussed. It was demonstrated that the Winkler medium can present a rough simulation about the real elastic foundations. To cover the limitations of the Winkler foundation, it is better to employ a more complicated spring system instead of the conventional Winkler model. One of these models is the Pasternak model which includes an additional shear spring as well as the former linear one to satisfy the float common motion of linear springs. The shear springs that are utilized in the Pasternak model are usually called Pasternak springs. It is worth mentioning that this type of elastic foundation is generally called Winkler–Pasternak model because of the presence of both linear and shear springs. In the following parts, we will try to show the effect of adding the Pasternak coefficient to the previously implemented Winkler one and import this term in the governing equations of nanosized beams and plates. The effect of this type of medium must be included in the motion equations by calculating the variation of work done by springs as mentioned in Eq. (9.1).

9.2.1 Analysis of Nanobeams Embedded on Winkler Pasternak Foundation

Here, the work done by Winkler–Pasternak medium must be defined in the integral form. Thereafter, the variation of this work must be calculated to apply the effect of the elastic foundation on the motion equations of the beams. The displacement vector of beam, which is required for calculating external work, is expressed in Eq. (9.3). The force vector of the Winkler–Pasternak medium can be expressed as:

$$\mathbf{F} = \left(0, 0, k_w u - k_p \frac{\partial^2 u}{\partial x^2}\right) \tag{9.36}$$

Hence, the work done by Winkler–Pasternak foundation can be expressed as:

$$V = \frac{1}{2} \int_V \left(k_w u^2 - k_p \frac{\partial^2 u}{\partial x^2} u\right) dV \tag{9.37}$$

Once Eq. (4.10) is employed for the u of the Euler–Bernoulli beam and the effect of the variation operator is applied on Eq. (9.37), the variation of work done by Winkler–Pasternak medium can be expressed in the following form:

$$\delta V = \int_V \left(k_w w - k_p \frac{\partial^2 w}{\partial x^2}\right) \delta w \, dV \tag{9.38}$$

where k_p denotes the Pasternak coefficient of the two-parameter elastic foundation. Utilizing Eq. (4.15) instead of Eq. (4.10) for the u and going through the previous procedure, the variation of work done by Winkler–Pasternak medium for a refined sinusoidal beam can be written as:

$$\delta V = \int_V \left(k_w (w_b + w_s) - k_p \frac{\partial^2 (w_b + w_s)}{\partial x^2}\right) \delta(w_b + w_s) dV \tag{9.39}$$

Here, the Euler–Lagrange equations of beams rested on Winkler–Pasternak foundation can be achieved. For an Euler–Bernoulli beam, Eqs. (4.24), (4.26) and (9.38) must be substituted in Eq. (4.19). Therefore, the Euler–Lagrange equations of a classical beam can be expressed as:

$$\frac{\partial N}{\partial x} = I_0 \frac{\partial^2 u}{\partial t^2} - I_1 \frac{\partial^3 w}{\partial x \partial t^2} \tag{9.40}$$

$$\frac{\partial^2 M}{\partial x^2} - k_w w + k_p \frac{\partial^2 w}{\partial x^2} = I_0 \frac{\partial^2 w}{\partial t^2} + I_1 \frac{\partial^3 u}{\partial x \partial t^2} - I_2 \frac{\partial^4 w}{\partial x^2 \partial t^2} \tag{9.41}$$

Moreover, one can find the Euler–Lagrange equations of refined sinusoidal beams rested on a two-parameter Pasternak substrate by inserting Eqs. (4.31), (4.33) and (9.39) in Eq. (4.19). The obtained equations are:

$$\frac{\partial N}{\partial x} = I_0 \frac{\partial^2 u}{\partial t^2} - I_1 \frac{\partial^3 w_b}{\partial x \partial t^2} - J_1 \frac{\partial^3 w_s}{\partial x \partial t^2} \tag{9.42}$$

$$\frac{\partial^2 M^b}{\partial x^2} - k_w\left(w_b + w_s\right) + k_p\frac{\partial^2\left(w_b + w_s\right)}{\partial x^2} = I_0\frac{\partial^2\left(w_b + w_s\right)}{\partial t^2} + I_1\frac{\partial^3 u}{\partial x\partial t^2} - I_2\frac{\partial^4 w_b}{\partial x^2\partial t^2} - J_2\frac{\partial^4 w_s}{\partial x^2\partial t^2}$$
$$\tag{9.43}$$

$$\frac{\partial^2 M^s}{\partial x^2} + \frac{\partial Q_x}{\partial x} - k_w\left(w_b + w_s\right) + k_p\frac{\partial^2\left(w_b + w_s\right)}{\partial x^2} = I_0\frac{\partial^2\left(w_b + w_s\right)}{\partial t^2} + J_1\frac{\partial^3 u}{\partial x\partial t^2}$$
$$- J_2\frac{\partial^4 w_b}{\partial x^2\partial t^2} - K_2\frac{\partial^4 w_s}{\partial x^2\partial t^2} \tag{9.44}$$

Moreover, the governing equations of the nanobeams rested on Winkler–Pasternak medium can be achieved by inserting the constitutive equations of nanobeams in the Euler–Lagrange equations obtained above. For the Euler–Bernoulli nanobeams, the governing equations can be achieved by substituting Eqs. (4.38) and (4.39) in Eqs. (9.40) and (9.41) as follows:

$$\left(1 - \lambda^2\nabla^2\right)\left[A_{xx}\frac{\partial^2 u}{\partial x^2} - B_{xx}\frac{\partial^3 w}{\partial x^3}\right] + \left(1 - \mu^2\nabla^2\right)\left[-I_0\frac{\partial^2 u}{\partial t^2} + I_1\frac{\partial^3 w}{\partial x\partial t^2}\right] = 0 \tag{9.45}$$

$$\left(1 - \lambda^2\nabla^2\right)\left[B_{xx}\frac{\partial^3 u}{\partial x^3} - D_{xx}\frac{\partial^4 w}{\partial x^4}\right] + \left(1 - \mu^2\nabla^2\right)\left[k_w w - k_p\frac{\partial^2 w}{\partial x^2} - I_0\frac{\partial^2 w}{\partial t^2}\right.$$
$$\left. - I_1\frac{\partial^3 u}{\partial x\partial t^2} + I_2\frac{\partial^4 w}{\partial x^2\partial t^2}\right] = 0 \tag{9.46}$$

Further, the governing equations of refined sinusoidal nanobeams, rested on a Winkler–Pasternak medium, can be derived when Eqs. (4.40)–(4.43) are substituted in Eqs. (9.42)–(9.44)

$$\left(1 - \lambda^2\nabla^2\right)\left[A_{xx}\frac{\partial^2 u}{\partial x^2} - B_{xx}\frac{\partial^3 w_b}{\partial x^3} - B_{xx}^s\frac{\partial^3 w_s}{\partial x^3}\right] + \left(1 - \mu^2\nabla^2\right)$$
$$\left[-I_0\frac{\partial^2 u}{\partial t^2} + I_1\frac{\partial^3 w_b}{\partial x\partial t^2} + J_1\frac{\partial^3 w_s}{\partial x\partial t^2}\right] = 0 \tag{9.47}$$

$$\left(1 - \lambda^2\nabla^2\right)\left[B_{xx}\frac{\partial^3 u}{\partial x^3} - D_{xx}\frac{\partial^4 w_b}{\partial x^4} - D_{xx}^s\frac{\partial^4 w_s}{\partial x^4}\right] + \left(1 - \mu^2\nabla^2\right)\left[k_w\left(w_b + w_s\right)\right.$$
$$\left. - k_p\frac{\partial^2\left(w_b + w_s\right)}{\partial x^2} - I_0\frac{\partial^2\left(w_b + w_s\right)}{\partial t^2} - I_1\frac{\partial^3 u}{\partial x\partial t^2} + I_2\frac{\partial^4 w_b}{\partial x^2\partial t^2} + J_2\frac{\partial^4 w_s}{\partial x^2\partial t^2}\right] = 0 \tag{9.48}$$

$$\left(1 - \lambda^2\nabla^2\right)\left[B_{xx}^s\frac{\partial^3 u}{\partial x^3} - D_{xx}^s\frac{\partial^4 w_b}{\partial x^4} - H_{xx}^s\frac{\partial^4 w_s}{\partial x^4} + A_x^s\frac{\partial^2 w_s}{\partial x^2}\right] + \left(1 - \mu^2\nabla^2\right)\left[k_w\left(w_b + w_s\right)\right.$$
$$\left. - k_p\frac{\partial^2\left(w_b + w_s\right)}{\partial x^2} - I_0\frac{\partial^2\left(w_b + w_s\right)}{\partial t^2} - J_1\frac{\partial^3 u}{\partial x\partial t^2} + J_2\frac{\partial^4 w_b}{\partial x^2\partial t^2} + K_2\frac{\partial^4 w_s}{\partial x^2\partial t^2}\right] = 0 \tag{9.49}$$

Now, solving the governing equations can help obtain the wave frequency and phase velocity responses of nanobeams. As stated before, the differences between solving the problem with and without the effect of a resting foundation can be observed in some of the arrays of the stiffness matrix involved in modeling the foundation. Hence, for investigating an Euler–Bernoulli nanobeam, the component k_{22} of the stiffness matrix defined in Eq. (4.51) must be modified. Therefore, this component must be in the following form once the plate is rested on a Winkler–Pasternak foundation:

$$k_{22} = -\left(1 - \lambda^2\nabla^2\right)D_{xx}\beta^4 - \left(1 - \mu^2\nabla^2\right)\left[k_w + k_p\beta^2\right] \tag{9.50}$$

In addition, for investigating a refined nanobeam embedded on a Winkler–Pasternak foundation, the arrays k_{22}, k_{23} and k_{33} must be modified. Thus, the following arrays must be employed when analyzing a refined nanobeam seated on a Winkler–Pasternak substrate:

$$k_{22} = -\left(1 - \lambda^2 \nabla^2\right) D_{xx} \beta^4 - \left(1 - \mu^2 \nabla^2\right) \left[k_w + k_p \beta^2\right],$$
$$k_{23} = -\left(1 - \lambda^2 \nabla^2\right) D_{xx}^s \beta^4 - \left(1 - \mu^2 \nabla^2\right) \left[k_w + k_p \beta^2\right],$$
$$k_{33} = -\left(1 - \lambda^2 \nabla^2\right) \left[H_{xx}^s \beta^4 + A_x^s \ \beta^2\right] - \left(1 - \mu^2 \nabla^2\right) \left[k_w + k_p \beta^2\right] \tag{9.51}$$

Again, it must be noticed that the other components of stiffness and mass matrices for both Euler–Bernoulli and refined sinusoidal nanobeams are the same as those reported in Section 4.3.5 of Chapter 4.

9.2.2 Analysis of Nanoplates Embedded on Winkler Pasternak Foundation

In this section, the effect of resting the nanoplate on the Winkler–Pasternak foundation will be considered, and the force vector of this type of foundation will be presented for plate-type elements. Indeed, the effect of Pasternak coefficient is modeled via the Laplacian operator. This operator possesses one dimension when probing a beam, whereas the effect of the transverse side of the structure cannot be dismissed while investigating a plate.

The displacement vector of a plate was introduced in Eq. (9.19). The force vector of the Winkler–Pasternak medium for a plate can be written in the following form:

$$\mathbf{F} = \left(0, 0, k_w u \ - k_p \left(\frac{\partial^2 u}{\partial x^2} + \frac{\partial^2 u}{\partial y^2}\right)\right) \tag{9.52}$$

Thus, the work done by the Winkler–Pasternak medium is:

$$V = \frac{1}{2} \int_V \left(k_w u^2 - k_p \left(\frac{\partial^2 u}{\partial x^2} + \frac{\partial^2 u}{\partial y^2}\right) u \ \right) dV \tag{9.53}$$

Now, the variation of the work done by the foundation for a Kirchhoff–Love plate can be obtained when Eq. (4.58) is inserted as the u and the variation operator is applied on the above equation. The variation of the work done by the medium can be written in the following form:

$$\delta V = \int_V \left(k_w w - k_p \left(\frac{\partial^2 w}{\partial x^2} + \frac{\partial^2 w}{\partial y^2}\right)\right) \delta w \, dV \tag{9.54}$$

Also, if Eq. (4.65) is substituted instead of the u , the variation of the work done by the Winkler–Pasternak substrate can be expressed as:

$$\delta V = \int_V \left(k_w (w_b + w_s) - k_p \left(\frac{\partial^2 (w_b + w_s)}{\partial x^2} + \frac{\partial^2 (w_b + w_s)}{\partial y^2}\right)\right) \delta (w_b + w_s) dV \tag{9.55}$$

Furthermore, the effect of resting the plate on a Winkler–Pasternak foundation must be applied on the motion equations of the plate. In other words, the Euler–Lagrange equations of the plate must be written again to cover the effect of the elastic substrate on the motion equations of the plate. When using the relations of the Kirchhoff–Love plates, the Euler–Lagrange equations can be obtained when Eqs. (4.72), (4.74) and (9.54) are substituted in Eq. (4.19). Thus, the obtained equations can be expressed as:

$$\frac{\partial N_{xx}}{\partial x} + \frac{\partial N_{xy}}{\partial y} = I_0 \frac{\partial^2 u}{\partial t^2} - I_1 \frac{\partial^3 w}{\partial x \partial t^2} \tag{9.56}$$

$$\frac{\partial N_{xy}}{\partial x} + \frac{\partial N_{yy}}{\partial y} = I_0 \frac{\partial^2 v}{\partial t^2} - I_1 \frac{\partial^3 w}{\partial y \partial t^2} \tag{9.57}$$

$$\frac{\partial^2 M_{xx}}{\partial x^2} + 2\frac{\partial^2 M_{xy}}{\partial x \partial y} + \frac{\partial^2 M_{yy}}{\partial y^2} - k_w w + k_p \left(\frac{\partial^2 w}{\partial x^2} + \frac{\partial^2 w}{\partial y^2}\right)$$

$$= I_0 \frac{\partial^2 w}{\partial t^2} + I_1 \left(\frac{\partial^3 u}{\partial x \partial t^2} + \frac{\partial^3 v}{\partial y \partial t^2}\right) - I_2 \left(\frac{\partial^4 w}{\partial x^2 \partial t^2} + \frac{\partial^4 w}{\partial y^2 \partial t^2}\right) \tag{9.58}$$

The Euler–Lagrange equations of higher-order plates must also be obtained. To obtain these equations, Eqs. (4.80), (4.83) and (9.55) should be substituted in Eq. (4.19). Finally, the Euler–Lagrange relations of a refined sinusoidal plate can be expressed as:

$$\frac{\partial N_{xx}}{\partial x} + \frac{\partial N_{xy}}{\partial y} = I_0 \frac{\partial^2 u}{\partial t^2} - I_1 \frac{\partial^3 w_b}{\partial x \partial t^2} - J_1 \frac{\partial^3 w_s}{\partial x \partial t^2} \tag{9.59}$$

$$\frac{\partial N_{xy}}{\partial x} + \frac{\partial N_{yy}}{\partial y} = I_0 \frac{\partial^2 v}{\partial t^2} - I_1 \frac{\partial^3 w_b}{\partial y \partial t^2} - J_1 \frac{\partial^3 w_s}{\partial y \partial t^2} \tag{9.60}$$

$$\frac{\partial^2 M_{xx}^b}{\partial x^2} + 2\frac{\partial^2 M_{xy}^b}{\partial x \partial y} + \frac{\partial^2 M_{yy}^b}{\partial y^2} - k_w \left(w_b + w_s\right) + k_p \left(\frac{\partial^2 (w_b + w_s)}{\partial x^2} + \frac{\partial^2 (w_b + w_s)}{\partial y^2}\right)$$

$$= I_0 \frac{\partial^2 (w_b + w_s)}{\partial t^2} + I_1 \left(\frac{\partial^3 u}{\partial x \partial t^2} + \frac{\partial^3 v}{\partial y \partial t^2}\right) - I_2 \left(\frac{\partial^4 w_b}{\partial x^2 \partial t^2} + \frac{\partial^4 w_b}{\partial y^2 \partial t^2}\right)$$

$$- J_2 \left(\frac{\partial^4 w_s}{\partial x^2 \partial t^2} + \frac{\partial^4 w_s}{\partial y^2 \partial t^2}\right) \tag{9.61}$$

$$\frac{\partial^2 M_{xx}^s}{\partial x^2} + 2\frac{\partial^2 M_{xy}^s}{\partial x \partial y} + \frac{\partial^2 M_{yy}^s}{\partial y^2} + \frac{\partial Q_x}{\partial x} + \frac{\partial Q_y}{\partial y} - k_w \left(w_b + w_s\right)$$

$$+ k_p \left(\frac{\partial^2 (w_b + w_s)}{\partial x^2} + \frac{\partial^2 (w_b + w_s)}{\partial y^2}\right) = I_0 \frac{\partial^2 (w_b + w_s)}{\partial t^2} + J_1 \left(\frac{\partial^3 u}{\partial x \partial t^2} + \frac{\partial^3 v}{\partial y \partial t^2}\right)$$

$$- J_2 \left(\frac{\partial^4 w_b}{\partial x^2 \partial t^2} + \frac{\partial^4 w_b}{\partial y^2 \partial t^2}\right) - K_2 \left(\frac{\partial^4 w_s}{\partial x^2 \partial t^2} + \frac{\partial^4 w_s}{\partial y^2 \partial t^2}\right) \tag{9.62}$$

Next, the governing equations of nanoplates rested on two-parameter elastic seat must be achieved. For a Kirchhoff–Love nanoplate, Eqs. (4.89) and (4.90) must be inserted in Eqs. (9.56)–(9.58). The governing equations of classical plates can be expressed as:

$$\left(1 - \lambda^2 \nabla^2\right) \left[A_{11}\frac{\partial^2 u}{\partial x^2} + \left(A_{12} + A_{66}\right)\frac{\partial^2 v}{\partial x \partial y} + A_{66}\frac{\partial^2 u}{\partial y^2} - B_{11}\frac{\partial^3 w}{\partial x^3} - \left(B_{12} + 2B_{66}\right)\frac{\partial^3 w}{\partial x \partial y^2}\right]$$

$$+ \left(1 - \mu^2 \nabla^2\right) \left[-I_0 \frac{\partial^2 u}{\partial t^2} + I_1 \frac{\partial^3 w}{\partial x \partial t^2}\right] = 0 \tag{9.63}$$

$$\left(1 - \lambda^2 \nabla^2\right) \left[A_{22}\frac{\partial^2 v}{\partial y^2} + \left(A_{12} + A_{66}\right)\frac{\partial^2 u}{\partial x \partial y} + A_{66}\frac{\partial^2 v}{\partial x^2} - B_{22}\frac{\partial^3 w}{\partial y^3} - \left(B_{12} + 2B_{66}\right)\frac{\partial^3 w}{\partial x^2 \partial y}\right]$$

$$+ \left(1 - \mu^2 \nabla^2\right) \left[-I_0 \frac{\partial^2 v}{\partial t^2} + I_1 \frac{\partial^3 w}{\partial y \partial t^2}\right] = 0 \tag{9.64}$$

$$\left(1 - \lambda^2 \nabla^2\right) \left[B_{11} \frac{\partial^3 u}{\partial x^3} + \left(B_{12} + 2B_{66}\right) \frac{\partial^3 u}{\partial x \partial y^2} + B_{22} \frac{\partial^3 v}{\partial y^3} + \left(B_{12} + 2B_{66}\right) \frac{\partial^3 v}{\partial x^2 \partial y} - D_{11} \frac{\partial^4 w}{\partial x^4} \right.$$

$$- 2\left(D_{12} + 2D_{66}\right) \frac{\partial^4 w}{\partial x^2 \partial y^2} - D_{22} \frac{\partial^4 w}{\partial y^4} \right] + \left(1 - \mu^2 \nabla^2\right) \left[k_w w - k_p \left(\frac{\partial^2 w}{\partial x^2} + \frac{\partial^2 w}{\partial y^2} \right) - I_0 \frac{\partial^2 w}{\partial t^2} \right.$$

$$- I_1 \left(\frac{\partial^3 u}{\partial x \partial t^2} + \frac{\partial^3 v}{\partial y \partial t^2} \right) + I_2 \left(\frac{\partial^4 w}{\partial x^2 \partial t^2} + \frac{\partial^4 w}{\partial y^2 \partial t^2} \right) \right] = 0 \tag{9.65}$$

Furthermore, the governing equations of refined shear deformable nanoplates can be derived by inserting Eqs. (4.93)–(4.96) in Eqs. (9.59)–(9.62). Once the aforementioned substitution is procured, the governing equations become:

$$\left(1 - \lambda^2 \nabla^2\right) \left[A_{11} \frac{\partial^2 u}{\partial x^2} + \left(A_{12} + A_{66}\right) \frac{\partial^2 v}{\partial x \partial y} + A_{66} \frac{\partial^2 u}{\partial y^2} - B_{11} \frac{\partial^3 w_b}{\partial x^3} - \left(B_{12} + 2B_{66}\right) \frac{\partial^3 w_b}{\partial x \partial y^2} \right.$$

$$- B_{11}^s \frac{\partial^3 w_s}{\partial x^3} - \left(B_{12}^s + 2B_{66}^s\right) \frac{\partial^3 w_s}{\partial x \partial y^2} \right] + \left(1 - \mu^2 \nabla^2\right) \left[-I_0 \frac{\partial^2 u}{\partial t^2} + I_1 \frac{\partial^3 w_b}{\partial x \partial t^2} + J_1 \frac{\partial^3 w_s}{\partial x \partial t^2} \right] = 0 \tag{9.66}$$

$$\left(1 - \lambda^2 \nabla^2\right) \left[A_{22} \frac{\partial^2 v}{\partial y^2} + \left(A_{12} + A_{66}\right) \frac{\partial^2 u}{\partial x \partial y} + A_{66} \frac{\partial^2 v}{\partial x^2} - B_{22} \frac{\partial^3 w_b}{\partial y^3} - \left(B_{12} + 2B_{66}\right) \frac{\partial^3 w_b}{\partial x^2 \partial y} \right.$$

$$- B_{22}^s \frac{\partial^3 w_s}{\partial y^3} - \left(B_{12}^s + 2B_{66}^s\right) \frac{\partial^3 w_s}{\partial x^2 \partial y} \right] + \left(1 - \mu^2 \nabla^2\right) \left[-I_0 \frac{\partial^2 v}{\partial t^2} + I_1 \frac{\partial^3 w_b}{\partial y \partial t^2} + J_1 \frac{\partial^3 w_s}{\partial y \partial t^2} \right] = 0 \tag{9.67}$$

$$\left(1 - \lambda^2 \nabla^2\right) \left[B_{11} \frac{\partial^3 u}{\partial x^3} + \left(B_{12} + 2B_{66}\right) \frac{\partial^3 u}{\partial x \partial y^2} + B_{22} \frac{\partial^3 v}{\partial y^3} + \left(B_{12} + 2B_{66}\right) \frac{\partial^3 v}{\partial x^2 \partial y} - D_{11} \frac{\partial^4 w_b}{\partial x^4} \right.$$

$$- 2\left(D_{12} + 2D_{66}\right) \frac{\partial^4 w_b}{\partial x^2 \partial y^2} - D_{22} \frac{\partial^4 w_b}{\partial y^4} - D_{11}^s \frac{\partial^4 w_s}{\partial x^4} - 2\left(D_{12}^s + 2D_{66}^s\right) \frac{\partial^4 w_s}{\partial x^2 \partial y^2} - D_{22}^s \frac{\partial^4 w_s}{\partial y^4} \right)$$

$$+ \left(1 - \mu^2 \nabla^2\right) \left[k_w \left(w_b + w_s\right) - k_p \left(\frac{\partial^2 \left(w_b + w_s\right)}{\partial x^2} + \frac{\partial^2 \left(w_b + w_s\right)}{\partial y^2} \right) - I_0 \frac{\partial^2 \left(w_b + w_s\right)}{\partial t^2} \right.$$

$$- I_1 \left(\frac{\partial^3 u}{\partial x \partial t^2} + \frac{\partial^3 v}{\partial y \partial t^2} \right) + I_2 \left(\frac{\partial^4 w_b}{\partial x^2 \partial t^2} + \frac{\partial^4 w_b}{\partial y^2 \partial t^2} \right) + J_2 \left(\frac{\partial^4 w_s}{\partial x^2 \partial t^2} + \frac{\partial^4 w_s}{\partial y^2 \partial t^2} \right) \right] = 0 \tag{9.68}$$

$$\left(1 - \lambda^2 \nabla^2\right) \left[B_{11}^s \frac{\partial^3 u}{\partial x^3} + \left(B_{12}^s + 2B_{66}^s\right) \frac{\partial^3 u}{\partial x \partial y^2} + B_{22}^s \frac{\partial^3 v}{\partial y^3} + \left(B_{12}^s + 2B_{66}^s\right) \frac{\partial^3 v}{\partial x^2 \partial y} - D_{11}^s \frac{\partial^4 w_b}{\partial x^4} \right.$$

$$- 2\left(D_{12}^s + 2D_{66}^s\right) \frac{\partial^4 w_b}{\partial x^2 \partial y^2} - D_{22}^s \frac{\partial^4 w_b}{\partial y^4} - H_{11}^s \frac{\partial^4 w_s}{\partial x^4} - 2\left(H_{12}^s + 2H_{66}^s\right) \frac{\partial^4 w_s}{\partial x^2 \partial y^2} - H_{22}^s \frac{\partial^4 w_s}{\partial y^4}$$

$$+ A_{44}^s \left(\frac{\partial^2 \left(w_b + w_s\right)}{\partial x^2} + \frac{\partial^2 \left(w_b + w_s\right)}{\partial y^2} \right) \right] + \left(1 - \mu^2 \nabla^2\right) \left[k_w \left(w_b + w_s\right) \right.$$

$$- \left(\frac{\partial^2 \left(w_b + w_s\right)}{\partial x^2} + \frac{\partial^2 \left(w_b + w_s\right)}{\partial y^2} \right) - I_0 \frac{\partial^2 \left(w_b + w_s\right)}{\partial t^2} - J_1 \left(\frac{\partial^3 u}{\partial x \partial t^2} + \frac{\partial^3 v}{\partial y \partial t^2} \right)$$

$$+ J_2 \left(\frac{\partial^4 w_b}{\partial x^2 \partial t^2} + \frac{\partial^4 w_b}{\partial y^2 \partial t^2} \right) + K_2 \left(\frac{\partial^4 w_s}{\partial x^2 \partial t^2} + \frac{\partial^4 w_s}{\partial y^2 \partial t^2} \right) \right] = 0 \tag{9.69}$$

At the end of this section, the solution of the wave propagation problem of nanoplates rested on Winkler–Pasternak substrate must be surveyed. It was discussed in previous paragraphs that the difference between the solutions of a nanostructure with and without a medium can be summarized in those arrays of the stiffness matrix, which are in contact with the foundation parameters. For the case of a Kirchhoff–Love nanoplate, all the arrays of stiffness matrix can be achieved from Eq. (4.107) except the following one:

$$
\begin{aligned}
k_{33} = &- \left(1 + \lambda^2\left(\beta_1^2 + \beta_2^2\right)\right)\left[D_{11}\beta_1^4 + 2\left(D_{12} + 2D_{66}\right)\beta_1^2\beta_2^2 + D_{22}\beta_2^4\right] \\
&- \left(1 + \mu^2\left(\beta_1^2 + \beta_2^2\right)\right)\left[k_w + k_p\left(\beta_1^2 + \beta_2^2\right)\right]
\end{aligned}
\tag{9.70}
$$

It is clear that the arrays of mass matrix are the same as those reported in Eq. (4.108). Moreover, when a refined sinusoidal nanoplate is assumed to be analyzed, the components of the stiffness matrix are exactly the same as those presented in Eq. (4.109). However, the components that must be changed are presented below:

$$
\begin{aligned}
k_{33} = &- \left(1 + \lambda^2\left(\beta_1^2 + \beta_2^2\right)\right)\left[D_{11}\beta_1^4 + 2\left(D_{12} + 2D_{66}\right)\beta_1^2\beta_2^2 + D_{22}\beta_2^4\right] \\
&- \left(1 + \mu^2\left(\beta_1^2 + \beta_2^2\right)\right)\left[k_w + k_p\left(\beta_1^2 + \beta_2^2\right)\right], \\
k_{34} = &- \left(1 + \lambda^2\left(\beta_1^2 + \beta_2^2\right)\right)\left[D_{11}^s\beta_1^4 + 2\left(D_{12}^s + 2D_{66}^s\right)\beta_1^2\beta_2^2 + D_{22}^s\beta_2^4\right] \\
&- \left(1 + \mu^2\left(\beta_1^2 + \beta_2^2\right)\right)\left[k_w + k_p\left(\beta_1^2 + \beta_2^2\right)\right], \\
k_{44} = &- \left(1 + \lambda^2\left(\beta_1^2 + \beta_2^2\right)\right)\left[H_{11}^s\beta_1^4 + 2\left(H_{12}^s + 2H_{66}^s\right)\beta_1^2\beta_2^2 + H_{22}^s\beta_2^4 \right.\\
&\left. + A_{44}^s\left(\beta_1^2 + \beta_2^2\right)\right] - \left(1 + \mu^2\left(\beta_1^2 + \beta_2^2\right)\right)\left[k_w + k_p\left(\beta_1^2 + \beta_2^2\right)\right]
\end{aligned}
\tag{9.71}
$$

Similar to Kirchhoff–Love nanoplates, it must be declared that the arrays of mass matrix are completely the same as the expressions mentioned in Eq. (4.110).

9.3   visco-Pasternak Foundation

Now, we present a mathematical formulation for the wave dispersion problem of nanostructures rested on a viscoelastic substrate. It was discussed in the previous sections that the Winkler and Pasternak models can present stiff models that can be utilized as the seat for the structural elements like beams and plates. In some practical applications, the dynamic output of the system might be required to be attenuated. Thus, the damping effect must be added to the medium as well as to the previously introduced stiff elements. For this, a linear damper will be added to the Pasternak elastic model to generate the well-known visco-Pasternak model. In what follows, effect of adding a damper on the wave dispersion responses of the nanobeams and nanoplates will be discussed on the basis of both classical and shear deformable kinematic models.

9.3.1   Analysis of Nanobeams Embedded on visco-Pasternak Foundation

Similar to previous sections, here, a force vector must be introduced for obtaining the motion equations of a beam-type element embedded on a visco–Pasternak foundation. The displacement vector will not be presented because it was formerly introduced in Eq. (9.3). The force vector of a visco–Pasternak substrate can be expressed in the following form:

$$
\mathbf{F} = \left(0, 0, k_w u \ - k_p \frac{\partial^2 u}{\partial x^2} - c_d \frac{\partial u}{\partial t}\right)
\tag{9.72}
$$

where c_d is the damping coefficient of the viscoelastic medium, and k_w and k_p correspond with the previously introduced Winkler and Pasternak coefficients, respectively.

Implementing the above equation and the definition of the displacement vector from Eq. (9.3), the work done by a viscoelastic substrate can be formulated in the following form:

$$V = \frac{1}{2} \int_V \left(k_w u^2 - k_p \frac{\partial^2 u}{\partial x^2} u - c_d \frac{\partial u}{\partial t} u \right) dV \qquad (9.73)$$

Now, the variation of the work done by the visco-Pasternak medium for an Euler–Bernoulli beam can be enriched by applying the effect of the variation operator on the above equation as well as using the displacement field of the aforementioned beam model, as defined in Eq. (4.10). After mathematical simplifications, the variation of work done by the medium can be expressed as:

$$\delta V = \int_V \left(k_w w - k_p \frac{\partial^2 w}{\partial x^2} - c_d \frac{\partial w}{\partial t} \right) \delta w \, dV \qquad (9.74)$$

Besides, whenever the displacement field of a refined higher-order beam, defined in Eq. (4.15), is selected, the variation of the work done by the viscoelastic medium can be written in the following form:

$$\delta V = \int_V \left(k_w \left(w_b + w_s \right) - k_p \frac{\partial^2 (w_b + w_s)}{\partial x^2} - c_d \frac{\partial (w_b + w_s)}{\partial t} \right) \delta \left(w_b + w_s \right) dV \qquad (9.75)$$

Now, the motion equations of beams rested on a three-parameter viscoelastic medium can be achieved by using the variation of the work done by the foundation in the definition of the Hamilton's principle. For a classical beam, Eqs. (4.24), (4.26) and (9.74) should be substituted in Eq. (4.19) to obtain the Euler–Lagrange equations. Once the aforementioned substitution is performed, the Euler–Lagrange equations of the Euler–Bernoulli beams embedded on a visco-Pasternak substrate can be expressed as below:

$$\frac{\partial N}{\partial x} = I_0 \frac{\partial^2 u}{\partial t^2} - I_1 \frac{\partial^3 w}{\partial x \partial t^2} \qquad (9.76)$$

$$\frac{\partial^2 M}{\partial x^2} - k_w w + k_p \frac{\partial^2 w}{\partial x^2} + c_d \frac{\partial w}{\partial t} = I_0 \frac{\partial^2 w}{\partial t^2} + I_1 \frac{\partial^3 u}{\partial x \partial t^2} - I_2 \frac{\partial^4 w}{\partial x^2 \partial t^2} \qquad (9.77)$$

Similarly, the Euler–Lagrange equations of a refined beam can be achieved by inserting Eqs. (4.31), (4.33) and (9.75) in Eq. (4.19). Once this mathematical substitution is completed, the final Euler–Lagrange equations for a shear deformable beam, embedded on a viscoelastic medium, can be written as:

$$\frac{\partial N}{\partial x} = I_0 \frac{\partial^2 u}{\partial t^2} - I_1 \frac{\partial^3 w_b}{\partial x \partial t^2} - J_1 \frac{\partial^3 w_s}{\partial x \partial t^2} \qquad (9.78)$$

$$\frac{\partial^2 M^b}{\partial x^2} - k_w \left(w_b + w_s \right) + k_p \frac{\partial^2 (w_b + w_s)}{\partial x^2} + c_d \frac{\partial (w_b + w_s)}{\partial t}$$
$$= I_0 \frac{\partial^2 (w_b + w_s)}{\partial t^2} + I_1 \frac{\partial^3 u}{\partial x \partial t^2} - I_2 \frac{\partial^4 w_b}{\partial x^2 \partial t^2} - J_2 \frac{\partial^4 w_s}{\partial x^2 \partial t^2} \qquad (9.79)$$

$$\frac{\partial^2 M^s}{\partial x^2} + \frac{\partial Q_x}{\partial x} - k_w \left(w_b + w_s \right) + k_p \frac{\partial^2 (w_b + w_s)}{\partial x^2} + c_d \frac{\partial (w_b + w_s)}{\partial t}$$
$$= I_0 \frac{\partial^2 (w_b + w_s)}{\partial t^2} + J_1 \frac{\partial^3 u}{\partial x \partial t^2} - J_2 \frac{\partial^4 w_b}{\partial x^2 \partial t^2} - K_2 \frac{\partial^4 w_s}{\partial x^2 \partial t^2} \qquad (9.80)$$

Now, the motion equations are derived for a beam considered to be resting on a three-parameter foundation. In the next step, the influences of small scale must be added to obtain the governing equations for nanobeams rested on a visco-Pasternak medium. For this, the constitutive equations of nonlocal strain gradient elasticity will be inserted in the above Euler–Lagrange equations. First, for an Euler–Bernoulli nanobeam, Eqs. (4.38) and (4.39) must be substituted in Eqs. (9.76) and (9.77). By this substitution, the final partial differential governing equations of a classical nanobeam can be expressed in the following form:

$$\left(1 - \lambda^2 \nabla^2\right) \left[A_{xx} \frac{\partial^2 u}{\partial x^2} - B_{xx} \frac{\partial^3 w}{\partial x^3}\right] + \left(1 - \mu^2 \nabla^2\right) \left[-I_0 \frac{\partial^2 u}{\partial t^2} + I_1 \frac{\partial^3 w}{\partial x \partial t^2}\right] = 0 \qquad (9.81)$$

$$\left(1 - \lambda^2 \nabla^2\right) \left[B_{xx} \frac{\partial^3 u}{\partial x^3} - D_{xx} \frac{\partial^4 w}{\partial x^4}\right] + \left(1 - \mu^2 \nabla^2\right) \left[k_w w - k_p \frac{\partial^2 w}{\partial x^2} - c_d \frac{\partial w}{\partial t}\right.$$
$$\left. - I_0 \frac{\partial^2 w}{\partial t^2} - I_1 \frac{\partial^3 u}{\partial x \partial t^2} + I_2 \frac{\partial^4 w}{\partial x^2 \partial t^2}\right] = 0 \qquad (9.82)$$

Furthermore, the governing equations of refined shear deformable nanobeams can be obtained by inserting Eqs. (4.40)–(4.43) in Eqs. (9.78)–(9.80). Once the aforementioned procedure is completed, the governing equations of a refined higher-order nanobeam embedded on a viscoelastic substrate can be written in the following form:

$$\left(1 - \lambda^2 \nabla^2\right) \left[A_{xx} \frac{\partial^2 u}{\partial x^2} - B_{xx} \frac{\partial^3 w_b}{\partial x^3} - B_{xx}^s \frac{\partial^3 w_s}{\partial x^3}\right] + \left(1 - \mu^2 \nabla^2\right)$$
$$\times \left[-I_0 \frac{\partial^2 u}{\partial t^2} + I_1 \frac{\partial^3 w_b}{\partial x \partial t^2} + J_1 \frac{\partial^3 w_s}{\partial x \partial t^2}\right] = 0 \qquad (9.83)$$

$$\left(1 - \lambda^2 \nabla^2\right) \left[B_{xx} \frac{\partial^3 u}{\partial x^3} - D_{xx} \frac{\partial^4 w_b}{\partial x^4} - D_{xx}^s \frac{\partial^4 w_s}{\partial x^4}\right] + \left(1 - \mu^2 \nabla^2\right) \left[k_w \left(w_b + w_s\right)\right.$$
$$\left. - k_p \frac{\partial^2 \left(w_b + w_s\right)}{\partial x^2} - c_d \frac{\partial \left(w_b + w_s\right)}{\partial t} - I_0 \frac{\partial^2 \left(w_b + w_s\right)}{\partial t^2} - I_1 \frac{\partial^3 u}{\partial x \partial t^2}\right.$$
$$\left. + I_2 \frac{\partial^4 w_b}{\partial x^2 \partial t^2} + J_2 \frac{\partial^4 w_s}{\partial x^2 \partial t^2}\right] = 0 \qquad (9.84)$$

$$\left(1 - \lambda^2 \nabla^2\right) \left[B_{xx}^s \frac{\partial^3 u}{\partial x^3} - D_{xx}^s \frac{\partial^4 w_b}{\partial x^4} - H_{xx}^s \frac{\partial^4 w_s}{\partial x^4} + A_x^s \frac{\partial^2 w_s}{\partial x^2}\right] + \left(1 - \mu^2 \nabla^2\right)$$
$$\times \left[k_w \left(w_b + w_s\right) - k_p \frac{\partial^2 \left(w_b + w_s\right)}{\partial x^2} - c_d \frac{\partial \left(w_b + w_s\right)}{\partial t} - I_0 \frac{\partial^2 \left(w_b + w_s\right)}{\partial t^2}\right.$$
$$\left. - J_1 \frac{\partial^3 u}{\partial x \partial t^2} + J_2 \frac{\partial^4 w_b}{\partial x^2 \partial t^2} + K_2 \frac{\partial^4 w_s}{\partial x^2 \partial t^2}\right] = 0 \qquad (9.85)$$

Now, the governing equations must be solved to achieve the wave frequency and phase velocity of the dispersed waves. In this case (i.e., when the structure is rested on a foundation with a viscous term), the classical dynamic equation previously defined in Eq. (1.3) will not be obtained. In fact, in a dynamic analysis without the effects of damping, only differentiations with respect to time exist, whereas in a problem where effects of damping parameters are included, we will see odd differentiations with respect to the time domain.

In a very famous type of damped dynamic problems, the final equation can be expressed in the following matrix form:

$$\left([K] + \omega[C] - \omega^2[M]\right)[\Delta] = 0 \tag{9.86}$$

In such problems, first and second order differentiations with respect to time occur in the final governing equations of the problem. Thus, the final response of the problem can be achieved by setting the determinant of the above equation equal to zero. For analyzing a nanostructure rested on a viscoelastic substrate, solving the governing equations will result in an eigenvalue problem similar to that reported in Eq. (9.86). It is worth mentioning that the mass matrix is the same as those reported in the previous sections. Also, the stiffness matrix for a nanostructure rested on a visco-Pasternak medium is the same as that of a nanostructure embedded on a Winkler–Pasternak foundation. Hence, the only difference is in the presence of the damping matrix C. In what follows, we will describe the damping matrix's components for both Euler–Bernoulli and refined shear deformable nanobeams. For an Euler–Bernoulli nanobeam, the only nonzero array of the damping matrix can be written as:

$$c_{22} = \left(1 + \mu^2 \beta^2\right) i c_d \tag{9.87}$$

where i denotes the imaginary unit. It is worth mentioning that the damping matrix of a classical nanobeam is of 2×2 order. Now, the solution of the refined sinusoidal nanobeams rested on a visco-Pasternak medium should be completed. In this case, the stiffness, mass and damping matrices are of order 3×3 because of the participation of three dependent variables, namely, u, w_b and w_s, in construction of the governing equations. Similar to the Euler–Bernoulli nanobeams, here, the mass and stiffness matrices are identical with those stated for the nanobeams rested on a Winkler–Pasternak medium. In this case, the nonzero components of the damping matrix are:

$$c_{22} = c_{23} = c_{33} = \left(1 + \mu^2 \beta^2\right) i c_d \tag{9.88}$$

It is worth mentioning that the symmetric form of the mass and stiffness matrices can also be observed in the damping matrix.

9.3.2 Analysis of Nanoplates Embedded on visco-Pasternak Foundation

In this section, we will complete the presented formulations in this chapter by formulating the wave propagation problem for the nanoplates that are embedded on a visco-Pasternak medium. Similar to previous sections on nanoplates, the first part of the formulations relates to the introduction of the force vector of the visco-Pasternak substrate that affects the nanoplates' motion equations. This vector can be expressed in the following form:

$$\mathbf{F} = \left(0, 0, k_w u - k_p \left(\frac{\partial^2 u}{\partial x^2} + \frac{\partial^2 u}{\partial y^2}\right) - c_d \frac{\partial u}{\partial t}\right) \tag{9.89}$$

Employing the above equation, the work done by the viscoelastic medium can be written as:

$$V = \frac{1}{2} \int_V \left(k_w u^2 - k_p \left(\frac{\partial^2 u}{\partial x^2} + \frac{\partial^2 u}{\partial y^2}\right) u - c_d \frac{\partial u}{\partial t} u\right) dV \tag{9.90}$$

Now, the variation of the above equation must be calculated to be used in the primary definition of the Hamilton's principle for a nanoplate. First, using the displacement field of the Kirchhoff–Love plate hypothesis (see Eq. (4.58)) and applying the effect of the variation operator, the work done by the visco-Pasternak foundation on a classical plate can be formulated in the following form:

$$\delta V = \int_V \left(k_w w - k_p \left(\frac{\partial^2 w}{\partial x^2} + \frac{\partial^2 w}{\partial y^2} \right) - c_d \frac{\partial w}{\partial t} \right) \delta w \, dV \tag{9.91}$$

Using the same mathematical operations and implementing Eq. (4.65) as the displacement field of the refined plates, the variation of the work done by the visco-Pasternak substrate can be expressed as:

$$\delta V = \int_V \left(k_w \left(w_b + w_s \right) - k_p \left(\frac{\partial^2 \left(w_b + w_s \right)}{\partial x^2} + \frac{\partial^2 \left(w_b + w_s \right)}{\partial y^2} \right) - c_d \frac{\partial \left(w_b + w_s \right)}{\partial t} \right) \delta \left(w_b + w_s \right) dV \tag{9.92}$$

Now, the Euler–Lagrange equations of the plates rested on a visco-Pasternak foundation can be enriched. First, this procedure will be adopted for a classical plate. For this, one should substitute Eqs. (4.72), (4.74) and (9.91) in Eq. (4.19). Hence, the Euler–Lagrange equations for a Kirchhoff–Love plate can be expressed as:

$$\frac{\partial N_{xx}}{\partial x} + \frac{\partial N_{xy}}{\partial y} = I_0 \frac{\partial^2 u}{\partial t^2} - I_1 \frac{\partial^3 w}{\partial x \partial t^2} \tag{9.93}$$

$$\frac{\partial N_{xy}}{\partial x} + \frac{\partial N_{yy}}{\partial y} = I_0 \frac{\partial^2 v}{\partial t^2} - I_1 \frac{\partial^3 w}{\partial y \partial t^2} \tag{9.94}$$

$$\frac{\partial^2 M_{xx}}{\partial x^2} + 2 \frac{\partial^2 M_{xy}}{\partial x \partial y} + \frac{\partial^2 M_{yy}}{\partial y^2} - k_w w + k_p \left(\frac{\partial^2 w}{\partial x^2} + \frac{\partial^2 w}{\partial y^2} \right) + c_d \frac{\partial w}{\partial t}$$
$$= I_0 \frac{\partial^2 w}{\partial t^2} + I_1 \left(\frac{\partial^3 u}{\partial x \partial t^2} + \frac{\partial^3 v}{\partial y \partial t^2} \right) - I_2 \left(\frac{\partial^4 w}{\partial x^2 \partial t^2} + \frac{\partial^4 w}{\partial y^2 \partial t^2} \right) \tag{9.95}$$

Afterward, the same mathematical operations must be performed for the refined shear deformable plates. Again, the definition of the Hamilton's principle must be employed (see Eq. (4.19)). For the refined sinusoidal plates, Eqs. (4.80), (4.83) and (9.92) must be utilized and substituted in the Hamilton's principle. Whenever the substitution procedure is completed, the Euler–Lagrange equations of the refined plates rested on a viscoelastic medium can be written as:

$$\frac{\partial N_{xx}}{\partial x} + \frac{\partial N_{xy}}{\partial y} = I_0 \frac{\partial^2 u}{\partial t^2} - I_1 \frac{\partial^3 w_b}{\partial x \partial t^2} - J_1 \frac{\partial^3 w_s}{\partial x \partial t^2} \tag{9.96}$$

$$\frac{\partial N_{xy}}{\partial x} + \frac{\partial N_{yy}}{\partial y} = I_0 \frac{\partial^2 v}{\partial t^2} - I_1 \frac{\partial^3 w_b}{\partial y \partial t^2} - J_1 \frac{\partial^3 w_s}{\partial y \partial t^2} \tag{9.97}$$

$$\frac{\partial^2 M_{xx}^b}{\partial x^2} + 2 \frac{\partial^2 M_{xy}^b}{\partial x \partial y} + \frac{\partial^2 M_{yy}^b}{\partial y^2} - k_w \left(w_b + w_s \right) + k_p \left(\frac{\partial^2 \left(w_b + w_s \right)}{\partial x^2} + \frac{\partial^2 \left(w_b + w_s \right)}{\partial y^2} \right)$$
$$+ c_d \frac{\partial \left(w_b + w_s \right)}{\partial t} = I_0 \frac{\partial^2 \left(w_b + w_s \right)}{\partial t^2} + I_1 \left(\frac{\partial^3 u}{\partial x \partial t^2} + \frac{\partial^3 v}{\partial y \partial t^2} \right)$$
$$- I_2 \left(\frac{\partial^4 w_b}{\partial x^2 \partial t^2} + \frac{\partial^4 w_b}{\partial y^2 \partial t^2} \right) - J_2 \left(\frac{\partial^4 w_s}{\partial x^2 \partial t^2} + \frac{\partial^4 w_s}{\partial y^2 \partial t^2} \right) \tag{9.98}$$

$$\frac{\partial^2 M_{xx}^s}{\partial x^2} + 2\frac{\partial^2 M_{xy}^s}{\partial x \partial y} + \frac{\partial^2 M_{yy}^s}{\partial y^2} + \frac{\partial Q_x}{\partial x} + \frac{\partial Q_y}{\partial y} - k_w(w_b + w_s)$$

$$+ k_p \left(\frac{\partial^2 (w_b + w_s)}{\partial x^2} + \frac{\partial^2 (w_b + w_s)}{\partial y^2} \right) + c_d \frac{\partial (w_b + w_s)}{\partial t} = I_0 \frac{\partial^2 (w_b + w_s)}{\partial t^2}$$

$$+ J_1 \left(\frac{\partial^3 u}{\partial x \partial t^2} + \frac{\partial^3 v}{\partial y \partial t^2} \right) - J_2 \left(\frac{\partial^4 w_b}{\partial x^2 \partial t^2} + \frac{\partial^4 w_b}{\partial y^2 \partial t^2} \right) - K_2 \left(\frac{\partial^4 w_s}{\partial x^2 \partial t^2} + \frac{\partial^4 w_s}{\partial y^2 \partial t^2} \right) \quad (9.99)$$

Thereafter, we should seek for the governing equations of the nanoplate when the effects of the viscoelastic substrate are included. First, the constitutive equations of Kirchhoff–Love nanoplates will be inserted in the motion equations of such plates to obtain the governing equations of such nanostructures. For this, Eqs. (4.89) and (4.90) must be substituted in Eqs. (9.93)–(9.95). Thus, the governing equations can be expressed as:

$$\left(1 - \lambda^2 \nabla^2\right) \left[A_{11}\frac{\partial^2 u}{\partial x^2} + (A_{12} + A_{66})\frac{\partial^2 v}{\partial x \partial y} + A_{66}\frac{\partial^2 u}{\partial y^2} - B_{11}\frac{\partial^3 w}{\partial x^3} \right.$$

$$\left. - (B_{12} + 2B_{66})\frac{\partial^3 w}{\partial x \partial y^2} \right] + \left(1 - \mu^2 \nabla^2\right) \left[-I_0 \frac{\partial^2 u}{\partial t^2} + I_1 \frac{\partial^3 w}{\partial x \partial t^2} \right] = 0 \quad (9.100)$$

$$\left(1 - \lambda^2 \nabla^2\right) \left[A_{22}\frac{\partial^2 v}{\partial y^2} + (A_{12} + A_{66})\frac{\partial^2 u}{\partial x \partial y} + A_{66}\frac{\partial^2 v}{\partial x^2} - B_{22}\frac{\partial^3 w}{\partial y^3} \right.$$

$$\left. - (B_{12} + 2B_{66})\frac{\partial^3 w}{\partial x^2 \partial y} \right] + \left(1 - \mu^2 \nabla^2\right) \left[-I_0 \frac{\partial^2 v}{\partial t^2} + I_1 \frac{\partial^3 w}{\partial y \partial t^2} \right] = 0 \quad (9.101)$$

$$\left(1 - \lambda^2 \nabla^2\right) \left[B_{11}\frac{\partial^3 u}{\partial x^3} + (B_{12} + 2B_{66})\frac{\partial^3 u}{\partial x \partial y^2} + B_{22}\frac{\partial^3 v}{\partial y^3} + (B_{12} + 2B_{66})\frac{\partial^3 v}{\partial x^2 \partial y} - D_{11}\frac{\partial^4 w}{\partial x^4} \right.$$

$$\left. - 2(D_{12} + 2D_{66})\frac{\partial^4 w}{\partial x^2 \partial y^2} - D_{22}\frac{\partial^4 w}{\partial y^4} \right] + \left(1 - \mu^2 \nabla^2\right) \left[k_w w - k_p \left(\frac{\partial^2 w}{\partial x^2} + \frac{\partial^2 w}{\partial y^2} \right) \right.$$

$$\left. + c_d \frac{\partial w}{\partial t} - I_0 \frac{\partial^2 w}{\partial t^2} - I_1 \left(\frac{\partial^3 u}{\partial x \partial t^2} + \frac{\partial^3 v}{\partial y \partial t^2} \right) + I_2 \left(\frac{\partial^4 w}{\partial x^2 \partial t^2} + \frac{\partial^4 w}{\partial y^2 \partial t^2} \right) \right] = 0 \quad (9.102)$$

Next, Eqs. (4.93) and (4.96) must be inserted in Eqs. (9.96)–(9.99) to obtain the governing equations of nanoplates embedded on a visco-Pasternak foundation in the framework of the nonlocal strain gradient theory of elasticity. The governing equations of such a nanostructure can be expressed in the following form:

$$\left(1 - \lambda^2 \nabla^2\right) \left[A_{11}\frac{\partial^2 u}{\partial x^2} + (A_{12} + A_{66})\frac{\partial^2 v}{\partial x \partial y} + A_{66}\frac{\partial^2 u}{\partial y^2} - B_{11}\frac{\partial^3 w_b}{\partial x^3} \right.$$

$$\left. - (B_{12} + 2B_{66})\frac{\partial^3 w_b}{\partial x \partial y^2} - B_{11}^s\frac{\partial^3 w_s}{\partial x^3} - (B_{12}^s + 2B_{66}^s)\frac{\partial^3 w_s}{\partial x \partial y^2} \right]$$

$$+ \left(1 - \mu^2 \nabla^2\right) \left[-I_0 \frac{\partial^2 u}{\partial t^2} + I_1 \frac{\partial^3 w_b}{\partial x \partial t^2} + J_1 \frac{\partial^3 w_s}{\partial x \partial t^2} \right] = 0 \quad (9.103)$$

$$\left(1 - \lambda^2 \nabla^2\right) \left[A_{22}\frac{\partial^2 v}{\partial y^2} + (A_{12} + A_{66})\frac{\partial^2 u}{\partial x \partial y} + A_{66}\frac{\partial^2 v}{\partial x^2} - B_{22}\frac{\partial^3 w_b}{\partial y^3} \right.$$

$$\left. - (B_{12} + 2B_{66})\frac{\partial^3 w_b}{\partial x^2 \partial y} - B_{22}^s\frac{\partial^3 w_s}{\partial y^3} - (B_{12}^s + 2B_{66}^s)\frac{\partial^3 w_s}{\partial x^2 \partial y} \right]$$

$$+ \left(1 - \mu^2 \nabla^2\right) \left[-I_0 \frac{\partial^2 v}{\partial t^2} + I_1 \frac{\partial^3 w_b}{\partial y \partial t^2} + J_1 \frac{\partial^3 w_s}{\partial y \partial t^2} \right] = 0 \quad (9.104)$$

$$\left(1 - \lambda^2 \nabla^2\right)\left[B_{11}\frac{\partial^3 u}{\partial x^3} + \left(B_{12} + 2B_{66}\right)\frac{\partial^3 u}{\partial x \partial y^2} + B_{22}\frac{\partial^3 v}{\partial y^3} + \left(B_{12} + 2B_{66}\right)\frac{\partial^3 v}{\partial x^2 \partial y} - D_{11}\frac{\partial^4 w_b}{\partial x^4}\right.$$

$$- 2\left(D_{12} + 2D_{66}\right)\frac{\partial^4 w_b}{\partial x^2 \partial y^2} - D_{22}\frac{\partial^4 w_b}{\partial y^4} - D_{11}^s\frac{\partial^4 w_s}{\partial x^4} - 2\left(D_{12}^s + 2D_{66}^s\right)\frac{\partial^4 w_s}{\partial x^2 \partial y^2}$$

$$\left. - D_{22}^s\frac{\partial^4 w_s}{\partial y^4}\right] + \left(1 - \mu^2 \nabla^2\right)\left[k_w\left(w_b + w_s\right) - k_p\left(\frac{\partial^2\left(w_b + w_s\right)}{\partial x^2} + \frac{\partial^2\left(w_b + w_s\right)}{\partial y^2}\right)\right.$$

$$- c_d\frac{\partial\left(w_b + w_s\right)}{\partial t} - I_0\frac{\partial^2\left(w_b + w_s\right)}{\partial t^2} - I_1\left(\frac{\partial^3 u}{\partial x \partial t^2} + \frac{\partial^3 v}{\partial y \partial t^2}\right)$$

$$\left. + I_2\left(\frac{\partial^4 w_b}{\partial x^2 \partial t^2} + \frac{\partial^4 w_b}{\partial y^2 \partial t^2}\right) + J_2\left(\frac{\partial^4 w_s}{\partial x^2 \partial t^2} + \frac{\partial^4 w_s}{\partial y^2 \partial t^2}\right)\right] = 0 \tag{9.105}$$

$$\left(1 - \lambda^2 \nabla^2\right)\left[B_{11}^s\frac{\partial^3 u}{\partial x^3} + \left(B_{12}^s + 2B_{66}^s\right)\frac{\partial^3 u}{\partial x \partial y^2} + B_{22}^s\frac{\partial^3 v}{\partial y^3} + \left(B_{12}^s + 2B_{66}^s\right)\frac{\partial^3 v}{\partial x^2 \partial y} - D_{11}^s\frac{\partial^4 w_b}{\partial x^4}\right.$$

$$- 2\left(D_{12}^s + 2D_{66}^s\right)\frac{\partial^4 w_b}{\partial x^2 \partial y^2} - D_{22}^s\frac{\partial^4 w_b}{\partial y^4} - H_{11}^s\frac{\partial^4 w_s}{\partial x^4} - 2\left(H_{12}^s + 2H_{66}^s\right)\frac{\partial^4 w_s}{\partial x^2 \partial y^2}$$

$$\left. - H_{22}^s\frac{\partial^4 w_s}{\partial y^4} + A_{44}^s\left(\frac{\partial^2\left(w_b + w_s\right)}{\partial x^2} + \frac{\partial^2\left(w_b + w_s\right)}{\partial y^2}\right)\right] + \left(1 - \mu^2 \nabla^2\right)\left[k_w\left(w_b + w_s\right)\right.$$

$$- \left(\frac{\partial^2\left(w_b + w_s\right)}{\partial x^2} + \frac{\partial^2\left(w_b + w_s\right)}{\partial y^2}\right) - c_d\frac{\partial\left(w_b + w_s\right)}{\partial t} - I_0\frac{\partial^2\left(w_b + w_s\right)}{\partial t^2}$$

$$\left. - J_1\left(\frac{\partial^3 u}{\partial x \partial t^2} + \frac{\partial^3 v}{\partial y \partial t^2}\right) + J_2\left(\frac{\partial^4 w_b}{\partial x^2 \partial t^2} + \frac{\partial^4 w_b}{\partial y^2 \partial t^2}\right) + K_2\left(\frac{\partial^4 w_s}{\partial x^2 \partial t^2} + \frac{\partial^4 w_s}{\partial y^2 \partial t^2}\right)\right] = 0$$

$$\tag{9.106}$$

Finally, the obtained governing equations must be solved via the introduced method to enrich the wave dispersion responses of the nanostructures rested on a three-parameter visco-Pasternak medium. As expressed in Eq. (9.86), the solution of the dynamic problem of a continuous system considering the damping's effect includes an additional damping matrix in addition to the ordinary stiffness and mass matrices of the system. For the case of analyzing a nanoplate, it should be considered that the dimensions of the obtained eigenvalue problem are 3×3 for a Kirchhoff–Love nanoplate and 4×4 for a refined shear deformable one because of the involved degrees of freedom that are employed in the derivation of the governing equations. Similar to the previous section on nanobeams, it is clear that the mass and stiffness matrices for both Kirchhoff–Love and refined nanoplates are the same as those reported for a nanoplate embedded on two-parameter elastic Winkler–Pasternak foundation. Hence, in the following relations, we will only present the nonzero components of the damping matrix. For the Kirchhoff–Love nanoplates, the only nonzero term of the damping matrix is:

$$c_{33} = \left(1 + \mu^2\left(\beta_1^2 + \beta_2^2\right)\right)ic_d \tag{9.107}$$

In addition, for a refined shear deformable nanoplate, the nonzero components of the symmetric damping matrix can be written in the following form:

$$c_{22} = c_{23} = c_{33} = \left(1 + \mu^2\left(\beta_1^2 + \beta_2^2\right)\right)ic_d \tag{9.108}$$

9.4 Numerical Results and Discussion

In this section, graphical results will be presented to emphasize the role of each of the foundation parameters on the wave propagation curves of FG nanobeams and nanoplates. It is proven that the effects of various terms can be better observed when the dimensionless form of the substrate parameters are utilized. Therefore, the dimensionless form of Winkler, Pasternak and damping coefficients can be expressed in the following form:

$$K_w = \frac{k_w a^4}{10^{-16} D^*}, K_p = \frac{k_p a^2}{10^{-1} D^*}, C_d = c_d \frac{a^2}{10^- \sqrt{\rho_c h D^*}}, D^* = \frac{E_c}{12(1 - \nu_c^2)} \qquad (9.109)$$

In the above equation, a is the length of the nanobeam or nanoplate and h denotes the thickness of these nanostructures. In what follows, the effects of both elastic and viscoelastic foundations on the phase velocity curves of FG nanobeams will be illustrated, and after finishing the mentioned procedure, we will investigate similar influences on the mechanical performance of waves traveling in FG nanosized plates.

Similar to the first case study, the effects of resting a nanobeam on various elastic substrates will be presented in Figure 9.1. In Figure 9.1, the phase velocity curves of FG nanobeams are drawn as a function of the wave number for different amounts of stiffness terms of two-parameter elastic medium. It can be realized that the mechanical response of the nanobeam can be only affected in a limited range of wave numbers smaller than $\beta = 0.04$ (1/nm). In such wave numbers, the phase velocity of the propagated waves can be intensified when a greater Winkler or Pasternak coefficient is implemented. It is clear that the effect of the Pasternak term is more sensible than that of the Winkler parameter. The physical reason of the velocity enhancement due to increase in the stiffness of the substrate is that,

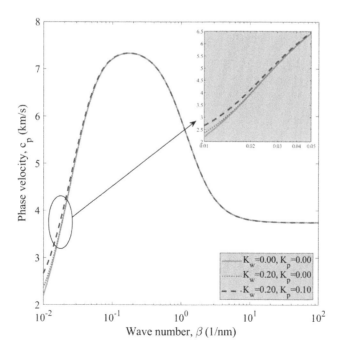

FIGURE 9.1
Variation of phase velocity versus wave number for FG nanobeams rested on an elastic medium with various Winkler and Pasternak stiffnesses ($c < 1$, $p = 0.5$).

by increasing the medium's stiffness, the structural stiffness of the nanobeam grows, and therefore, the equivalent stiffness of the nanobeam increases resulting in higher mechanical responses.

In addition, the effect of the gradient index of the employed functionally graded material (FGM) on the phase velocity curves of the FG nanobeams is shown in Figure 9.2. This study is performed in wave number of $\beta = 0.01$ (1/nm) to cover the effects of the elastic foundation on the mechanical response of the nanobeam. As studied in previous chapters, it is acceptable to see a decrease in the phase velocity of the nanobeam while a greater value is assigned to the gradient index. Indeed, the volume fraction of the ceramic phase of the FGM becomes greater than that of the metalic phase, which results in a reduction in the stiffness of the FGM across the thickness direction. Again, it is observable that the phase velocity increases as either Winkler or Pasternak coefficients are intensified. The physical reason of this issue is the previously mentioned improvement that can be found in the equivalent stiffness of the nanostructure by adding the springs to the medium.

After numerical investigation of the effects of elastic coefficients of the Winkler–Pasternak substrate, it is time to survey how the damping coefficient of the visco-Pasternak medium can affect the phase velocity characteristics of FG nanobeams. For this, the conventional phase velocity versus wave number plot is employed for various damping coefficients of the viscoelastic medium in certain values of both Winkler and Pasternak coefficients. Similar to Figure 9.1, it is observed that only in small wave numbers the effect of the damping coefficient of the viscoelastic substrate can be seen and this term cannot change the phase velocity of the nanobeam in high wave numbers. However, the quality of the effect of damping coefficient on the phase velocity of waves dispersed in nanobeams is completely different from those reported in former illustrations for the elastic terms. In fact, the damping effects can attenuate the mechanical responses of a desired structure. Based on this physical point of view, it is not strange to see a decrease in the mechanical response of FG nanobeams as the damping coefficient is added (Figure 9.3).

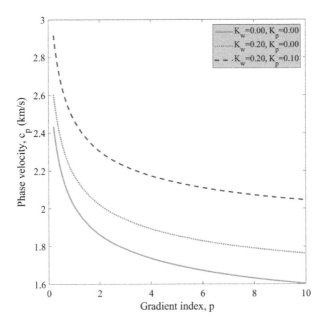

FIGURE 9.2

Variation of phase velocity versus gradient index for FG nanobeams rested on an elastic medium with various Winkler and Pasternak stiffnesses ($c < 1$, $\beta = 0.01$ (1/nm)).

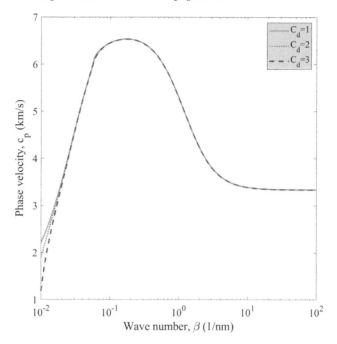

FIGURE 9.3
Variation of phase velocity versus wave number for FG nanobeams rested on a visco-Pasternak medium with various damping coefficients ($c < 1$, $p = 2$, $K_w = 0.2$, $K_p = 0.1$).

In previous illustrations, it was proven that the effects of both elastic and viscoelastic foundations can be observed in small wave numbers. In the next case study, authors mainly aim to highlight the effects of wave number on the damped velocity of the propagated waves in constant elastic stiff coefficients of the medium. It is clear from Figure 9.4 that the increase in the damping coefficient can finally result in the complete damping of the dynamic response of the nanobeam. However, depending on the employed wave number, the critical value that can completely damp the phase velocity of the nanobeam differs. It can be perceived that, by changing the wave number from $\beta = 0.01$ (1/nm) to $\beta = 0.02$ (1/nm), a remarkable value must be added to the damping coefficient to strengthen the damping effect in a way that can defeat the phase velocity of the nanobeam. So, it can be concluded that in the applications designers aim to damp the dynamic wave response of a nanobeam, implementation of large wave numbers can result in employing a damper with large viscosity which may not be desired.

As the final illustration for the numerical analysis of FG nanobeams rested on a desired substrate, the phase velocity curves of the mentioned nanostructure is plotted against the damping coefficient of the viscoelastic medium while various values are assigned to the stiff terms of the medium (see Figure 9.5). Similar to former illustrations, it can be again declared that the phase velocity of the dispersed waves goes through a diminishing path, resulting in the final value of zero for the mechanical response. Also, it can be observed that the greater are the stiffnesses of the employed springs in the foundation of the nanostructure, the higher can be the dynamic response of the nanobeam that can be related to the stiffness enhancement that can be made in the nanostructure while stiffer springs are selected. It is worth mentioning that the influence of the Pasternak parameter can be sensed easier than that of the Winkler parameter.

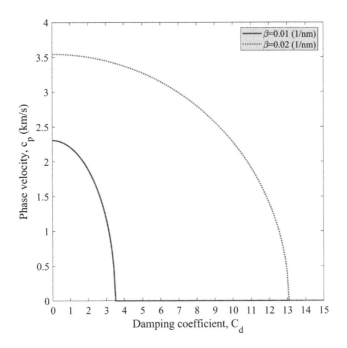

FIGURE 9.4
Effect of the wave number on the phase velocity curves of FG nanobeams on varying
damping coefficients ($c < 1$, $p = 2$, $K_w = 0.2$, $K_p = 0.1$).

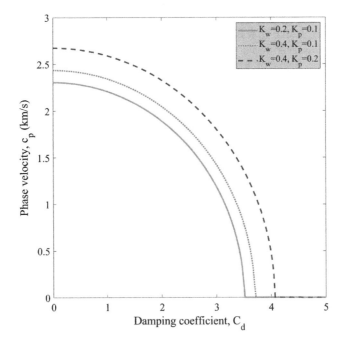

FIGURE 9.5
Coupled influences of the viscoelastic foundation parameters on the phase velocity behaviors
of FG nanobeams ($c < 1$, $p = 2$).

Now, it is time to survey the same problem for FG nanoplates. It can be estimated that, because of the similarities in the behaviors of various continuous systems, similar trends can be predicted for wave propagation in the FG nanoplates. Entirely, the same illustrations will be depicted for demonstrating the effects of different types of substrate on the wave propagation curves of FG nanoplates. However, some tiny changes are provided for the results of plates, which will be explained in the following paragraphs.

Figure 9.6 shows the change in the phase velocity curves of FG nanoplates rested on an elastic medium. It is observed that an increase in the value of the stiff springs of the foundation can intensify the phase speed of the elastic waves because of the increase in the equivalent stiffness of the continuous system. As explained in Figure 9.1, one can see that the wave response of the FG nanoplate can be aggrandized by adding the foundation parameters only in small wave numbers, and when utilizing a large wave number, no specific enhancement can be seen in the dispersion curves. Also, it is interesting to point that, in comparison with FG nanobeams, here stiffer springs are required to enrich the mechanical response of the nanoplate. In our estimations, the Pasternak coefficient possesses higher effect on the wave speed in the FG nanoplate compared with the Winkler term.

Further, the influence of using different gradient indices on the phase speed of the dispersed waves in the FG nanoplates is shown in Figure 9.7 for various substrate parameters. It is clear that adding each of the Winkler or Pasternak coefficients can result in reaching a stiffer nanoplate that can provide higher velocities. Furthermore, the phase speed of the waves will be reduced when a greater value is assigned to the gradient index. The reason for this is the decrease in the stiffness of the FGM on adding the gradient index. Again it can be figured out that a higher stiffness enhancement can be observed by adding the Pasternak parameter while a similar enhancement is imposed on each of the elastic springs of the foundation.

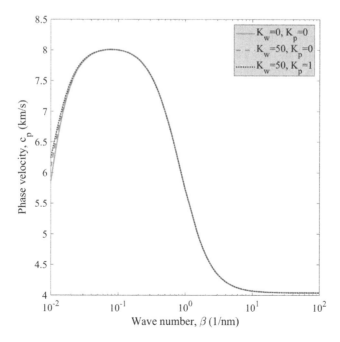

FIGURE 9.6
Variation of phase velocity versus wave number for FG nanoplates rested on an elastic medium with various Winkler and Pasternak stiffnesses ($c < 1$, $p = 2$).

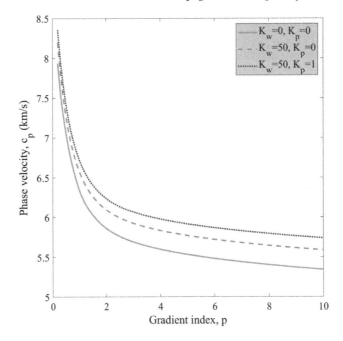

FIGURE 9.7
Variation of phase velocity versus gradient index for FG nanoplates rested on an elastic medium with various Winkler and Pasternak stiffnesses ($c < 1$, $\beta = 0.01$ (1/nm)).

Next, focusing on the impact of the damping coefficient of the visco-Pasternak substrate, the wave speed variations of FG nanoplates are plotted versus wave number for different values of the damping coefficient at constant values of the elastic terms of the medium. Obviously, the mechanical response of the system will be reduced while the dimensionless damping coefficient of the medium is increased. It is worth mentioning that the decrease in the wave response can be sensed whenever damping coefficient is changed from $C_d = 50$ to $C_d = 75$ instead of changing from $C_d = 25$ to $C_d = 50$. Similar to Winkler and Pasternak coefficients, damping coefficient can affect the wave velocity in a limited range of wave numbers, specially $\beta < 0.02$ (1/nm). The physical reason for this decreasing trend is the viscosity of the employed dampers, which reduces the stiffness of the system and generates compliance in the system, leading to the aforementioned change (Figure 9.8).

In another numerical example, the combined effects of damping phenomenon and material's distribution on the dynamic responses of the system are depicted in Figure 9.9. It can be observed that, by adding the value of the damping coefficient, a continuous attenuation occurs and the phase speed of the dispersed waves will finally reach zero. The reason for this reducing phenomenon was discussed previously. Also, it can be realized that the influence of the gradient index of the FGM depends on the value of the damping parameter. Indeed, nanoplates with smaller FGM gradient index possess greater responses unless damping coefficient exceeds $C_d = 75$. After this certain value of damping coefficient of the viscoelastic substrate, one can observe an aggrandation in the phase speed of the scattered waves while a higher gradient index is selected. This dual behavior occurs due to the presence of damping phenomenon and cannot be observed in cases where damping effects are dismissed. It can be concluded that, once the attenuation process is completed, no wave can propagate in the media because the damping effect of the viscoelastic foundation defeats the stiffness of the FG nanoplate.

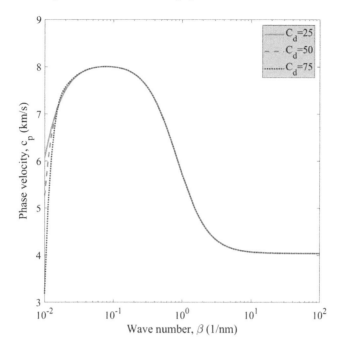

FIGURE 9.8

Variation of phase velocity versus wave number for FG nanoplates rested on a visco-Pasternak medium with various damping coefficients ($c < 1$, $p = 2$, $K_w = 50$, $K_p = 1$).

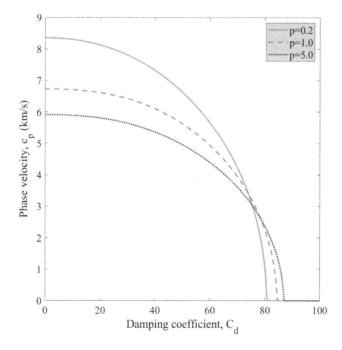

FIGURE 9.9

Illustration of the influences of gradient index and damping coefficient on the phase velocity of FG nanoplates ($c < 1$, $\beta = 0.01$ (1/nm), $K_w = 50$, $K_p = 1$).

Finally, different values of both Winkler and Pasternak coefficients are assigned in Figure 9.10, presenting the variation of phase velocity of FG nanoplates against damping coefficient. Once again, the damped behaviors of the nanoplate can be seen in this figure. It is clear that the wave velocity will diminish to zero as the damping coefficient grows in a continuous manner. Meanwhile, it is observed that the dynamic response of the system experiences an increase when using a viscoelastic medium with stiffer springs. According to the figure, it can be found that, if a similar change in the stiffness of Winkler and Pasternak coefficients is generated, Pasternak coefficient can affect the dispersion curves of the nanostructure more than the Winkler one. In fact, Pasternak springs are stiffer than Winkler ones in an identical value of their dimensionless parameters.

It must be mentioned that the influences of some of the involved parameters on the wave propagation responses of FG nanostructures embedded on visco-Pasternak foundation were not included in the above case studies. For example, the effects of nonlocal and length-scale terms, which were responsible for the consideration of the small scale of the continuous system, were not studied. The major reason was presented above. In other words, the effects of both elastic and viscous components of the foundation on the mechanical responses of the FG nanostructures were observable in very small wave numbers. On the other hand, it was shown in Chapter 4 that the scale parameters cannot change the wave dispersion curves of nanostructures in small wave numbers, and that their influences can be sensed in the wave numbers greater than $\beta = 0.1$ (1/nm). Hence, authors decided not to include the mentioned parameters in the numerical investigations because their influences were small enough to be neglected. Moreover, when the damping effects of the visco-Pasternak substrate are considered, it can be observed that a small increase in the wave number will result in a remarkable increase in the damping coefficient, which can damp the wave velocity. Thus, other variants more than wave number were used to present the results of the studied issue.

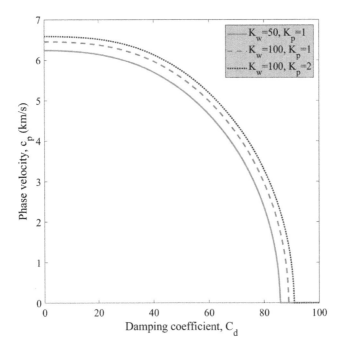

FIGURE 9.10
Coupled influences of the viscoelastic foundation parameters on the phase velocity behaviors of FG nanoplates ($c < 1$, $p = 2$).

10

Thermal Effects on Wave Propagation Characteristics of Smart Nanostructures

In this chapter, the influence of thermal environment on the wave propagation responses of functionally graded (FG) nanostructures will be surveyed for both classical and shear deformable beams and plates. It is worth mentioning that only the in-plane thermal loading will be considered here. In addition, different temperature profiles will be considered, and the material properties will be assumed to be a function of the temperature. The results demonstrate that, in this case, the stiffness of the nanostructure decreases dramatically, and due to this reduction in the equivalent stiffness matrix of the mentioned element, the wave propagation curves experience a drop down and the dynamic responses of the nanostructures reduce.

10.1   Introduction

As stated in the previous chapters, one of the most crucial superiorities of functionally graded materials (FGMs) in comparison with the conventional homogeneous or heterogeneous materials is the refined thermal resistance of such materials, which cannot be found in aforementioned materials [75]. In many engineering fields, FGMs are broadly used in applications where endurance of large thermal loading is required for the designed device. Thus, the thermomechanical analysis of the structures made from FGMs is of incredible importance. Further, it was mentioned in Chapter 4 that FG nanosized structures have attracted designers' attention to be used as structural devices in smart tiny systems. Due to the abovementioned importance of FG nanostructures in thermal environments, we will present a general and complete thermal analysis regarding the wave propagation characteristics of FG nanostructures.

In what follows, the mathematical formulation of the wave propagation problem for both beam-type and plate-type nanoscale elements will be updated to include the effects of the thermal environment. Afterward, different temperature profiles will be introduced to account for uniform, linear and nonlinear temperature gradients in numerical case studies.

10.2   Analysis of FG Nanobeams

Here, the effect of temperature change will be added to the previously introduced governing equations of nanobeams by presenting a modified version of motion equations, which covers the effects of thermal environment. Indeed, the effect of thermal loading appears in the variation of work done by external forces. First, this process will be completed for the

Euler–Bernoulli nanobeams, thereafter, the formulations of refined higher-order beams will be presented.

Once a beam is subjected to a temperature gradient, a thermal loading is applied to the beam due to the generated temperature change. This force can be expressed in the following mathematical form:

$$N^T = \int_A E(z)\alpha(z)\big(T(z) - T_0\big)dA \tag{10.1}$$

where $E(z)$, $\alpha(z)$, $T(z)$ and T_0 denote Young's moduli, coefficient of thermal expansion (CTE), local temperature profile and reference temperature, respectively. Now, the influence of this force must be added to the previously derived motion equations (see Chapter 4 for the derivation procedure). Now, it is better to introduce the way in which the effect of any external force can be applied on the equations of the beams' motion. The work done by the external axial force, F, applied on a beam can be expressed as:

$$V = \frac{1}{2}\int_0^L F\left(\frac{\partial u}{\partial x}\right)^2 dx \tag{10.2}$$

Now, the work done by thermal loading, N^T, applied on an Euler–Bernoulli beam can be achieved by substituting for the thermal force and displacement from Eqs. (10.1) and (4.10), respectively, in Eq. (10.2). By this substitution, the work done on the beam can be expressed in the following form:

$$V = \frac{1}{2}\int_0^L N^T\left(\frac{\partial w}{\partial x}\right)^2 dx \tag{10.3}$$

Using Eq. (4.15) instead of Eq. (4.10) for the deflection of the beam, the work done by thermal loading becomes:

$$V = \frac{1}{2}\int_0^L N^T\left(\frac{\partial(w_b + w_s)}{\partial x}\right)^2 dx \tag{10.4}$$

Now, the variation of the work done by thermal force on an Euler–Bernoulli beam can be calculated in the following form:

$$\delta V = \int_0^L N^T \frac{\partial w}{\partial x}\frac{\partial \delta w}{\partial x}dx \tag{10.5}$$

In addition, the variation of the external work done by thermal loading on the refined higher-order beams can be written as follows:

$$\delta V = \int_0^L N^T \frac{\partial(w_b + w_s)}{\partial x}\frac{\partial \delta(w_b + w_s)}{\partial x}dx \tag{10.6}$$

Next, the motion equations of the Euler–Bernoulli beam must be updated to include the thermal effects. For this, Eqs. (4.24), (4.26) and (10.5) should be substituted in the definition of the Hamilton's principle, stated in Eq. (4.19). Once the mathematical simplifications are performed, the Euler–Lagrange equations of the beam subjected to thermal loading on the basis of the Euler–Bernoulli beam hypothesis can be written as:

$$\frac{\partial N}{\partial x} = I_0\frac{\partial^2 u}{\partial t^2} - I_1\frac{\partial^3 w}{\partial x \partial t^2} \tag{10.7}$$

$$\frac{\partial^2 M}{\partial x^2} - N^T \frac{\partial^2 w}{\partial x^2} = I_0 \frac{\partial^2 w}{\partial t^2} + I_1 \frac{\partial^3 u}{\partial x \partial t^2} - I_2 \frac{\partial^4 w}{\partial x^2 \partial t^2} \tag{10.8}$$

Furthermore, the Euler–Lagrange equations of beams under thermal loading with respect to the influences of shear deformation can be obtained when Eqs. (4.31), (4.33) and (10.6) are inserted in Eq. (4.19) as follows:

$$\frac{\partial N}{\partial x} = I_0 \frac{\partial^2 u}{\partial t^2} - I_1 \frac{\partial^3 w_b}{\partial x \partial t^2} - J_1 \frac{\partial^3 w_s}{\partial x \partial t^2} \tag{10.9}$$

$$\frac{\partial^2 M^b}{\partial x^2} - N^T \frac{\partial^2 (w_b + w_s)}{\partial x^2} = I_0 \frac{\partial^2 (w_b + w_s)}{\partial t^2} + I_1 \frac{\partial^3 u}{\partial x \partial t^2} - I_2 \frac{\partial^4 w_b}{\partial x^2 \partial t^2} - J_2 \frac{\partial^4 w_s}{\partial x^2 \partial t^2} \tag{10.10}$$

$$\frac{\partial^2 M^s}{\partial x^2} + \frac{\partial Q_x}{\partial x} - N^T \frac{\partial^2 (w_b + w_s)}{\partial x^2} = I_0 \frac{\partial^2 (w_b + w_s)}{\partial t^2} + J_1 \frac{\partial^3 u}{\partial x \partial t^2} - J_2 \frac{\partial^4 w_b}{\partial x^2 \partial t^2} - K_2 \frac{\partial^4 w_s}{\partial x^2 \partial t^2} \tag{10.11}$$

Now, the governing equations of the beams under thermal loading can be obtained by substituting for the constitutive equations of the nonlocal strain gradient elasticity in the Euler–Lagrange equations of the beams. However, some modifications are needed on the constitutive equations of nanobeams to include the effects of thermal environment. Indeed, the fundamental Eq. (2.17) must be re-written in the following form to capture the effects of thermal strain:

$$\left(1 - \mu^2 \nabla^2\right) \sigma_{ij} = \left(1 - \lambda^2 \nabla^2\right) C_{ijkl} \left(\varepsilon_{kl} - \alpha \Delta T\right) \tag{10.12}$$

where α is the CTE and ΔT is the temperature gradient. By integrating from the above relation over the nanobeam's thickness for an Euler–Bernoulli nanobeams, the following constitutive equations can be achieved:

$$\left(1 - \mu^2 \nabla^2\right) N = \left(1 - \lambda^2 \nabla^2\right) \left(A_{xx} \frac{\partial u}{\partial x} - B_{xx} \frac{\partial^2 w}{\partial x^2}\right) - N^T \tag{10.13}$$

$$\left(1 - \mu^2 \nabla^2\right) M = \left(1 - \lambda^2 \nabla^2\right) \left(B_{xx} \frac{\partial u}{\partial x} - D_{xx} \frac{\partial^2 w}{\partial x^2}\right) - M^T \tag{10.14}$$

where A_{xx}, B_{xx} and D_{xx} are the cross-sectional rigidities that can be calculated using Eq. (4.44). The thermal resultants, N^T and M^T, can be defined in the following form:

$$\left[N^T, M^T\right] = \int_A E(z) \alpha(z) \left[1, z\right] dA \tag{10.15}$$

Using a similar procedure, the constitutive equations for the refined sinusoidal nanobeams can be expressed in the following form:

$$\left(1 - \mu^2 \nabla^2\right) N = \left(1 - \lambda^2 \nabla^2\right) \left(A_{xx} \frac{\partial u}{\partial x} - B_{xx} \frac{\partial^2 w_b}{\partial x^2} - B_{xx}^s \frac{\partial^2 w_s}{\partial x^2}\right) - N^T \tag{10.16}$$

$$\left(1 - \mu^2 \nabla^2\right) M^b = \left(1 - \lambda^2 \nabla^2\right) \left(B_{xx} \frac{\partial u}{\partial x} - D_{xx} \frac{\partial^2 w_b}{\partial x^2} - D_{xx}^s \frac{\partial^2 w_s}{\partial x^2}\right) - M_b^T \tag{10.17}$$

$$\left(1 - \mu^2 \nabla^2\right) M^s = \left(1 - \lambda^2 \nabla^2\right) \left(B_{xx}^s \frac{\partial u}{\partial x} - D_{xx}^s \frac{\partial^2 w_b}{\partial x^2} - H_{xx}^s \frac{\partial^2 w_s}{\partial x^2}\right) - M_s^T \tag{10.18}$$

where cross-sectional rigidities are identical with those defined in Eq. (4.44). In addition, the thermal resultants can be defined as bellow:

$$[N^T, M_b^T, M_s^T] = \int_A E(z)\alpha(z)[1, z, f(z)]dA \qquad (10.19)$$

It must be mentioned that Eq. (4.43) remains unchangeable. Now, one can obtain the governing equations of the problem when the effects of thermal loading are included. For an Euler–Bernoulli nanobeam, the governing equations can be achieved by substituting Eqs. (10.13) and (10.14) in Eqs. (10.7) and (10.8) as follows:

$$\left(1 - \lambda^2\nabla^2\right)\left[A_{xx}\frac{\partial^2 u}{\partial x^2} - B_{xx}\frac{\partial^3 w}{\partial x^3}\right] + \left(1 - \mu^2\nabla^2\right)\left[-I_0\frac{\partial^2 u}{\partial t^2} + I_1\frac{\partial^3 w}{\partial x\partial t^2}\right] = 0 \qquad (10.20)$$

$$\left(1 - \lambda^2\nabla^2\right)\left[B_{xx}\frac{\partial^3 u}{\partial x^3} - D_{xx}\frac{\partial^4 w}{\partial x^4}\right] + \left(1 - \mu^2\nabla^2\right)$$
$$\times \left[N^T\frac{\partial^2 w}{\partial x^2} - I_0\frac{\partial^2 w}{\partial t^2} - I_1\frac{\partial^3 u}{\partial x\partial t^2} + I_2\frac{\partial^4 w}{\partial x^2\partial t^2}\right] = 0 \qquad (10.21)$$

Moreover, the governing equations of refined higher-order nanobeams under thermal loading can be obtained by substituting Eqs. (10.16)–(10.18) and Eq. (4.43) in Eqs. (10.9)–(10.11) in the following form:

$$\left(1 - \lambda^2\nabla^2\right)\left[A_{xx}\frac{\partial^2 u}{\partial x^2} - B_{xx}\frac{\partial^3 w_b}{\partial x^3} - B_{xx}^s\frac{\partial^3 w_s}{\partial x^3}\right] + \left(1 - \mu^2\nabla^2\right)$$
$$\times \left[-I_0\frac{\partial^2 u}{\partial t^2} + I_1\frac{\partial^3 w_b}{\partial x\partial t^2} + J_1\frac{\partial^3 w_s}{\partial x\partial t^2}\right] = 0 \qquad (10.22)$$

$$\left(1 - \lambda^2\nabla^2\right)\left[B_{xx}\frac{\partial^3 u}{\partial x^3} - D_{xx}\frac{\partial^4 w_b}{\partial x^4} - D_{xx}^s\frac{\partial^4 w_s}{\partial x^4}\right] + \left(1 - \mu^2\nabla^2\right)$$
$$\times \left[N^T\frac{\partial^2(w_b + w_s)}{\partial x^2} - I_0\frac{\partial^2(w_b + w_s)}{\partial t^2} - I_1\frac{\partial^3 u}{\partial x\partial t^2} + I_2\frac{\partial^4 w_b}{\partial x^2\partial t^2} + J_2\frac{\partial^4 w_s}{\partial x^2\partial t^2}\right] = 0$$
$$(10.23)$$

$$\left(1 - \lambda^2\nabla^2\right)\left[B_{xx}^s\frac{\partial^3 u}{\partial x^3} - D_{xx}^s\frac{\partial^4 w_b}{\partial x^4} - H_{xx}^s\frac{\partial^4 w_s}{\partial x^4} + A_x^s\frac{\partial^2 w_s}{\partial x^2}\right] + \left(1 - \mu^2\nabla^2\right)$$
$$\times \left[N^T\frac{\partial^2(w_b + w_s)}{\partial x^2} - I_0\frac{\partial^2(w_b + w_s)}{\partial t^2} - J_1\frac{\partial^3 u}{\partial x\partial t^2} + J_2\frac{\partial^4 w_b}{\partial x^2\partial t^2} + K_2\frac{\partial^4 w_s}{\partial x^2\partial t^2}\right] = 0$$
$$(10.24)$$

Now, the governing equations are completely derived for both Euler–Bernoulli and refined shear deformable nanobeams. The next step is to seek for the solution of the problem when the nanobeam is subjected to temperature change. It is clear that the differences between the wave propagation analyses of FG nanobeams with and without the influences of thermal environment can be summarized as thermal loading. Hence, the general form of the solution for a thermally affected wave problem is the same as our previous solution expressed in Chapter 4. In fact, the arrays of the mass matrix of the nanobeam are completely the same as those reported in Eqs. (4.52) and (4.54) for the Euler–Bernoulli and refined sinusoidal nanobeams, respectively.

However, the components of the stiffness matrix are not completely the same as those presented in Eqs. (4.51) and (4.53). Indeed, the arrays involved with the deflection of the nanobeam must be modified. For investigating an Euler–Bernoulli nanobeam, the k_{22} array of the Eq. (4.51) must be re-written in the following form to account for the effects of thermal loading in the solution:

$$k_{22} = -\left(1 + \lambda^2 \beta^2\right) D_{xx} \beta^4 + \left(1 + \mu^2 \beta^2\right) N^T \tag{10.25}$$

Meanwhile, the k_{22}, k_{23} and k_{33} components of the Eq. (4.53) must be edited as follows to consider the thermal effects:

$$
\begin{aligned}
k_{22} &= -\left(1 + \lambda^2 \beta^2\right) D_{xx} \beta^4 + \left(1 + \mu^2 \beta^2\right) N^T, \\
k_{23} &= -\left(1 + \lambda^2 \beta^2\right) D^s_{xx} \beta^4 + \left(1 + \mu^2 \beta^2\right) N^T, \\
k_{33} &= -\left(1 + \lambda^2 \beta^2\right) \left[H^s_{xx} \beta^4 + A^s_x \ \beta^2\right] + \left(1 + \mu^2 \beta^2\right) N^T
\end{aligned}
\tag{10.26}
$$

10.3 Analysis of FG Nanoplates

In this section, a procedure similar to that presented in Section 10.2 will be presented to depict the mathematical formulation of the wave propagation problem within FG nanoplates when the nanoplate is considered to be subjected to thermal loading. Similar to all other analyses in the book, both classical and shear deformable theorems will be utilized to derive the partial differential governing equations for both cases. Similar to beams, subjecting a plate to a thermal environment produces in-plane forces that must be applied on the governing equations of the plate. Before presenting the energy-based formulations, the thermal loading in both longitudinal and transverse directions in an isotropic plate must be introduced. The thermal force can be expressed in the following form:

$$N^T = \int_{-\frac{h}{2}}^{\frac{h}{2}} \frac{E(z)}{1 - \nu(z)} \alpha(z) \left(T(z) - T_0\right) dz \tag{10.27}$$

where $E(z)$, $\nu(z)$ and $\alpha(z)$ correspond with Young's moduli, Poisson's ratio and CTE of the plate, respectively. Also, again $T(z)$ and T_0 are temperature profile and reference temperature, respectively. Now, the mathematical formulation for the work done by a biaxial in-plane load applied on the plate must be introduced. The work done by the aforementioned load can be expressed as follows:

$$V = \frac{1}{2} \int_A \left[F_x \left(\frac{\partial u}{\partial x}\right)^2 + F_y \left(\frac{\partial u}{\partial y}\right)^2 \right] dA \tag{10.28}$$

Using the above equation and the concept of isotropic materials (i.e., identity of the applied thermal forces in x and y directions), the work done by thermal force can be achieved for a FG plate with deflection u . Here, by selecting Eq. (4.58) as the deflection of the plate and setting $F_x = F_y = N^T$, the work done by thermal loading can be written in the following form:

$$V = \frac{1}{2} \int_A N^T \left[\left(\frac{\partial w}{\partial x}\right)^2 + \left(\frac{\partial w}{\partial y}\right)^2 \right] dA \tag{10.29}$$

Replacing Eq. (4.65) instead of Eq. (4.58) for the plate's deflection, the work done by thermal loading will be in the following form for a refined sinusoidal plate:

$$V = \frac{1}{2} \int_A N^T \left[\left(\frac{\partial (w_b + w_s)}{\partial x} \right)^2 + \left(\frac{\partial (w_b + w_s)}{\partial y} \right)^2 \right] dA \tag{10.30}$$

Next, the variation of the above equations must be computed to be entered in the definition of the Hamilton's principle. The variation of the work done by thermal loading for a Kirchhoff–Love plate can be expressed as:

$$\delta V = \int_A N^T \left[\frac{\partial w}{\partial x} \frac{\partial \delta w}{\partial x} + \frac{\partial w}{\partial y} \frac{\partial \delta w}{\partial y} \right] dA \tag{10.31}$$

Also, the variation of the external work done by thermal loading on the refined higher-order plates can be expressed as:

$$\delta V = \int_A N^T \left[\frac{\partial (w_b + w_s)}{\partial x} \frac{\partial \delta (w_b + w_s)}{\partial x} + \frac{\partial (w_b + w_s)}{\partial y} \frac{\partial \delta (w_b + w_s)}{\partial y} \right] dA \tag{10.32}$$

Now, the Euler–Lagrange equations of both Kirchhoff–Love and refined higher-order plates can be achieved with respect to the effects of the thermal environment. When a Kirchhoff–Love plate is considered, the Eqs. (4.72), (4.74) and (10.31) must be inserted in Eq. (4.19). When this substitution is completed, the Euler–Lagrange equations can be expressed in the following form:

$$\frac{\partial N_{xx}}{\partial x} + \frac{\partial N_{xy}}{\partial y} = I_0 \frac{\partial^2 u}{\partial t^2} - I_1 \frac{\partial^3 w}{\partial x \partial t^2} \tag{10.33}$$

$$\frac{\partial N_{xy}}{\partial x} + \frac{\partial N_{yy}}{\partial y} = I_0 \frac{\partial^2 v}{\partial t^2} - I_1 \frac{\partial^3 w}{\partial y \partial t^2} \tag{10.34}$$

$$\frac{\partial^2 M_{xx}}{\partial x^2} + 2 \frac{\partial^2 M_{xy}}{\partial x \partial y} + \frac{\partial^2 M_{yy}}{\partial y^2} - N^T \left(\frac{\partial^2 w}{\partial x^2} + \frac{\partial^2 w}{\partial y^2} \right)$$
$$= I_0 \frac{\partial^2 w}{\partial t^2} + I_1 \left(\frac{\partial^3 u}{\partial x \partial t^2} + \frac{\partial^3 v}{\partial y \partial t^2} \right) - I_2 \left(\frac{\partial^4 w}{\partial x^2 \partial t^2} + \frac{\partial^4 w}{\partial y^2 \partial t^2} \right) \tag{10.35}$$

Moreover, inserting Eqs. (4.80), (4.83) and (10.32) in Eq. (4.19), the Euler–Lagrange equations of the plates with respect to the effects of the thermal environment can be obtained in the following form:

$$\frac{\partial N_{xx}}{\partial x} + \frac{\partial N_{xy}}{\partial y} = I_0 \frac{\partial^2 u}{\partial t^2} - I_1 \frac{\partial^3 w_b}{\partial x \partial t^2} - J_1 \frac{\partial^3 w_s}{\partial x \partial t^2} \tag{10.36}$$

$$\frac{\partial N_{xy}}{\partial x} + \frac{\partial N_{yy}}{\partial y} = I_0 \frac{\partial^2 v}{\partial t^2} - I_1 \frac{\partial^3 w_b}{\partial y \partial t^2} - J_1 \frac{\partial^3 w_s}{\partial y \partial t^2} \tag{10.37}$$

$$\frac{\partial^2 M_{xx}^b}{\partial x^2} + 2 \frac{\partial^2 M_{xy}^b}{\partial x \partial y} + \frac{\partial^2 M_{yy}^b}{\partial y^2} - N^T \left(\frac{\partial^2 (w_b + w_s)}{\partial x^2} + \frac{\partial^2 (w_b + w_s)}{\partial y^2} \right) = I_0 \frac{\partial^2 (w_b + w_s)}{\partial t^2}$$
$$+ I_1 \left(\frac{\partial^3 u}{\partial x \partial t^2} + \frac{\partial^3 v}{\partial y \partial t^2} \right) - I_2 \left(\frac{\partial^4 w_b}{\partial x^2 \partial t^2} + \frac{\partial^4 w_b}{\partial y^2 \partial t^2} \right) - J_2 \left(\frac{\partial^4 w_s}{\partial x^2 \partial t^2} + \frac{\partial^4 w_s}{\partial y^2 \partial t^2} \right) \tag{10.38}$$

$$
\frac{\partial^2 M_{xx}^s}{\partial x^2} + 2\frac{\partial^2 M_{xy}^s}{\partial x \partial y} + \frac{\partial^2 M_{yy}^s}{\partial y^2} + \frac{\partial Q_x}{\partial x} + \frac{\partial Q_y}{\partial y} - N^T \left(\frac{\partial^2 (w_b + w_s)}{\partial x^2} + \frac{\partial^2 (w_b + w_s)}{\partial y^2} \right)
$$

$$
= I_0 \frac{\partial^2 (w_b + w_s)}{\partial t^2} + J_1 \left(\frac{\partial^3 u}{\partial x \partial t^2} + \frac{\partial^3 v}{\partial y \partial t^2} \right) - J_2 \left(\frac{\partial^4 w_b}{\partial x^2 \partial t^2} + \frac{\partial^4 w_b}{\partial y^2 \partial t^2} \right)
$$

$$
- K_2 \left(\frac{\partial^4 w_s}{\partial x^2 \partial t^2} + \frac{\partial^4 w_s}{\partial y^2 \partial t^2} \right)
\tag{10.39}
$$

Now, the governing equations of the FG nanoplate can be obtained when the constitutive equations of the thermally affected nanoplates, achieved from the nonlocal strain gradient elasticity, are substituted in the obtained Euler–Lagrange equations. Due to the existence of temperature gradient, the constitutive equations of the nanoplate must be modified to cover the thermal loading. Indeed, instead of Eq. (2.17), Eq. (10.12) should be implemented to derive the constitutive equations of the nanoplates considering the thermal effects. For a Kirchhoff–Love nanoplate, Eqs. (4.89) and (4.90) can be re-written in the following form integrating from Eq. (10.12) across the nanoplate's thickness:

$$
(1 - \mu^2 \nabla^2) \begin{bmatrix} N_{xx} \\ N_{yy} \\ N_{xy} \end{bmatrix} = (1 - \lambda^2 \nabla^2) \left(\begin{bmatrix} A_{11} & A_{12} & 0 \\ A_{12} & A_{22} & 0 \\ 0 & 0 & A_{66} \end{bmatrix} \begin{bmatrix} \frac{\partial u}{\partial x} \\ \frac{\partial v}{\partial y} \\ \frac{\partial u}{\partial y} + \frac{\partial v}{\partial x} \end{bmatrix} \right.
$$

$$
\left. + \begin{bmatrix} B_{11} & B_{12} & 0 \\ B_{12} & B_{22} & 0 \\ 0 & 0 & B_{66} \end{bmatrix} \begin{bmatrix} -\frac{\partial^2 w}{\partial x^2} \\ -\frac{\partial^2 w}{\partial y^2} \\ -2\frac{\partial^2 w}{\partial x \partial y} \end{bmatrix} - \begin{bmatrix} N^T \\ N^T \\ 0 \end{bmatrix} \right)
\tag{10.40}
$$

$$
(1 - \mu^2 \nabla^2) \begin{bmatrix} M_{xx} \\ M_{yy} \\ M_{xy} \end{bmatrix} = (1 - \lambda^2 \nabla^2) \left(\begin{bmatrix} B_{11} & B_{12} & 0 \\ B_{12} & B_{22} & 0 \\ 0 & 0 & B_{66} \end{bmatrix} \begin{bmatrix} \frac{\partial u}{\partial x} \\ \frac{\partial v}{\partial y} \\ \frac{\partial u}{\partial y} + \frac{\partial v}{\partial x} \end{bmatrix} \right.
$$

$$
\left. + \begin{bmatrix} D_{11} & D_{12} & 0 \\ D_{12} & D_{22} & 0 \\ 0 & 0 & D_{66} \end{bmatrix} \begin{bmatrix} -\frac{\partial^2 w}{\partial x^2} \\ -\frac{\partial^2 w}{\partial y^2} \\ -2\frac{\partial^2 w}{\partial x \partial y} \end{bmatrix} - \begin{bmatrix} M^T \\ M^T \\ 0 \end{bmatrix} \right)
\tag{10.41}
$$

where N^T and M^T stand for thermal force and moments, respectively, which can be calculated in the following form:

$$
[N^T, M^T] = \int_{-\frac{h}{2}}^{\frac{h}{2}} [1, z] \frac{E(z)}{1 - \nu(z)} \alpha(z) (T(z) - T_0) dz
\tag{10.42}
$$

One can obtain the cross-sectional rigidities from Eqs. (4.91) and (4.92). Also, for the refined sinusoidal nanoplates, the constitutive equations derived in Eqs. (4.93)–(4.95) can be expressed in the following new form:

$$
(1 - \mu^2 \nabla^2) \begin{bmatrix} N_{xx} \\ N_{yy} \\ N_{xy} \end{bmatrix} = (1 - \lambda^2 \nabla^2) \left(\begin{bmatrix} A_{11} & A_{12} & 0 \\ A_{12} & A_{22} & 0 \\ 0 & 0 & A_{66} \end{bmatrix} \begin{bmatrix} \frac{\partial u}{\partial x} \\ \frac{\partial v}{\partial y} \\ \frac{\partial u}{\partial y} + \frac{\partial v}{\partial x} \end{bmatrix} + \begin{bmatrix} B_{11} & B_{12} & 0 \\ B_{12} & B_{22} & 0 \\ 0 & 0 & B_{66} \end{bmatrix} \right.
$$

$$
\left. \times \begin{bmatrix} -\frac{\partial^2 w_b}{\partial x^2} \\ -\frac{\partial^2 w_b}{\partial y^2} \\ -2\frac{\partial^2 w_b}{\partial x \partial y} \end{bmatrix} + \begin{bmatrix} B_{11}^s & B_{12}^s & 0 \\ B_{12}^s & B_{22}^s & 0 \\ 0 & 0 & B_{66}^s \end{bmatrix} \begin{bmatrix} -\frac{\partial^2 w_s}{\partial x^2} \\ -\frac{\partial^2 w_s}{\partial y^2} \\ -2\frac{\partial^2 w_s}{\partial x \partial y} \end{bmatrix} - \begin{bmatrix} N^T \\ N^T \\ 0 \end{bmatrix} \right)
\tag{10.43}
$$

$$(1 - \mu^2 \nabla^2) \begin{bmatrix} M_{xx}^b \\ M_{yy}^b \\ M_{xy}^b \end{bmatrix} = (1 - \lambda^2 \nabla^2) \left(\begin{bmatrix} B_{11} & B_{12} & 0 \\ B_{12} & B_{22} & 0 \\ 0 & 0 & B_{66} \end{bmatrix} \begin{bmatrix} \frac{\partial u}{\partial x} \\ \frac{\partial v}{\partial y} \\ \frac{\partial u}{\partial y} + \frac{\partial v}{\partial x} \end{bmatrix} + \begin{bmatrix} D_{11} & D_{12} & 0 \\ D_{12} & D_{22} & 0 \\ 0 & 0 & D_{66} \end{bmatrix} \right.$$

$$\left. \times \begin{bmatrix} -\frac{\partial^2 w_b}{\partial x^2} \\ -\frac{\partial^2 w_b}{\partial y^2} \\ -2\frac{\partial^2 w_b}{\partial x \partial y} \end{bmatrix} + \begin{bmatrix} D_{11}^s & D_{12}^s & 0 \\ D_{12}^s & D_{22}^s & 0 \\ 0 & 0 & D_{66}^s \end{bmatrix} \begin{bmatrix} -\frac{\partial^2 w_s}{\partial x^2} \\ -\frac{\partial^2 w_s}{\partial y^2} \\ -2\frac{\partial^2 w_s}{\partial x \partial y} \end{bmatrix} - \begin{bmatrix} M_b^T \\ M_b^T \\ 0 \end{bmatrix} \right) \tag{10.44}$$

$$(1 - \mu^2 \nabla^2) \begin{bmatrix} M_{xx}^s \\ M_{yy}^s \\ M_{xy}^s \end{bmatrix} = (1 - \lambda^2 \nabla^2) \left(\begin{bmatrix} B_{11}^s & B_{12}^s & 0 \\ B_{12}^s & B_{22}^s & 0 \\ 0 & 0 & B_{66}^s \end{bmatrix} \begin{bmatrix} \frac{\partial u}{\partial x} \\ \frac{\partial v}{\partial y} \\ \frac{\partial u}{\partial y} + \frac{\partial v}{\partial x} \end{bmatrix} + \begin{bmatrix} D_{11}^s & D_{12}^s & 0 \\ D_{12}^s & D_{22}^s & 0 \\ 0 & 0 & D_{66}^s \end{bmatrix} \right.$$

$$\left. \times \begin{bmatrix} -\frac{\partial^2 w_b}{\partial x^2} \\ -\frac{\partial^2 w_b}{\partial y^2} \\ -2\frac{\partial^2 w_b}{\partial x \partial y} \end{bmatrix} + \begin{bmatrix} H_{11}^s & H_{12}^s & 0 \\ H_{12}^s & H_{22}^s & 0 \\ 0 & 0 & H_{66}^s \end{bmatrix} \begin{bmatrix} -\frac{\partial^2 w_s}{\partial x^2} \\ -\frac{\partial^2 w_s}{\partial y^2} \\ -2\frac{\partial^2 w_s}{\partial x \partial y} \end{bmatrix} - \begin{bmatrix} M_s^T \\ M_s^T \\ 0 \end{bmatrix} \right) \tag{10.45}$$

It is worth mentioning that Eq. (4.96) remains unchangeable. The cross-sectional rigidities used in Eqs. (10.43)–(10.45) can be obtained from Eqs. (4.97)–(4.99). Moreover, the thermal resultants can be calculated via the below relations:

$$[N^T, M_b^T, M_s^T] = \int_{-\frac{h}{2}}^{\frac{h}{2}} [1, z, f(z)] \frac{E(z)}{1 - \nu(z)} \alpha(z) (T(z) - T_0) dz \tag{10.46}$$

Next, the governing equations of FG nanoplates in a thermal environment can be obtained by substituting for the stress resulting from the newly derived constitutive equations in the Euler–Lagrange equations of the nanoplate. For a Kirchhoff–Love FG nanoplate, Eqs. (10.40) and (10.41) must be inserted in Eqs. (10.33)–(10.35). Once the aforementioned substitution is carried out, the governing equations of the FG nanoplates can be written as follows:

$$(1 - \lambda^2 \nabla^2) \left[A_{11} \frac{\partial^2 u}{\partial x^2} + (A_{12} + A_{66}) \frac{\partial^2 v}{\partial x \partial y} + A_{66} \frac{\partial^2 u}{\partial y^2} - B_{11} \frac{\partial^3 w}{\partial x^3} - (B_{12} + 2B_{66}) \frac{\partial^3 w}{\partial x \partial y^2} \right]$$
$$+ (1 - \mu^2 \nabla^2) \left[-I_0 \frac{\partial^2 u}{\partial t^2} + I_1 \frac{\partial^3 w}{\partial x \partial t^2} \right] = 0 \tag{10.47}$$

$$(1 - \lambda^2 \nabla^2) \left[A_{22} \frac{\partial^2 v}{\partial y^2} + (A_{12} + A_{66}) \frac{\partial^2 u}{\partial x \partial y} + A_{66} \frac{\partial^2 v}{\partial x^2} - B_{22} \frac{\partial^3 w}{\partial y^3} - (B_{12} + 2B_{66}) \frac{\partial^3 w}{\partial x^2 \partial y} \right]$$
$$+ (1 - \mu^2 \nabla^2) \left[-I_0 \frac{\partial^2 v}{\partial t^2} + I_1 \frac{\partial^3 w}{\partial y \partial t^2} \right] = 0 \tag{10.48}$$

$$(1 - \lambda^2 \nabla^2) \left[B_{11} \frac{\partial^3 u}{\partial x^3} + (B_{12} + 2B_{66}) \frac{\partial^3 u}{\partial x \partial y^2} + B_{22} \frac{\partial^3 v}{\partial y^3} + (B_{12} + 2B_{66}) \frac{\partial^3 v}{\partial x^2 \partial y} - D_{11} \frac{\partial^4 w}{\partial x^4} \right.$$
$$\left. - 2(D_{12} + 2D_{66}) \frac{\partial^4 w}{\partial x^2 \partial y^2} - D_{22} \frac{\partial^4 w}{\partial y^4} \right] + (1 - \mu^2 \nabla^2) \left[N^T \left(\frac{\partial^2 w}{\partial x^2} + \frac{\partial^2 w}{\partial y^2} \right) \right.$$
$$\left. - I_0 \frac{\partial^2 w}{\partial t^2} - I_1 \left(\frac{\partial^3 u}{\partial x \partial t^2} + \frac{\partial^3 v}{\partial y \partial t^2} \right) + I_2 \left(\frac{\partial^4 w}{\partial x^2 \partial t^2} + \frac{\partial^4 w}{\partial y^2 \partial t^2} \right) \right] = 0 \tag{10.49}$$

Also, once Eqs. (10.43)–(10.45) and (4.100) are inserted in Eqs. (10.36)–(10.39), the governing equations of FG nanoplates in thermal environment with respect to the effects of shear deformation can be expressed by:

$$
\left(1 - \lambda^2 \nabla^2\right)\left[A_{11}\frac{\partial^2 u}{\partial x^2} + \left(A_{12} + A_{66}\right)\frac{\partial^2 v}{\partial x \partial y} + A_{66}\frac{\partial^2 u}{\partial y^2} - B_{11}\frac{\partial^3 w_b}{\partial x^3} - \left(B_{12} + 2B_{66}\right)\frac{\partial^3 w_b}{\partial x \partial y^2}\right.
$$
$$
\left. - B_{11}^s\frac{\partial^3 w_s}{\partial x^3} - \left(B_{12}^s + 2B_{66}^s\right)\frac{\partial^3 w_s}{\partial x \partial y^2}\right] + \left(1 - \mu^2 \nabla^2\right)\left[-I_0\frac{\partial^2 u}{\partial t^2} + I_1\frac{\partial^3 w_b}{\partial x \partial t^2} + J_1\frac{\partial^3 w_s}{\partial x \partial t^2}\right] = 0
$$

$$(10.50)$$

$$
\left(1 - \lambda^2 \nabla^2\right)\left[A_{22}\frac{\partial^2 v}{\partial y^2} + \left(A_{12} + A_{66}\right)\frac{\partial^2 u}{\partial x \partial y} + A_{66}\frac{\partial^2 v}{\partial x^2} - B_{22}\frac{\partial^3 w_b}{\partial y^3} - \left(B_{12} + 2B_{66}\right)\frac{\partial^3 w_b}{\partial x^2 \partial y}\right.
$$
$$
\left. - B_{22}^s\frac{\partial^3 w_s}{\partial y^3} - \left(B_{12}^s + 2B_{66}^s\right)\frac{\partial^3 w_s}{\partial x^2 \partial y}\right] + \left(1 - \mu^2 \nabla^2\right)\left[-I_0\frac{\partial^2 v}{\partial t^2} + I_1\frac{\partial^3 w_b}{\partial y \partial t^2} + J_1\frac{\partial^3 w_s}{\partial y \partial t^2}\right] = 0
$$

$$(10.51)$$

$$
\left(1 - \lambda^2 \nabla^2\right)\left[B_{11}\frac{\partial^3 u}{\partial x^3} + \left(B_{12} + 2B_{66}\right)\frac{\partial^3 u}{\partial x \partial y^2} + B_{22}\frac{\partial^3 v}{\partial y^3} + \left(B_{12} + 2B_{66}\right)\frac{\partial^3 v}{\partial x^2 \partial y} - D_{11}\frac{\partial^4 w_b}{\partial x^4}\right.
$$
$$
- 2\left(D_{12} + 2D_{66}\right)\frac{\partial^4 w_b}{\partial x^2 \partial y^2} - D_{22}\frac{\partial^4 w_b}{\partial y^4} - D_{11}^s\frac{\partial^4 w_s}{\partial x^4} - 2\left(D_{12}^s + 2D_{66}^s\right)\frac{\partial^4 w_s}{\partial x^2 \partial y^2}
$$
$$
\left. - D_{22}^s\frac{\partial^4 w_s}{\partial y^4}\right) + \left(1 - \mu^2 \nabla^2\right)\left[N^T\left(\frac{\partial^2\left(w_b + w_s\right)}{\partial x^2} + \frac{\partial^2\left(w_b + w_s\right)}{\partial y^2}\right) - I_0\frac{\partial^2\left(w_b + w_s\right)}{\partial t^2}\right.
$$
$$
\left. - I_1\left(\frac{\partial^3 u}{\partial x \partial t^2} + \frac{\partial^3 v}{\partial y \partial t^2}\right) + I_2\left(\frac{\partial^4 w_b}{\partial x^2 \partial t^2} + \frac{\partial^4 w_b}{\partial y^2 \partial t^2}\right) + J_2\left(\frac{\partial^4 w_s}{\partial x^2 \partial t^2} + \frac{\partial^4 w_s}{\partial y^2 \partial t^2}\right)\right] = 0
$$

$$(10.52)$$

$$
\left(1 - \lambda^2 \nabla^2\right)\left[B_{11}^s\frac{\partial^3 u}{\partial x^3} + \left(B_{12}^s + 2B_{66}^s\right)\frac{\partial^3 u}{\partial x \partial y^2} + B_{22}^s\frac{\partial^3 v}{\partial y^3} + \left(B_{12}^s + 2B_{66}^s\right)\frac{\partial^3 v}{\partial x^2 \partial y}\right.
$$
$$
- D_{11}^s\frac{\partial^4 w_b}{\partial x^4} - 2\left(D_{12}^s + 2D_{66}^s\right)\frac{\partial^4 w_b}{\partial x^2 \partial y^2} - D_{22}^s\frac{\partial^4 w_b}{\partial y^4} - H_{11}^s\frac{\partial^4 w_s}{\partial x^4}
$$
$$
\left. - 2\left(H_{12}^s + 2H_{66}^s\right)\frac{\partial^4 w_s}{\partial x^2 \partial y^2} - H_{22}^s\frac{\partial^4 w_s}{\partial y^4} + A_{44}^s\left(\frac{\partial^2\left(w_b + w_s\right)}{\partial x^2} + \frac{\partial^2\left(w_b + w_s\right)}{\partial y^2}\right)\right]
$$
$$
+ \left(1 - \mu^2 \nabla^2\right)\left[N^T\left(\frac{\partial^2\left(w_b + w_s\right)}{\partial x^2} + \frac{\partial^2\left(w_b + w_s\right)}{\partial y^2}\right) - I_0\frac{\partial^2\left(w_b + w_s\right)}{\partial t^2}\right.
$$
$$
\left. - J_1\left(\frac{\partial^3 u}{\partial x \partial t^2} + \frac{\partial^3 v}{\partial y \partial t^2}\right) + J_2\left(\frac{\partial^4 w_b}{\partial x^2 \partial t^2} + \frac{\partial^4 w_b}{\partial y^2 \partial t^2}\right) + K_2\left(\frac{\partial^4 w_s}{\partial x^2 \partial t^2} + \frac{\partial^4 w_s}{\partial y^2 \partial t^2}\right)\right] = 0
$$

$$(10.53)$$

Now, the obtained governing equations must be solved using the analytical wave solution to enrich the phase velocity and wave frequency of the dispersed waves. Again, it must be pointed that the inertia-based part of the solution is completely the same as those presented in Chapter 4 because thermal loading does not affect this part of the formulations. In other words, the arrays of mass matrix for the Kirchhoff–Love and refined shear deformable nanoplates are identical with those reported in Eqs. (4.108) and (4.110), respectively.

Moreover, the changes in stiffness matrix belong to those arrays that are related to the deflection of the nanoplate, and those related to the longitudinal and transverse components of the nanoplate's displacement are the same as those reported in Eqs. (4.107) and (4.109) for Kirchhoff–Love and refined sinusoidal nanoplates, respectively. The modified version of the changed array of the stiffness matrix of a Kirchhoff–Love nanoplate, namely k_{33} array, can be written in the following form:

$$k_{33} = -\left(1 + \lambda^2\left(\beta_1^2 + \beta_2^2\right)\right)\left[D_{11}\beta_1^4 + 2\left(D_{12} + 2D_{66}\right)\beta_1^2\beta_2^2 + D_{22}\beta_2^4\right]$$
$$+ \left(1 + \mu^2\left(\beta_1^2 + \beta_2^2\right)\right)N^T\left(\beta_1^2 + \beta_2^2\right) \tag{10.54}$$

Also, the new form of the modified arrays of stiffness matrix for a refined shear deformable nanoplate which is subjected to a thermal loading can be expressed as:

$$k_{33} = -\left(1 + \lambda^2\left(\beta_1^2 + \beta_2^2\right)\right)\left[D_{11}\beta_1^4 + 2\left(D_{12} + 2D_{66}\right)\beta_1^2\beta_2^2 + D_{22}\beta_2^4\right]$$
$$+ \left(1 + \mu^2\left(\beta_1^2 + \beta_2^2\right)\right)N^T\left(\beta_1^2 + \beta_2^2\right),$$
$$k_{34} = -\left(1 + \lambda^2\left(\beta_1^2 + \beta_2^2\right)\right)\left[D_{11}^s\beta_1^4 + 2\left(D_{12}^s + 2D_{66}^s\right)\beta_1^2\beta_2^2 + D_{22}^s\beta_2^4\right]$$
$$+ \left(1 + \mu^2\left(\beta_1^2 + \beta_2^2\right)\right)N^T\left(\beta_1^2 + \beta_2^2\right),$$
$$k_{44} = -\left(1 + \lambda^2\left(\beta_1^2 + \beta_2^2\right)\right)\left[H_{11}^s\beta_1^4 + 2\left(H_{12}^s + 2H_{66}^s\right)\beta_1^2\beta_2^2 + H_{22}^s\beta_2^4\right.$$
$$\left. + A_{44}^s\left(\beta_1^2 + \beta_2^2\right)\right] + \left(1 + \mu^2\left(\beta_1^2 + \beta_2^2\right)\right)N^T\left(\beta_1^2 + \beta_2^2\right) \tag{10.55}$$

10.4 Different Types of Temperature Raise

In this section, the various types of temperature raises will be presented to be employed in the following numerical examples. In this chapter, four types of thermal loading will be studied, namely, uniform temperature raise (UTR), linear temperature raise (LTR), sinusoidal temperature raise (STR) and nonlinear temperature raise (NLTR). The difference between the aforementioned types of thermal loading is in the local temperature profile across the thickness of the nanoplate. In the following, these types of thermal loading will be presented in the mathematical form.

10.4.1 Uniform Temperature Raise (UTR)

Here UTR will be presented to clarify how the final form of this type of thermal loading, N^T, appears in the governing equations and solutions. In this type of temperature raise, the spatial conditions cannot affect the temperature profile, and the local temperature possesses a certain value of T everywhere across the thickness of the nanoplate. Hence, the temperature change will be $T - T_0$, and due to this fact, the thermal loading for beam and plate will be in the following form:

$$N_{\text{nanobeam}}^T = \int_A E(z)\alpha(z)(T - T_0)dA,$$

$$N_{\text{nanoplate}}^T = \int_{-\frac{h}{2}}^{\frac{h}{2}} \frac{E(z)}{1 - \nu(z)}\alpha(z)(T - T_0)dz \tag{10.56}$$

10.4.2 Linear Temperature Raise (LTR)

In the LTR, the local temperature in every point across the thickness is a linear function of the dimensionless thickness, z/h. The mathematical representation of $T(z)$ is as follows:

$$T(z) = T_0 + \Delta T\left(\frac{z}{h} + \frac{1}{2}\right) \tag{10.57}$$

where T_0 is the reference temperature and ΔT is the applied temperature gradient. Following the formula for the thermal loading applied on the beam (see Eq. (10.1)) and plate (see Eq. (10.27)), the expression $T(z) - T_0$ must be obtained. The mentioned term can be obtained in the following form:

$$T(z) - T_0 = \Delta T\left(\frac{z}{h} + \frac{1}{2}\right) \tag{10.58}$$

Therefore, the thermal loadings applied on the nanobeam and nanoplate can be re-written as below:

$$N^T_{\text{nanobeam}} = \int_A E(z)\alpha(z)\Delta T\left(\frac{z}{h} + \frac{1}{2}\right)dA,$$

$$N^T_{\text{nanoplate}} = \int_{-\frac{h}{2}}^{\frac{h}{2}} \frac{E(z)}{1 - \nu(z)}\alpha(z)\Delta T\left(\frac{z}{h} + \frac{1}{2}\right)dz \tag{10.59}$$

10.4.3 Sinusoidal Temperature Raise (STR)

Similar to LTR, in the STR-type thermal loading, the local temperature profile is a function of the dimensionless thickness. In this case, the local temperature possesses a trigonometric term which is a linear function of the dimensionless thickness. In this case, the temperature profile can be expressed as:

$$T(z) = T_0 + \Delta T\left[1 - \cos\left(\frac{z}{h} + \frac{1}{2}\right)\right] \tag{10.60}$$

Using Eq. (10.60), the term $T(z) - T_0$ form for the STR can be derived in the following form:

$$T(z) - T_0 = \Delta T\left[1 - \cos\left(\frac{z}{h} + \frac{1}{2}\right)\right] \tag{10.61}$$

Therefore, the thermal force applied on the nanostructures can be achieved as below:

$$N^T_{\text{nanobeam}} = \int_A E(z)\alpha(z)\Delta T\left[1 - \cos\left(\frac{z}{h} + \frac{1}{2}\right)\right]dA,$$

$$N^T_{\text{nanoplate}} = \int_{-\frac{h}{2}}^{\frac{h}{2}} \frac{E(z)}{1 - \nu(z)}\alpha(z)\Delta T\left[1 - \cos\left(\frac{z}{h} + \frac{1}{2}\right)\right]dz \tag{10.62}$$

10.4.4 Nonlinear Temperature Raise (NLTR)

In this section, we are about to find an expression for the spatial distribution of the local temperature across the thickness of the nanostructures. The nanobeam or nanoplate is assumed to be subjected to a steady-state nonlinear heat conduction. The initial form of the heat conduction problem in the thickness direction can be expressed as:

$$-\frac{d}{dz}\left(\kappa(z, T_0)\frac{dT}{dz}\right) = 0 \tag{10.63}$$

where κ is the thermal conductivity and T_0 is the reference temperature in each desired thickness. The boundary conditions at the top and bottom surfaces of the nanostructures can be written as:

$$T\left(\frac{h}{2}\right) = T_c, T\left(-\frac{h}{2}\right) = T_m \tag{10.64}$$

By solving Eq. (10.63) with boundary values denoted in Eq. (10.64), the following expression can be presented for the local temperature profile in terms of thickness:

$$T(z) = T_m + \Delta T \frac{\int_{-\frac{h}{2}} \frac{1}{\kappa(\,,T_0)} dz}{\int_{-\frac{h}{2}}^{\frac{h}{2}} \frac{1}{\kappa(\,,T_0)} dz} \tag{10.65}$$

where ΔT is equal to $T_c - T_m$. Also, the temperature T_c is presumed to be identical with the reference temperature, T_0. Therefore, the expression $T(z) - T_0$ can be written as:

$$T(z) - T_0 = \Delta T \frac{\int_{-\frac{h}{2}} \frac{1}{\kappa(\,,T_0)} dz}{\int_{-\frac{h}{2}}^{\frac{h}{2}} \frac{1}{\kappa(\,,T_0)} dz} \tag{10.66}$$

Now, the CTE and thermal conductivity of the FGM are required. Thus, these must be reached using following relations according to the Mori-Tanaka homogenization model for the FGMs:

$$\frac{\alpha_e - \alpha_m}{\alpha_c - \alpha_m} = \frac{\frac{1}{\kappa_e} - \frac{1}{\kappa_m}}{\frac{1}{\kappa_c} - \frac{1}{\kappa_m}} \tag{10.67}$$

$$\frac{\kappa_e - \kappa_m}{\kappa_c - \kappa_m} = \frac{V_c}{1 + V_m \frac{(\kappa_c - \kappa_m)}{3\kappa_m}} \tag{10.68}$$

Now, the thermal loading applied on nanobeam and nanoplate can be re-written as follows:

$$N_{\text{nanobeam}}^T = \int_A \left[E(z)\alpha(z)\Delta T \frac{\int_{-\frac{h}{2}} \frac{1}{\kappa(\,,T_0)} dz}{\int_{-\frac{h}{2}}^{\frac{h}{2}} \frac{1}{\kappa(\,,T_0)} dz} \right] dA,$$

$$N_{\text{nanoplate}}^T = \int_{-\frac{h}{2}}^{\frac{h}{2}} \left[\frac{E(z)}{1-\nu(z)} \alpha(z) \Delta T \frac{\int_{-\frac{h}{2}} \frac{1}{\kappa(\,,T_0)} dz}{\int_{-\frac{h}{2}}^{\frac{h}{2}} \frac{1}{\kappa(\,,T_0)} dz} \right] dz \tag{10.69}$$

10.5   Numerical Results and Discussion

In this section, numerical examples will be presented for both nanobeams and nanoplates to illustrate the effects of the thermal environment on the wave dispersion characteristics of FG nanostructures. In the following numerical studies, material properties of the FGM are assumed to be functions of the reference temperature for presenting a more reliable thermomechanical analysis. The material properties of the FGM are the same as those reported in the Ref. [58]. First, the illustrations of FG nanobeams will be studied, and thereafter, similar examples will be depicted for FG nanoplates. In Figure 10.1, the effect of thermal environment on dynamic responses of FG nanobeams is presented plotting the phase velocity curve versus the wave number. The phase velocity of the composite nanobeam experiences a decreasing trend as the temperature increase is assumed to be a nonzero value. Naturally, the higher the temperature raise, the more will be the speed decrease for the waves traveling inside the FG nanobeam. The major reason of this trend is the softening impact of temperature on the stiffness of the constituent material. Indeed, the modulus of the FGM will be decreased while a nonzero value is assigned to the temperature increase. Here, this phenomenon can be better sensed due to the temperature dependency of the material properties. So, it is acceptable to observe such a decreasing behavior once the temperature raise is intensified because of the stiffness reduction, which results in the consequent decrease in the equivalent stiffness of the nanosized beam.

Moreover, the difference between the dispersion curve of FG nanobeams subjected to various types of thermal loading is shown in the Figure 10.2. Among all the types of thermal loading, UTR is more powerful than the others in generating decreases in the wave propagation responses of FG nanobeams. However, setting a comparison between the effect of different types of thermal loading requires paying attention to the range of the wave number. In fact, in small wave numbers, the lowest phase speed belongs to UTR followed by NLTR, LTR and STR, respectively. This trend can be approximately observed in other wave numbers, however, some exceptions can be seen in the Figure 10.2. For instance, in wave numbers greater than $\beta = 0.1$ (1/nm), the difference between the phase velocities under UTR- and NLTR-type thermal loadings is small enough to be dismissed. Similar phenomenon can be observed in the wave numbers greater than $\beta = 2$ (1/nm)

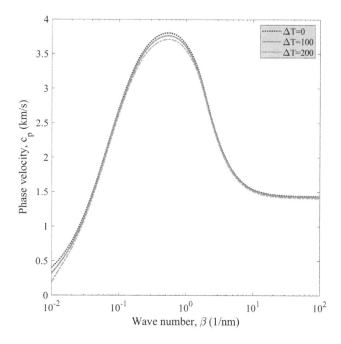

FIGURE 10.1
Variation of phase velocity versus wave number for FG nanobeams subjected to different temperature increases using UTR ($c < 1$, $p = 1$).

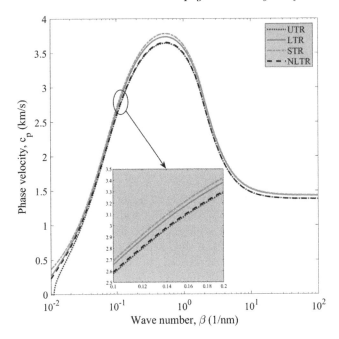

FIGURE 10.2
Illustration of the influence of applying various types of temperature increase on FG nanobeams ($c < 1$, $p = 1$).

for the nanobeams subjected to STR- and LTR-type thermal loadings. Therefore, it can be concluded that once one is tended to reduce the speed of the propagated waves in an FG nanobeam with a UTR, it is better to choose small wave numbers to gain as much as possible from this type of thermal loading.

In Figure 10.3, the effect of thermal environment on the phase speed of the dispersed waves is depicted as well as that of the material distribution parameter. It is again illustrated that, in each value of the gradient index, phase velocity can be decreased when the nanobeam is presumed to be subjected to a thermal loading. This decreasing influence can be seen in the total wave number range and is not dedicated to particular wave numbers. As illustrated in Figure 10.2, the reason for this reduction is the reduction in the stiffness of the FG nanobeam on increasing the temperature raise. Meanwhile, employment of a bigger gradient index will result in decreasing the stiffness of the nanobeam, which causes a remarkable decrease in the equivalent stiffness of the nanobeam, and hence wave speed will be decreased. Similar to the effect of temperature, the influence of gradient index can also be observed in all wave numbers.

Another crucial case study aims to observe the effect of the selected wave number on the attenuation rate of phase speed once the temperature is increased in Figure 10.4. According to this figure, it can be well observed that the diminishing path of the phase velocity critically depends on the value of the wave number in a way that any tiny increase in the wave number results in a huge increment in temperature raise, which can make the speed of the propagated waves to be zero. In fact, as the wave number increases (up to wave number of $\beta = 0.5$ (1/nm)), the magnitude of the phase velocity becomes greater, and this reciprocity makes it harder to completely damp the velocity of the dispersed waves by adding the temperature increase. It must be declared that the presented sentences are valid for the cases where scale factor is assumed to be smaller than one ($c < 1$). In the case of using the Eringen's theory (i.e., zero scale factor or $c = 0$), very big wave numbers will

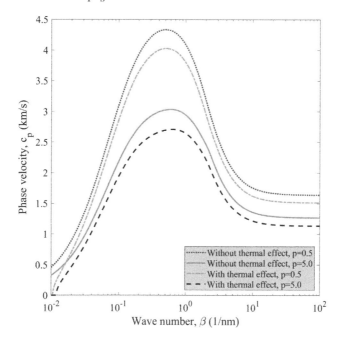

FIGURE 10.3
Coupled effects of thermal environment and gradient index on the phase speed curves of FG nanobeams ($c < 1$).

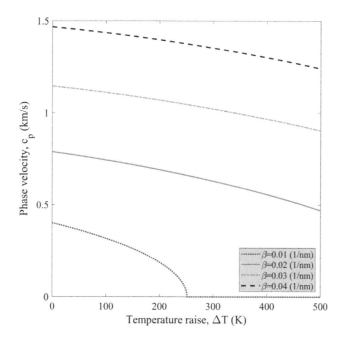

FIGURE 10.4
Variation of the phase velocity of FG nanobeams against temperature increase for various wave numbers ($c < 1$, $p = 1$).

behave the same as very small ones, and the attenuation at the high-range wave numbers will be possible.

On the other hand, the variation of phase speed against temperature increase is plotted in Figure 10.5 to demonstrate how various types of temperature raise can affect the phase velocity of the dispersed waves inside a nanobeam at the wave number of $\beta = 0.01$ (1/nm). In this figure, the findings of Figure 10.2 can be seen once again. It is clear that the UTR-type thermal loading is the most powerful thermal loading among all other types, which can reduce the phase velocity of the propagated waves with a higher rate compared with the others. A large temperature gradient is needed once it is required to reduce the phase speed of the dispersed waves by STR-type thermal loading. Thus, implementation of the STR is impossible in the real world applications due to the huge temperature gradients which is needed in this type of thermal loading to damp the propagated waves.

As the last case study about FG nanobeams, Figure 10.6 shows the effect of different types of thermal loading on the variation of the escape frequency of the nanobeams when the temperature increase is varied from 0 to 500° K. The results of this figure are in accordance with those illustrated in Figure 10.2, where both NLTR- and UTR-types predicted similar behaviors for the high-range wave numbers. It can be again concluded that the implementation of STR-type thermal loading is an improper choice for reducing the dynamic responses of the propagated waves. Because by choosing this type of thermal loading, an applicable temperature raise must be utilized to generate a reduction in the escape frequency. Besides, employing LTR will result in approximately similar behavior, and therefore, the implementation of this type of thermal loading is not recommended. Hence, one may be able to decrease the escape frequency either by applying an UTR or subjecting the nanobeam to a heat conduction-type thermal loading. Again it must be considered that the depicted results are valid for the condition of selecting a scale factor lesser than one, and the reduction rate differs from that of this case.

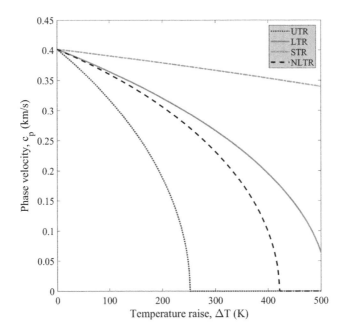

FIGURE 10.5
Variation of the phase velocity of FG nanobeams against temperature increase using different types of thermal loading ($c < 1$, $p = 1$, $\beta = 0.01$ (1/nm)).

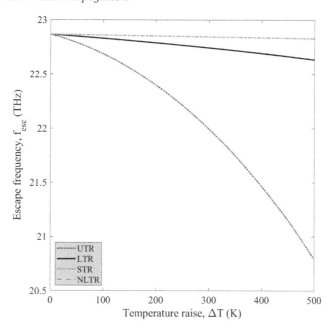

FIGURE 10.6
Variation of the escape frequency of FG nanobeams against temperature increase using different types of thermal loading ($c < 1$, $p = 1$).

Now, the illustrations of FG nanoplates will be depicted to show the influences of thermal environment on the wave dispersion curves of such composite nanostructures. In Figure 10.7, the effect of subjecting the FG nanoplate to a uniform temperature change on the phase speed curve of such nanostructure is depicted. The figure reveals that, in the case of applying a thermal loading on the nanoplate, the phase velocity of the nanoplate decreases. As stated in the previous discussions about the nanobeams, it can be again denoted that the main reason of this phenomenon is that the stiffness of the implemented FGM decreases while the nanoplate is placed in an environment which has a thermal gradient. Thus, it is natural to see a reduction in the speed of dispersed waves due to the direct relation between the material's stiffness with the dynamic responses of the continua. However, it must be considered that as same as nanobeams, the wave propagation speed is not independent from the wave number. In other words, the reduction in the stiffness of the nanoplate can affect the velocity of the elastic waves in a limited range of wave numbers approximately up to wave number of $\beta = 0.05$ (1/nm). Once waves with a very high wave number are employed, one cannot observe any decrease in the velocity of the waves scattered in the media by applying a temperature gradient on the nanosized plate.

Furthermore, the influence of different types of temperature distribution on the velocity curves of dispersed waves inside FG nanoplates is illustrated in Figure 10.8. According to the figure, it can be seen that the most critical reduction in the wave propagation responses of FG nanoplates occures in the UTR-type thermal loading followed by LTR, NLTR and STR. It is worth mentioning that there is no special difference between the LTR and NLTR type, and the dispersion curve branches corresponding with these two types are the same all over the wave number range. Again, it is clear that the wave propagation characteristics cannot be influenced by thermal environment in the case of utilizing high wave numbers. Hence, the designers of continuous nanostructures must consider the effects of UTR-type thermal loading more than the other types.

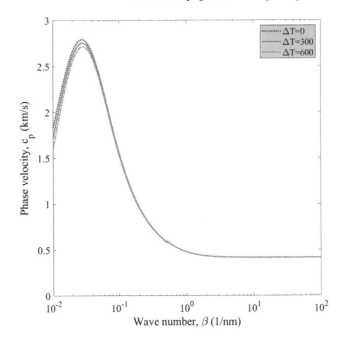

FIGURE 10.7
Variation of phase velocity versus wave number for FG nanoplates subjected to different temperature increases using UTR ($c < 1$, $p = 1$).

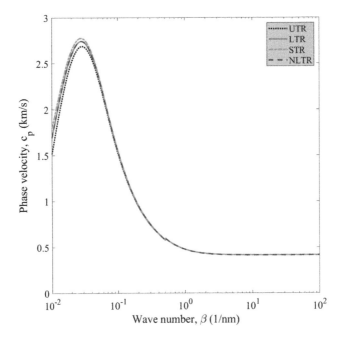

FIGURE 10.8
Illustration of the influence of applying various types of temperature increase on FG nanoplates ($c < 1$, $p = 1$).

Another case study refers to the investigation of the mixed impact of thermal environment and material's distribution parameter on the wave propagation curves of FG nanoplates by drawing the variation of phase speed against wave number in Figure 10.9. It can be concluded that the phase velocity curves will be shifted downward in the case of utilizing a high gradient index. The physical reason for this phenomenon is the lower stiffness of the constituent material in such cases in comparison to cases where small gradient indices are employed. Also, similar to the former illustrations, the dispersion responses of the FG nanoplates will be reduced once the effects of thermal gradient are included because of the predictable stiffness softening phenomenon, which happens in the FGM with temperature increase. Hence, it is better to disperse elastic waves in FG nanoplates isolated from a thermal environment to achieve a large wave velocity.

As expressed in the previous figures, the mechanical behaviors of the waves dispersed in the FG nanostructures depends on the range of wave number which is studied in a critical manner. Therefore, to emphasize on the incredible role of wave number on the wave propagation responses of FG nanoplates, Figure 10.10 is devoted to survey this issue in the framework of a plot drawing phase velocity versus temperature raise when the nanoplate is considered to be subjected to UTR. Obviously, it can be perceived that the phase velocity of FG nanoplates reduce with the increase in temperature continuously. It is clear that if wave number is assumed to be 0.02 (1/nm) instead of 0.01 (1/nm), the amount of temperature increase required for damping the velocity will be critically increased and reaches a value that cannot be applied in the real world applications. Thus, it is important to account for the impact of wave number once it tends to damp the wave dispersion characteristics of FG nanoplates by seating the nanoplate in a thermal environment.

Finally, the effect of various types of temperature gradient on the wave propagation speed inside FG nanoplates is presented in Figure 10.11 by plotting the variation of phase

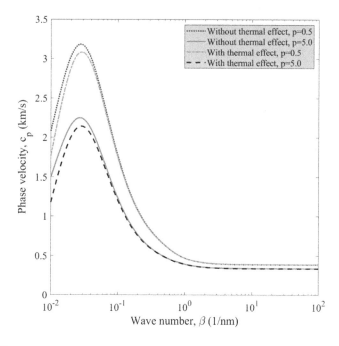

FIGURE 10.9
Coupled effects of thermal environment and gradient index on the phase speed curves of FG nanoplates ($c < 1$).

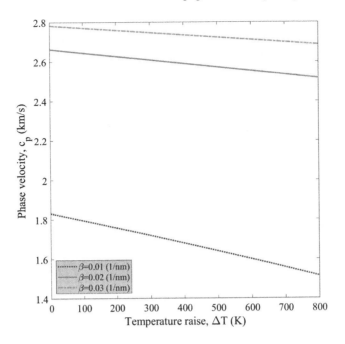

FIGURE 10.10
Variation of the phase velocity of FG nanoplates against temperature increase for various wave numbers ($c < 1$, $p = 1$).

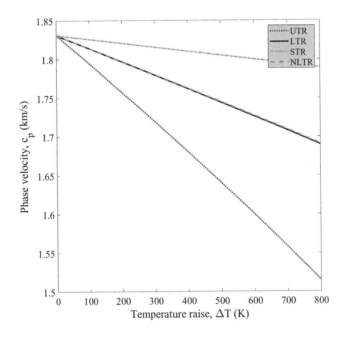

FIGURE 10.11
Variation of the phase velocity of FG nanoplates against temperature raise using different types of thermal loading ($c < 1$, $p = 1$, $\beta = 0.01$ (1/nm)).

speed against temperature increase in the wave number of $\beta = 0.01$ (1/nm). As stated earlier, the most critical condition can be reported when UTR distribution is selected. In other words, dependency of the propagation speed to the temperature can be most sensed when the nanoplate is subjected to a uniform distribution of temperature change among the nanostructure's thickness. Next, LTR and NLTR types of thermal loading can affect the wave velocity in a critical way followed by negligible influence of the STR. Indeed, the LTR and NLTR-type thermal loadings possess identical influences on the wave propagation responses of FG nanoplates. It is observable that, even by adding $800°$ K to the initial temperature of the environment, the nanoplate's wave velocity cannot be achieved to a certain zero value. Therefore, in an indirect way, the effect of wave number is stated in this diagram once again.

To conclude the presented results, it must be mentioned that the wave propagation characteristics of FG nanobeams and nanoplates are too sensitive to the variations of the wave number. In fact, the effects of involved variants, here the temperature raise can be seen only in a particular range of wave numbers, and after such a range, the previously observed trends cannot be seen anymore. In general, the wave propagation responses of FG nanostructures will be reduced while the effects of thermal environment are included. Also, it must be mentioned that the efficiency range of the wave number can be varied if the thickness of the nanostructure is changed. In fact, the higher the nanostructures' thickness is, the bigger will be the efficiency range of the wave numbers that the effect of thermal environment can be observed in it. Hence, all the abovementioned results must be considered as case studies obtained via a particular amount of the nanostructure's thickness. So, such an efficiency range may be invalid for another nanobeam or nanoplate with a different thickness from that implemented in the present study. However, the trends and qualitative results are the same for every thickness value.

11

Magnetic Field Effects on Wave Propagation Characteristics of Smart Nanostructures

In this chapter, the effects of the presence of external magnetic field on the wave propagation characteristics of functionally graded (FG) nanostructures will be investigated to show how the induced force generated by this field can intensify the dispersion curves of FG nanosized beams and plates. For this, again the dynamic form of the principle of virtual work will be extended for nanostructures in the presence of the external work done by the induced magnetic force. To consider the induced magnetic force in continuum Maxwell's relations will be expressed for beam-type and plate-type nanostructures. The size dependency issue will be covered in this chapter similar to previous ones on the basis of the nonlocal strain gradient elasticity. The presented linear dispersion analysis will be solved using the well-known analytical solution of waves scattered in the nanosized continuous systems. The presented numerical case studies reveal that the effect of external magnetic field can be observed in small wave numbers similar to that of the thermal environment.

11.1 Introduction

As mentioned in previous chapters, industrial and engineering designs have moved toward implementation of multi-task devices to satisfy various purposes. This trend can be better sensed in the field of aerospace, nuclear and mechanics [172,173]. In many industrial applications, mechanical devices are placed in circumstances where an external applied electric and/or magnetic field exists. On the other hand, it is known that an induced magnetic force will appear in any elastic media which is supposed to be in a magnetically influenced environment. Indeed, the Maxwell's relation can justify how this induced force can be generated in the structure. Because of this and the widespread application of mechanical macro and/or nanodevices in industries dealing with the magnetically affected environments, it is important to consider the mechanical behaviors of structural elements subjected to magnetic field. Therefore, some researchers devoted a portion of their scientific activities to analyze the magnetoelastic responses of macro- and nano-scale mechanical elements. In one of the studies, Lee et al. [174] presented a powerful finite element (FE) solution for the transient responses of conducting plates under the action of the well-known Lorentz magnetic force. They could derive the time-dependent responses of the simulated fusion reactor in the framework of the modified Newton-Raphson method. Zheng et al. [173] procured a finite element method (FEM)-based study on the mechanical deflection behaviors of conductive plates in the presence of induced magnetic Lorentz force. They assumed the plate to be subjected to a time-varying magnetic field and derived the relations for the induced force by capturing the interaction between the magnetic field and Eddy current in continuum. The influence of externally induced Lorentz force is included in a dynamic analysis performed by Kiani [175], as well as that of eddy current to study the mechanical responses of nanosize wires which can be employed as conductors. On the basis of nonlocal elasticity theory of Eringen, Murmu et al. [176] carried out a natural frequency analysis on the double-walled carbon nanotubes (CNTs) including the influences of the external magnetic field. In another investigation, Murmu et al. [177] studied the dynamic characteristics of nanosize plates fabricated from graphene when an induced Lorentz force is applied on the nanoplate. They considered effects of the small size of the nanostructure by extending the differential nonlocal relations of Eringen. In addition, Kiani [178] presented another study dealing with the effects of three-dimensionally applied magnetic field on the frequency and stability responses of CNTs. The same author solved the vibration problem of nanosize plates subjected to an external magnetic field in the framework of the nonlocal hypothesis incorporated with various plate theories [179]. In this paper, both in- and out-of-plane fluctuations of the nanoplate were considered as case studies. Focusing on the effects of external magnetic field on the dynamic responses of nanoplates, Karličić et al. [180] probed the vibrational characteristics of nanoscale plates subjected to a moving mass for showing their application as tiny mass sensors. In this study, they utilized mono-layered graphene as the nanostructure. Besides, another frequency analysis was procured by Karličić et al. [181] to survey the effect of a flexural external magnetic field on the longitudinal oscillations of a series of nanosize rods placed in a magnetically affected environment. The effects of clamped and free edge supports on the dynamic responses of the nanosize rod chain were included in this investigation. Furthermore, Ghorbanpour Arani et al. [182] mixed the nonlocal constitutive equations of Eringen with the Euler–Lagrange equations of Reddy plates to analyze the magnetically influenced vibrational responses of nanoplates with moving edge conditions. Incorporating the effects of the magnetic field with the concept of the optimized FG composites, Ebrahimi and Barati [183] investigated the wave propagation problem inside nanosize beams. In this research, the equivalent material properties of the nanobeam were obtained on the basis of a dual power-law model, called sigmoid model. Moreover, the influence of the externally induced Lorentz force is considered by Ebrahimi and Barati [184] in another study dealing with the dispersion responses of FG single- and double-layered nanosize beams employing the sigmoid homogenization scheme. In this analysis, nanobeams are considered to be connected via elastic springs when probing the dispersion responses of dual nanobeams. Impacts of surface-to-volume ratio of nanosize elements were included in an investigation by Ebrahimi and Barati [185] dealing with natural frequency behaviors of rotary nanostructures subjected to an axial magnetic field. Another study dealing with the magnetically influenced propagation of flexural waves in the nanostructures was performed by Karami et al. [186], combining the crucial influences of temperature and the existence of porosity in the media on the dispersion behaviors of waves scattering in nanoplates made from FGMs. On the other hand, the issue of post-buckling problem in a nanosize beam was studied by Dai et al. [187] by considering the effects of externally induced magnetic force applied on the nanosize element. In this paper, the effects of nonlinear strain-displacement relation and thickness-stretching were also covered. The nonlocal strain gradient microstructure-dependent elasticity hypothesis was implemented by Ebrahimi and Dabbagh [188] probing the wave dispersion responses of nanosize double-layered graphenes whenever the system is subjected to Lorentz force. The coupling between the two layers of the nanoplate was modeled by utilizing elastic springs. In an other study, Jalaei and Ghorbanpour Arani [189] utilized the nonlocal elasticity coupled with the well-known Kelvin-Voigt viscoelastic model to consider for the transient bending characteristics of graphene sheets rested on a viscoelastic substrate when the continuous system is subjected to an external magnetic field, which produces an induced Lorentz force in the nanostructure.

11.2 Maxwell's Relations

In this section, the magnetic induction relations proposed by Maxwell will be presented to show the effect of an external magnetic field on a desired structure. First, the general form of these relations will be depicted, and afterward, the application of such relations for beam-type and plate-type elements will be shown. It is better to define the involved variants before beginning the derivation procedure. In the what follows, ρ_b, E, H, J_e, B and D are static charge density, electric field vector, magnetic field intensity, Eddy current density, magnetic field density and displacement current density, respectively. Now, the Maxwell's relations can be expressed in the following mathematical form:

$$\nabla \times H = \dot{D} + J_e, \nabla \times E = -\dot{B},$$
$$\nabla D = \rho_b, \nabla B = 0,$$
$$D = \epsilon E, B = \eta H \tag{11.1}$$

In the above equation, ϵ and η denote the electric and magnetic permeability of the material, respectively. It must be considered that these parameters are two simple scalars whenever the material is assumed to be isotropic. Assuming the displacement current density (D) and its derivative with respect to time $(\dot{D})$ to be zero, the electric and magnetic fields can be expressed in the below simplified form:

$$E = E_0 + e, H = H_0 + h \tag{11.2}$$

where $e(u,t)$ and $h(u,t)$ are small disturbances of the primary electromagnetic fields, E_0 and H_0, respectively, applied on the material. It is worth mentioning that u stands for the displacement field vector utilized to model the spatial behaviors of the structure. In other words, this vector equals $u = (u_x, 0, u)$ for a beam and $u = (u_x, u_y, u)$ for a plate. In this chapter, we aim to enrich the induced Lorentz force applied on the nanosize structures, hence, the effect of initial electric field is dismissed $(E_0 = 0)$. Utilizing above assumptions and revising Eq. (11.1), the following relations can be obtained:

$$\nabla \times h = J_e, \nabla \times e = -\eta \frac{\partial h}{\partial t},$$
$$\nabla h = 0, e = -\eta \left(\frac{\partial u}{\partial t} \times H_0 \right),$$
$$h = \nabla \times (u \times H_0) \tag{11.3}$$

Now, h must be calculated from the above relations. As mentioned before, the displacement field vectors of beams and plates are different. Here, we will use the displacement field of the plates and derive the equations for plate-type elements and in the following steps, we will show how to transform such equations for beams. In this study, we consider the magnetic field to be longitudinal, hence, the vector H_0 can be expressed in the following form:

$$H_0 = (H_x, 0, 0) \tag{11.4}$$

where H_x is the only nonzero component of the magnetic field which is parallel to the $x-$axis. Once the magnetic field introduced in Eq. (11.4) and the displacement field vector (u_x, u_y, u) is inserted in Eq. 11.3, one can obtain the below relation:

$$h = -H_x \left(\frac{\partial u_y}{\partial y} + \frac{\partial u}{\partial z} \right) \hat{e}_i + H_x \frac{\partial u_y}{\partial x} \hat{e}_j + H_x \frac{\partial u}{\partial x} \hat{e}_k \tag{11.5}$$

In the above equation, $\hat{e}_i$, $\hat{e}_j$ and $\hat{e}_k$ correspond with the unit vectors in the x, y and z directions, respectively. Finally, the induced Lorentz force applied on the structure because of the presence of magnetic field can be enriched in the following form:

$$
\begin{aligned}
\boldsymbol{f} &= \eta\big(\boldsymbol{J}_e \times \boldsymbol{H}\big) \\
&= \eta\left(0\hat{e}_i + H_x^2\left[\frac{\partial^2 u_y}{\partial x^2} + \frac{\partial^2 u_y}{\partial y^2} + \frac{\partial^2 u}{\partial y \partial z}\right]\hat{e}_j + H_x^2\left[\frac{\partial^2 u}{\partial x^2} + \frac{\partial^2 u}{\partial y^2} + \frac{\partial^2 u_y}{\partial y \partial z}\right]\hat{e}_k\right)
\end{aligned} \tag{11.6}
$$

It must be mentioned that the aforementioned force is a body force and is obtained per volume; therefore, it must be integrated over the volume of the structure to be transferred to the real value of the Lorentz force applied on the element. The expression presented in Eq. (11.6) is the total force applied on a plate; however, in this study the effects of the in-plane components of this loading will be dismissed. Therefore, the equivalent Lorentz force applied on the plate in the z−axis can be written as:

$$
f_L = \eta H_x^2\left[\frac{\partial^2 u}{\partial x^2} + \frac{\partial^2 u}{\partial y^2} + \frac{\partial^2 u_y}{\partial y \partial z}\right] \tag{11.7}
$$

On the other hand, the differentiation with respect to the independent variable z are ignored. So, the Lorentz force can be re-written as:

$$
f_L = \eta H_x^2\left[\frac{\partial^2 u}{\partial x^2} + \frac{\partial^2 u}{\partial y^2}\right] \tag{11.8}
$$

The above equation reveals the Lorentz force applied on a plate-type element. The induced force applied on a beam-type element can be achieved by simplifying the above equation. In fact, the displacement field of a beam is independent from the y variable. Hence, the induced force for a beam can be expressed as:

$$
f_L = \eta H_x^2 \frac{\partial^2 u}{\partial x^2} \tag{11.9}
$$

11.3   Analysis of FG Nanobeams

In this section, the effects of the external Lorentz force induced in the nanosize FG beams will be discussed. Indeed, the influence of the applied Lorentz force will appear in the variation of work done by external loadings. Therefore, it is necessary to calculate this term to obtain the equations of motion for FG nanobeams when the nanobeam is assumed to be placed in a magnetic environment. In what follows, the variation of work done by Lorentz force will be derived for both Euler–Bernoulli and refined sinusoidal nanoscale beams.

As stated in section 11.2, the relation obtained in Eq. (11.9) is a body force and must be integrated over the volume of the nanobeam to be applied on the Euler–Lagrange equations of the beams. The integration over volume must be divided in two integrations over the length and cross-section area of the beam, respectively. The first one (i.e., integration over the longitudinal direction) will be satisfied in the variation of work done by external forces. So, the integration over the cross-section area of the beam must be performed before calculating the variation of work done by this loading. Thus, the total Lorentz force applied on the beam can be defined in the following integral form:

$$
F_L = \int_A f_L \, dA = \eta A H_x^2 \frac{\partial^2 u}{\partial x^2} \tag{11.10}
$$

Now, the variation of the work done by this force can be calculated as Eq. (6.6) for both Euler–Bernoulli and refined beams. Using the displacement field of Euler–Bernoulli beams, introduced in Eq. (4.10), the variation of work done by Lorentz force can be expressed by:

$$\delta V = \int_0^L F_L \frac{\partial w}{\partial x} \frac{\partial \delta w}{\partial x} dx \tag{11.11}$$

Replacing Eq. (4.10) with Eq. (4.15), the variation of work done by the induced Lorentz force applied on the refined shear deformable beams can be defined in the following form:

$$\delta V = \int_0^L F_L \frac{\partial (w_b + w_s)}{\partial x} \frac{\partial \delta (w_b + w_s)}{\partial x} dx \tag{11.12}$$

Now, it is time to derive the Euler–Lagrange equations of FG beams in the presence of the applied Lorentz force. First, the motion equations of FG Euler–Bernoulli beams will be derived. For this, Eqs. (4.24), (4.26) and (11.11) must be inserted in the definition of the Hamilton's principle stated in Eq. (4.19). Performing the aforementioned substitution, the Euler–Lagrange equations of FG beams can be presented in the following form:

$$\frac{\partial N}{\partial x} = I_0 \frac{\partial^2 u}{\partial t^2} - I_1 \frac{\partial^3 w}{\partial x \partial t^2} \tag{11.13}$$

$$\frac{\partial^2 M}{\partial x^2} - F_L \frac{\partial^2 w}{\partial x^2} = I_0 \frac{\partial^2 w}{\partial t^2} + I_1 \frac{\partial^3 u}{\partial x \partial t^2} - I_2 \frac{\partial^4 w}{\partial x^2 \partial t^2} \tag{11.14}$$

Now, the same substitution must be procured for FG refined shear deformable beams. In this case, Eqs. (4.31), (4.33) and (11.12) should be inserted in Eq. (4.19). After the substitution, the Euler–Lagrange equations of the beam can be expressed in the following form based on the refined sinusoidal beam hypothesis:

$$\frac{\partial N}{\partial x} = I_0 \frac{\partial^2 u}{\partial t^2} - I_1 \frac{\partial^3 w_b}{\partial x \partial t^2} - J_1 \frac{\partial^3 w_s}{\partial x \partial t^2} \tag{11.15}$$

$$\frac{\partial^2 M^b}{\partial x^2} - F_L \frac{\partial^2 (w_b + w_s)}{\partial x^2} = I_0 \frac{\partial^2 (w_b + w_s)}{\partial t^2} + I_1 \frac{\partial^3 u}{\partial x \partial t^2} - I_2 \frac{\partial^4 w_b}{\partial x^2 \partial t^2} - J_2 \frac{\partial^4 w_s}{\partial x^2 \partial t^2} \tag{11.16}$$

$$\frac{\partial^2 M^s}{\partial x^2} + \frac{\partial Q_x}{\partial x} - F_L \frac{\partial^2 (w_b + w_s)}{\partial x^2} = I_0 \frac{\partial^2 (w_b + w_s)}{\partial t^2} + J_1 \frac{\partial^3 u}{\partial x \partial t^2} - J_2 \frac{\partial^4 w_b}{\partial x^2 \partial t^2} - K_2 \frac{\partial^4 w_s}{\partial x^2 \partial t^2} \tag{11.17}$$

Next, the influence of small scale on the Euler–Lagrange equations of the beam must be applied. For investigating classical nanobeams, Eqs. (4.38) and (4.39) must be inserted in Eqs. (11.13) and (11.14). Once the aforementioned substitution is performed, the governing equations of FG nanobeams subjected to Lorentz force can be written in the following form:

$$\left(1 - \lambda^2 \nabla^2\right) \left[A_{xx} \frac{\partial^2 u}{\partial x^2} - B_{xx} \frac{\partial^3 w}{\partial x^3} \right] + \left(1 - \mu^2 \nabla^2\right) \left[-I_0 \frac{\partial^2 u}{\partial t^2} + I_1 \frac{\partial^3 w}{\partial x \partial t^2} \right] = 0 \tag{11.18}$$

$$\left(1 - \lambda^2 \nabla^2\right) \left[B_{xx} \frac{\partial^3 u}{\partial x^3} - D_{xx} \frac{\partial^4 w}{\partial x^4} \right] + \left(1 - \mu^2 \nabla^2\right)$$
$$\times \left[F_L \frac{\partial^2 w}{\partial x^2} - I_0 \frac{\partial^2 w}{\partial t^2} - I_1 \frac{\partial^3 u}{\partial x \partial t^2} + I_2 \frac{\partial^4 w}{\partial x^2 \partial t^2} \right] = 0 \tag{11.19}$$

Doing the same mathematical operations, the governing equations of higher-order FG nanobeams can be achieved by substituting for N, M^b, M^s and Q_x from Eqs. (4.40)–(4.43) in Eqs. (11.15)–(11.17) in the following form:

$$
\left(1 - \lambda^2 \nabla^2\right) \left[A_{xx} \frac{\partial^2 u}{\partial x^2} - B_{xx} \frac{\partial^3 w_b}{\partial x^3} - B_{xx}^s \frac{\partial^3 w_s}{\partial x^3} \right] + \left(1 - \mu^2 \nabla^2\right)
$$

$$
\times \left[-I_0 \frac{\partial^2 u}{\partial t^2} + I_1 \frac{\partial^3 w_b}{\partial x \partial t^2} + J_1 \frac{\partial^3 w_s}{\partial x \partial t^2} \right] = 0
\tag{11.20}
$$

$$
\left(1 - \lambda^2 \nabla^2\right) \left[B_{xx} \frac{\partial^3 u}{\partial x^3} - D_{xx} \frac{\partial^4 w_b}{\partial x^4} - D_{xx}^s \frac{\partial^4 w_s}{\partial x^4} \right] + \left(1 - \mu^2 \nabla^2\right)
$$

$$
\times \left[F_L \frac{\partial^2 (w_b + w_s)}{\partial x^2} - I_0 \frac{\partial^2 (w_b + w_s)}{\partial t^2} - I_1 \frac{\partial^3 u}{\partial x \partial t^2} + I_2 \frac{\partial^4 w_b}{\partial x^2 \partial t^2} + J_2 \frac{\partial^4 w_s}{\partial x^2 \partial t^2} \right] = 0
\tag{11.21}
$$

$$
\left(1 - \lambda^2 \nabla^2\right) \left[B_{xx}^s \frac{\partial^3 u}{\partial x^3} - D_{xx}^s \frac{\partial^4 w_b}{\partial x^4} - H_{xx}^s \frac{\partial^4 w_s}{\partial x^4} + A_x^s \frac{\partial^2 w_s}{\partial x^2} \right] + \left(1 - \mu^2 \nabla^2\right)
$$

$$
\times \left[F_L \frac{\partial^2 (w_b + w_s)}{\partial x^2} - I_0 \frac{\partial^2 (w_b + w_s)}{\partial t^2} - J_1 \frac{\partial^3 u}{\partial x \partial t^2} + J_2 \frac{\partial^4 w_b}{\partial x^2 \partial t^2} + K_2 \frac{\partial^4 w_s}{\partial x^2 \partial t^2} \right] = 0
\tag{11.22}
$$

The next step is to solve the obtained governing equations. As former chapters, the well-known analytical solution, introduced in Chapter 1, will be inserted in the above set of equations to determine the mechanical frequency and phase speed of the propagated waves in the presence of Lorentz force. As can be figured out from the governing equations for both Euler–Bernoulli and higher-order beams, the components of mass matrix are the same as those reported in Eqs. (4.52) and (4.54) for Euler–Bernoulli and sinusoidal nanobeams, respectively. The only difference belongs to those components of the stiffness matrix which are totally involved with the deflection of the nanobeam. As stated in previous chapters, such components are k_{22} for Euler–Bernoulli nanobeam and k_{22}, k_{23} and k_{33} for higher-order nanobeams. Thus, the k_{22} array of Eq. (4.51) must be modified in the following form for a classical nanobeam:

$$
k_{22} = -\left(1 + \lambda^2 \beta^2\right) D_{xx} \beta^4 - \left(1 + \mu^2 \beta^2\right) F_L \ \beta^2
\tag{11.23}
$$

Moreover, the k_{22}, k_{23} and k_{33} arrays of the Eq. (4.53) must be re-written as:

$$
k_{22} = -\left(1 + \lambda^2 \beta^2\right) D_{xx} \beta^4 - \left(1 + \mu^2 \beta^2\right) F_L \ \beta^2,
$$

$$
k_{23} = -\left(1 + \lambda^2 \beta^2\right) D_{xx}^s \beta^4 - \left(1 + \mu^2 \beta^2\right) F_L \ \beta^2,
$$

$$
k_{33} = -\left(1 + \lambda^2 \beta^2\right) \left[H_{xx}^s \beta^4 + A_x^s \ \beta^2 \right] - \left(1 + \mu^2 \beta^2\right) F_L \ \beta^2
\tag{11.24}
$$

Once the aforementioned modifications were applied on the stiffness matrices of Euler–Bernoulli and higher-order nanobeams, the dispersion responses of the mentioned nanostructure can be obtained by solving the equivalent eigenvalue equation of the problem. In the next section, we will derive and solve the wave propagation problem of FG nanoplates, and after that the numerical results of both nanobeam- and nanoplate-type elements will be discussed.

11.4 Analysis of FG Nanoplates

The present section will present a mathematical framework for the wave propagation problem of FG nanoplates when the nanoplate is assumed to be subjected to Lorentz body force. The formulations for both Kirchhoff–Love and refined higher-order plates will be derived followed by applying the effects of small scale on the basis of the nonlocal strain gradient theory of elasticity. Again it must be mentioned that the Lorentz force derived in Section 11.2 is a body force and must be integrated over the element's volume to be applied on the nanoplate. However, in this section, type of integration differs from that presented in the former section because of the different nature of plate-type elements in comparison with the beam-type ones. In fact, for obtaining the motion equations of plates, this integration must be decomposed in two integrals, first on the thickness of the plate (i.e., integration with respect to z independent variable) and second on the area (i.e., dual integration with respect to x and y independent variables).

Using the primary definition stated in Eq. (11.8), and considering the displacement field of classical plates (see Eq. (4.58)), the Lorentz force which must be included in the computations for a Kirchhoff–Love plate can be expressed in the following form:

$$F_L = \int_{-\frac{h}{2}}^{\frac{h}{2}} f_L \, dz = \eta h H_x^2 \left(\frac{\partial^2 w}{\partial x^2} + \frac{\partial^2 w}{\partial y^2} \right) \tag{11.25}$$

Replacing Eq. (4.58) with Eq. (4.65), the Lorentz force applied on refined sinusoidal plates can be formulated in the following form:

$$F_L = \int_{-\frac{h}{2}}^{\frac{h}{2}} f_L \, dz = \eta h H_x^2 \left(\frac{\partial^2 (w_b + w_s)}{\partial x^2} + \frac{\partial^2 (w_b + w_s)}{\partial y^2} \right) \tag{11.26}$$

Now, the variation of work done by external Lorentz force for a classical plate can be expressed in the following form:

$$\delta V = \int_0^a \int_0^b F_L \left(\frac{\partial w}{\partial x} \frac{\partial \delta w}{\partial x} + \frac{\partial w}{\partial y} \frac{\partial \delta w}{\partial y} \right) dy dx \tag{11.27}$$

For refined sinusoidal plates, we have:

$$\delta V = \int_0^a \int_0^b F_L \left(\frac{\partial (w_b + w_s)}{\partial x} \frac{\partial \delta (w_b + w_s)}{\partial x} + \frac{\partial (w_b + w_s)}{\partial y} \frac{\partial \delta (w_b + w_s)}{\partial y} \right) dy dx \tag{11.28}$$

Now, the motion equations of FG plates can be derived using the dynamic form of the principle of virtual work. Considering the Kirchhoff–Love plates, the Euler–Lagrange equations can be achieved utilizing Eqs. (4.72), (4.74) and (11.27) in association with Eq. (4.19). Completing the substitution process, the following Euler–Lagrange equations can be enriched for classical plates:

$$\frac{\partial N_{xx}}{\partial x} + \frac{\partial N_{xy}}{\partial y} = I_0 \frac{\partial^2 u}{\partial t^2} - I_1 \frac{\partial^3 w}{\partial x \partial t^2} \tag{11.29}$$

$$\frac{\partial N_{xy}}{\partial x} + \frac{\partial N_{yy}}{\partial y} = I_0 \frac{\partial^2 v}{\partial t^2} - I_1 \frac{\partial^3 w}{\partial y \partial t^2} \tag{11.30}$$

$$\frac{\partial^2 M_{xx}}{\partial x^2} + 2 \frac{\partial^2 M_{xy}}{\partial x \partial y} + \frac{\partial^2 M_{yy}}{\partial y^2} - F_L \left(\frac{\partial^2 w}{\partial x^2} + \frac{\partial^2 w}{\partial y^2} \right)$$
$$= I_0 \frac{\partial^2 w}{\partial t^2} + I_1 \left(\frac{\partial^3 u}{\partial x \partial t^2} + \frac{\partial^3 v}{\partial y \partial t^2} \right) - I_2 \left(\frac{\partial^4 w}{\partial x^2 \partial t^2} + \frac{\partial^4 w}{\partial y^2 \partial t^2} \right) \tag{11.31}$$

Furthermore, the Euler–Lagrange equations of refined sinusoidal plates can be obtained once Eqs. (4.80), (4.83) and (11.28) are inserted in the definition of the Hamilton's principle, stated in Eq. (4.19). Accordingly, the motion equations of FG plates in the presence of the effects of shear deflection and magnetic loading can be expressed as:

$$\frac{\partial N_{xx}}{\partial x} + \frac{\partial N_{xy}}{\partial y} = I_0 \frac{\partial^2 u}{\partial t^2} - I_1 \frac{\partial^3 w_b}{\partial x \partial t^2} - J_1 \frac{\partial^3 w_s}{\partial x \partial t^2} \tag{11.32}$$

$$\frac{\partial N_{xy}}{\partial x} + \frac{\partial N_{yy}}{\partial y} = I_0 \frac{\partial^2 v}{\partial t^2} - I_1 \frac{\partial^3 w_b}{\partial y \partial t^2} - J_1 \frac{\partial^3 w_s}{\partial y \partial t^2} \tag{11.33}$$

$$\frac{\partial^2 M_{xx}^b}{\partial x^2} + 2\frac{\partial^2 M_{xy}^b}{\partial x \partial y} + \frac{\partial^2 M_{yy}^b}{\partial y^2} - F_L \left(\frac{\partial^2 (w_b + w_s)}{\partial x^2} + \frac{\partial^2 (w_b + w_s)}{\partial y^2} \right) = I_0 \frac{\partial^2 (w_b + w_s)}{\partial t^2}$$
$$+ I_1 \left(\frac{\partial^3 u}{\partial x \partial t^2} + \frac{\partial^3 v}{\partial y \partial t^2} \right) - I_2 \left(\frac{\partial^4 w_b}{\partial x^2 \partial t^2} + \frac{\partial^4 w_b}{\partial y^2 \partial t^2} \right) - J_2 \left(\frac{\partial^4 w_s}{\partial x^2 \partial t^2} + \frac{\partial^4 w_s}{\partial y^2 \partial t^2} \right) \tag{11.34}$$

$$\frac{\partial^2 M_{xx}^s}{\partial x^2} + 2\frac{\partial^2 M_{xy}^s}{\partial x \partial y} + \frac{\partial^2 M_{yy}^s}{\partial y^2} + \frac{\partial Q_x}{\partial x} + \frac{\partial Q_y}{\partial y} - F_L \left(\frac{\partial^2 (w_b + w_s)}{\partial x^2} + \frac{\partial^2 (w_b + w_s)}{\partial y^2} \right)$$
$$= I_0 \frac{\partial^2 (w_b + w_s)}{\partial t^2} + J_1 \left(\frac{\partial^3 u}{\partial x \partial t^2} + \frac{\partial^3 v}{\partial y \partial t^2} \right) - J_2 \left(\frac{\partial^4 w_b}{\partial x^2 \partial t^2} + \frac{\partial^4 w_b}{\partial y^2 \partial t^2} \right)$$
$$- K_2 \left(\frac{\partial^4 w_s}{\partial x^2 \partial t^2} + \frac{\partial^4 w_s}{\partial y^2 \partial t^2} \right) \tag{11.35}$$

Afterward, the above equations must be replaced with the governing equations in terms of displacement field of the nanoplate to include the effects of small scale on the mechanical responses of waves scattered in FG nanoplates. For this, the stress resultants of nanoplates, achieved in the framework of the nonlocal stress-strain gradient elasticity, should be substituted in the obtained Euler–Lagrange relations. Hence, Eqs. (4.89) and (4.90) must be substituted in Eqs. (11.29)–(11.31) to derive the governing equations of FG classical nanoplates subjected to Lorentz force. Based upon the above procedure, the governing equations of FG nanoplates can be written as:

$$\left(1 - \lambda^2 \nabla^2\right) \left[A_{11} \frac{\partial^2 u}{\partial x^2} + (A_{12} + A_{66}) \frac{\partial^2 v}{\partial x \partial y} + A_{66} \frac{\partial^2 u}{\partial y^2} - B_{11} \frac{\partial^3 w}{\partial x^3} - (B_{12} + 2B_{66}) \frac{\partial^3 w}{\partial x \partial y^2} \right]$$
$$+ \left(1 - \mu^2 \nabla^2\right) \left[-I_0 \frac{\partial^2 u}{\partial t^2} + I_1 \frac{\partial^3 w}{\partial x \partial t^2} \right] = 0 \tag{11.36}$$

$$\left(1 - \lambda^2 \nabla^2\right) \left[A_{22} \frac{\partial^2 v}{\partial y^2} + (A_{12} + A_{66}) \frac{\partial^2 u}{\partial x \partial y} + A_{66} \frac{\partial^2 v}{\partial x^2} - B_{22} \frac{\partial^3 w}{\partial y^3} - (B_{12} + 2B_{66}) \frac{\partial^3 w}{\partial x^2 \partial y} \right]$$
$$+ \left(1 - \mu^2 \nabla^2\right) \left[-I_0 \frac{\partial^2 v}{\partial t^2} + I_1 \frac{\partial^3 w}{\partial y \partial t^2} \right] = 0 \tag{11.37}$$

$$\left(1 - \lambda^2 \nabla^2\right) \left[B_{11} \frac{\partial^3 u}{\partial x^3} + (B_{12} + 2B_{66}) \frac{\partial^3 u}{\partial x \partial y^2} + B_{22} \frac{\partial^3 v}{\partial y^3} + (B_{12} + 2B_{66}) \frac{\partial^3 v}{\partial x^2 \partial y} - D_{11} \frac{\partial^4 w}{\partial x^4} \right.$$
$$- 2(D_{12} + 2D_{66}) \frac{\partial^4 w}{\partial x^2 \partial y^2} - D_{22} \frac{\partial^4 w}{\partial y^4} \right] + \left(1 - \mu^2 \nabla^2\right) \left[F_L \left(\frac{\partial^2 w}{\partial x^2} + \frac{\partial^2 w}{\partial y^2} \right) \right.$$
$$- I_0 \frac{\partial^2 w}{\partial t^2} - I_1 \left(\frac{\partial^3 u}{\partial x \partial t^2} + \frac{\partial^3 v}{\partial y \partial t^2} \right) + I_2 \left(\frac{\partial^4 w}{\partial x^2 \partial t^2} + \frac{\partial^4 w}{\partial y^2 \partial t^2} \right) \right] = 0 \tag{11.38}$$

Moreover, the coupled governing equations of FG nanoplates under applied Lorentz force can be expressed in the following form when Eqs. (4.97)–(4.100) are inserted in Eqs. (11.32)–(11.35):

$$\left(1 - \lambda^2 \nabla^2\right) \left[A_{11} \frac{\partial^2 u}{\partial x^2} + \left(A_{12} + A_{66}\right) \frac{\partial^2 v}{\partial x \partial y} + A_{66} \frac{\partial^2 u}{\partial y^2} - B_{11} \frac{\partial^3 w_b}{\partial x^3} - \left(B_{12} + 2B_{66}\right) \frac{\partial^3 w_b}{\partial x \partial y^2} \right.$$
$$\left. - B_{11}^s \frac{\partial^3 w_s}{\partial x^3} - \left(B_{12}^s + 2B_{66}^s\right) \frac{\partial^3 w_s}{\partial x \partial y^2} \right] + \left(1 - \mu^2 \nabla^2\right) \left[-I_0 \frac{\partial^2 u}{\partial t^2} + I_1 \frac{\partial^3 w_b}{\partial x \partial t^2} + J_1 \frac{\partial^3 w_s}{\partial x \partial t^2} \right] = 0$$

$$(11.39)$$

$$\left(1 - \lambda^2 \nabla^2\right) \left[A_{22} \frac{\partial^2 v}{\partial y^2} + \left(A_{12} + A_{66}\right) \frac{\partial^2 u}{\partial x \partial y} + A_{66} \frac{\partial^2 v}{\partial x^2} - B_{22} \frac{\partial^3 w_b}{\partial y^3} - \left(B_{12} + 2B_{66}\right) \frac{\partial^3 w_b}{\partial x^2 \partial y} \right.$$
$$\left. - B_{22}^s \frac{\partial^3 w_s}{\partial y^3} - \left(B_{12}^s + 2B_{66}^s\right) \frac{\partial^3 w_s}{\partial x^2 \partial y} \right] + \left(1 - \mu^2 \nabla^2\right) \left[-I_0 \frac{\partial^2 v}{\partial t^2} + I_1 \frac{\partial^3 w_b}{\partial y \partial t^2} + J_1 \frac{\partial^3 w_s}{\partial y \partial t^2} \right] = 0$$

$$(11.40)$$

$$\left(1 - \lambda^2 \nabla^2\right) \left[B_{11} \frac{\partial^3 u}{\partial x^3} + \left(B_{12} + 2B_{66}\right) \frac{\partial^3 u}{\partial x \partial y^2} + B_{22} \frac{\partial^3 v}{\partial y^3} + \left(B_{12} + 2B_{66}\right) \frac{\partial^3 v}{\partial x^2 \partial y} - D_{11} \frac{\partial^4 w_b}{\partial x^4} \right.$$
$$- 2\left(D_{12} + 2D_{66}\right) \frac{\partial^4 w_b}{\partial x^2 \partial y^2} - D_{22} \frac{\partial^4 w_b}{\partial y^4} - D_{11}^s \frac{\partial^4 w_s}{\partial x^4} - 2\left(D_{12}^s + 2D_{66}^s\right) \frac{\partial^4 w_s}{\partial x^2 \partial y^2}$$
$$\left. - D_{22}^s \frac{\partial^4 w_s}{\partial y^4} \right) + \left(1 - \mu^2 \nabla^2\right) \left[F_L \left(\frac{\partial^2 \left(w_b + w_s\right)}{\partial x^2} + \frac{\partial^2 \left(w_b + w_s\right)}{\partial y^2} \right) - I_0 \frac{\partial^2 \left(w_b + w_s\right)}{\partial t^2} \right.$$
$$\left. - I_1 \left(\frac{\partial^3 u}{\partial x \partial t^2} + \frac{\partial^3 v}{\partial y \partial t^2} \right) + I_2 \left(\frac{\partial^4 w_b}{\partial x^2 \partial t^2} + \frac{\partial^4 w_b}{\partial y^2 \partial t^2} \right) + J_2 \left(\frac{\partial^4 w_s}{\partial x^2 \partial t^2} + \frac{\partial^4 w_s}{\partial y^2 \partial t^2} \right) \right] = 0$$

$$(11.41)$$

$$\left(1 - \lambda^2 \nabla^2\right) \left[B_{11}^s \frac{\partial^3 u}{\partial x^3} + \left(B_{12}^s + 2B_{66}^s\right) \frac{\partial^3 u}{\partial x \partial y^2} + B_{22}^s \frac{\partial^3 v}{\partial y^3} + \left(B_{12}^s + 2B_{66}^s\right) \frac{\partial^3 v}{\partial x^2 \partial y} - D_{11}^s \frac{\partial^4 w_b}{\partial x^4} \right.$$
$$- 2\left(D_{12}^s + 2D_{66}^s\right) \frac{\partial^4 w_b}{\partial x^2 \partial y^2} - D_{22}^s \frac{\partial^4 w_b}{\partial y^4} - H_{11}^s \frac{\partial^4 w_s}{\partial x^4} - 2\left(H_{12}^s + 2H_{66}^s\right) \frac{\partial^4 w_s}{\partial x^2 \partial y^2} - H_{22}^s \frac{\partial^4 w_s}{\partial y^4}$$
$$\left. + A_{44}^s \left(\frac{\partial^2 \left(w_b + w_s\right)}{\partial x^2} + \frac{\partial^2 \left(w_b + w_s\right)}{\partial y^2} \right) \right] + \left(1 - \mu^2 \nabla^2\right) \left[F_{Lz} \left(\frac{\partial^2 \left(w_b + w_s\right)}{\partial x^2} + \frac{\partial^2 \left(w_b + w_s\right)}{\partial y^2} \right) \right.$$
$$- I_0 \frac{\partial^2 \left(w_b + w_s\right)}{\partial t^2} - J_1 \left(\frac{\partial^3 u}{\partial x \partial t^2} + \frac{\partial^3 v}{\partial y \partial t^2} \right) + J_2 \left(\frac{\partial^4 w_b}{\partial x^2 \partial t^2} + \frac{\partial^4 w_b}{\partial y^2 \partial t^2} \right)$$
$$\left. + K_2 \left(\frac{\partial^4 w_s}{\partial x^2 \partial t^2} + \frac{\partial^4 w_s}{\partial y^2 \partial t^2} \right) \right] = 0$$

$$(11.42)$$

Finally, the above equations must be solved to enrich the dispersion responses of FG nanoplates. For this, the stiffness and mass matrices for both Kirchhoff–Love and sinusoidal nanoplates must be constructed to obtain the wave frequency and phase velocity of the propagated waves solving the obtained eigenvalue equation consisted of aforementioned matrices. However, it can be realized that the arrays of mass matrices for classical and shear deformable plates are the same as those expressed in Eqs. (4.108) and (4.110), respectively. Similar to the former section about nanobeams subjected to induced Lorentz force, here, only components of stiffness matrix related to the deflection of the nanoplate must be modified, and other arrays of stiffness matrix remain the same as those reported in Eqs. (4.107) and (4.109) for classical and higher-order nanoplates, respectively.

The k_{33} array of Eq. (4.107) must be modified in the following form to include the effects of the Lorentz force:

$$k_{33} = -\left(1 + \lambda^2\left(\beta_1^2 + \beta_2^2\right)\right)\left[D_{11}\beta_1^4 + 2\left(D_{12} + 2D_{66}\right)\beta_1^2\beta_2^2 + D_{22}\beta_2^4\right]$$
$$+ \left(1 + \mu^2\left(\beta_1^2 + \beta_2^2\right)\right)F_L\left(\beta_1^2 + \beta_2^2\right) \tag{11.43}$$

Moreover, the modified k_{22}, k_{23} and k_{33} arrays of the stiffness matrix for a sinusoidal higher-order nanoplate can be re-written in the following form to include the effects of applied Lorentz force:

$$k_{33} = -\left(1 + \lambda^2\left(\beta_1^2 + \beta_2^2\right)\right)\left[D_{11}\beta_1^4 + 2\left(D_{12} + 2D_{66}\right)\beta_1^2\beta_2^2 + D_{22}\beta_2^4\right]$$
$$+ \left(1 + \mu^2\left(\beta_1^2 + \beta_2^2\right)\right)F_L\left(\beta_1^2 + \beta_2^2\right),$$
$$k_{34} = -\left(1 + \lambda^2\left(\beta_1^2 + \beta_2^2\right)\right)\left[D_{11}^s\beta_1^4 + 2\left(D_{12}^s + 2D_{66}^s\right)\beta_1^2\beta_2^2 + D_{22}^s\beta_2^4\right]$$
$$+ \left(1 + \mu^2\left(\beta_1^2 + \beta_2^2\right)\right)F_L\left(\beta_1^2 + \beta_2^2\right),$$
$$k_{44} = -\left(1 + \lambda^2\left(\beta_1^2 + \beta_2^2\right)\right)\left[H_{11}^s\beta_1^4 + 2\left(H_{12}^s + 2H_{66}^s\right)\beta_1^2\beta_2^2 + H_{22}^s\beta_2^4\right]$$
$$+ A_{44}^s\left(\beta_1^2 + \beta_2^2\right)\right] + \left(1 + \mu^2\left(\beta_1^2 + \beta_2^2\right)\right)F_L\left(\beta_1^2 + \beta_2^2\right) \tag{11.44}$$

Implementing the modified components of the stiffness matrices of classical and shear deformable nanosize plates, the wave frequency and phase velocity of FG nanoplates once subjected to magnetic force can be achieved. In what follows, numerical illustrations will be depicted to show how can the Lorentz force affect the mechanical response of the system.

11.5   Numerical Results and Discussion

In this section, a group of numerical examples will be presented to reveal the effect of the induced magnetic force on the dispersion curves of FG nanostructures. In all examples, magnetic permeability was assumed to be $4\pi \times 10^7$. First, illustrations dealing with nanobeams will be presented and interpreted, followed by those related to nanoplates.

As the first example, Figure 11.1 shows the influence of the presence of magnetic field on the wave frequency curves of FG nanobeams. In this figure, the variation of wave frequency is plotted against wave number for various amounts of magnetic field intensity. In addition, the effects of small scale are also included in the figure, presenting the dispersion curves for two situations, in which zero and nonzero values are assigned to the scale factor. Hence, placing the nanosize beam in a magnetic environment with a desired magnetic field intensity cannot result in observing changes in the dispersion curves of FG nanobeams. Indeed, the efficiency range of the induced magnetic force is limited to small wave numbers, in particular lower than $\beta = 0.1$ (1/nm). Inside the aforementioned range, the induced Lorentz force acts in a way in which the stiffness of the nanobeam becomes greater and the nanostructure will possess higher frequency values. However, such a phenomenon cannot be observed in wave numbers larger than the specified value. On the other hand, it can be seen that the greater the scale factor, the higher will be the frequency of the propagated waves. This issue originates from the hardening impact of the length-scale parameter on the mechanical responses of the nanosize elements. Hence, the effect of magnetic field on the wave resonance frequency will reduce as the scale factor grows.

Similar to Figure 11.1, the effect of induced Lorentz force on the dispersion speed of the scattered waves versus wave number is investigated in the framework of Figure 11.2 for

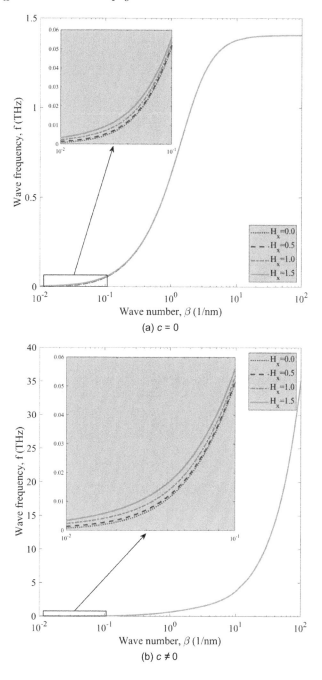

FIGURE 11.1

Variation of wave frequency of FG nanobeams versus wave number for various amounts of magnetic field intensity and scale factors ($p = 2$).

various scale factors to emphasize on the crucial role of both scale factor and magnetic field intensity on the phase speed of the propagated waves in FG nanobeams. First, our former findings about the size-dependent behaviors of nanostructures are showing themselves again in this figure. In fact, the phase velocity of the waves will be enlarged when the scale factor is increased. As stated in previous chapters, this trend is related to the increment in the

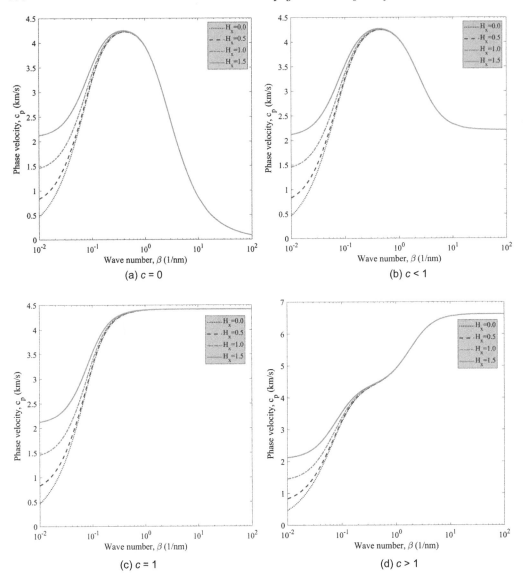

FIGURE 11.2
Variation of phase velocity of waves propagated in FG nanobeams versus wave number for various amounts of magnetic field intensity and scale factors ($p = 2$).

stiffness hardening in the nanosize element as the scale factor increase. On the other hand, the phase velocity can be intensified in wave numbers smaller than $\beta = 0.1$ (1/nm) once the magnetic field intensity is added. This hardening behavior cannot be observed in higher wave numbers anymore and this issue is in complete compatibility with Figure 11.1.

Further, the effect of magnetic field intensity on the phase velocity of the dispersed waves in the FG nanobeams can be seen in Figure 11.3 for various wave numbers. According to this figure, it can be realized that the phase speed of the scattered waves rises in a continuous nonlinear form as the magnitude of the magnetic field intensity increases. This trend is compatible with the results of previous figures. Also, it can be understood that the dynamic response of the nanostructure decreases as the gradient index becomes higher.

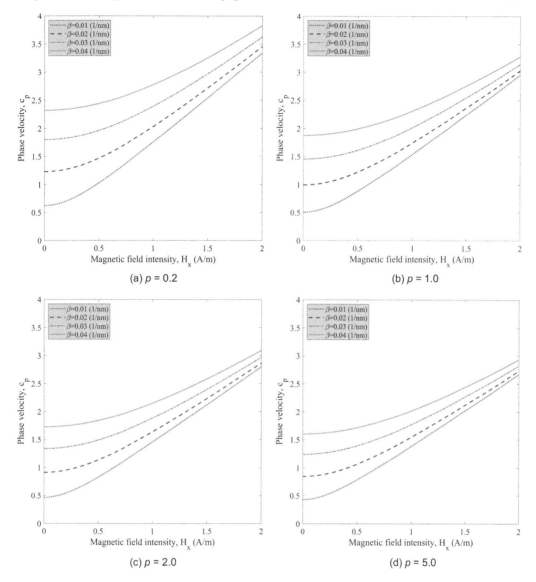

FIGURE 11.3

Variation of phase velocity of waves propagated in FG nanobeams versus magnetic field intensity for various wave numbers ($c < 1$).

The physical reason of this phenomenon is that, by adding this variant, the stiffness of the composite material becomes closer to that of the ceramic-rich phase which has lower stiffness in comparison with the metallic-rich one. In addition, as estimated before, the phase velocity grows when wave number is added within the range of $0.01 < \beta < 0.1$ (1/nm). This behavior can be verified by trends illustrated in Figure 11.2.

As the final illustration on FG nanobeam, combined influences of gradient index and magnetic field intensity on the phase speed of the scattered waves inside the media are shown in Figure 11.4. Based on this figure, the phase velocity will experience a gradual decreasing path as the gradient index increases. The physical reason for this phenomenon was explained in the interpretation of the previous figure and will not be explained again here. Also, another phenomenon discussed in the previous examples, i.e., increase in the

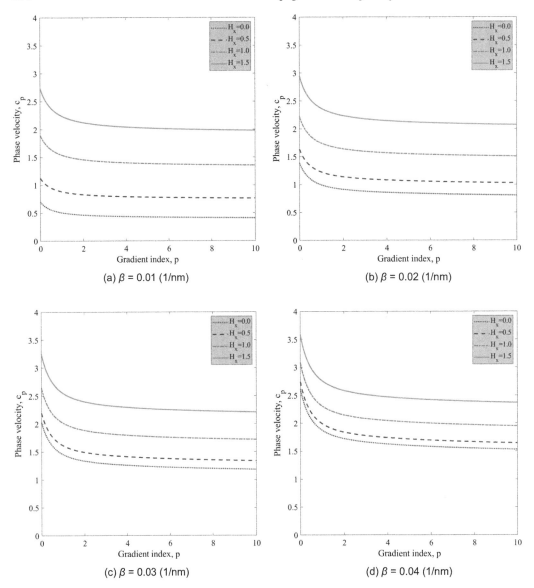

FIGURE 11.4
Variation of phase velocity of waves propagated in FG nanobeams versus gradient index for various magnetic field intensities $(c < 1)$.

phase velocity with increase in the amount of the selected wave number, can be observed here too. It is obvious that the dynamic response of the traveling waves becomes greater when the wave number is increased from $\beta = 0.01$ (1/nm) to a bigger one in the range of $\beta < 0.1$ (1/nm). On the other hand, it can be seen that the velocity of the dispersed waves grows as the magnetic field intensity enlarges, which is the main effect of the induced Lorentz force on the wave propagation responses of FG nanobeams.

Now, the influence of magnetically affected environment on the wave dispersion behaviors of FG nanoplates will be reviewed to show how can the dynamic responses of the continuous nanoplate-type system can be influenced by tuning the magnetic field intensity as well as other involved parameters. As the first illustration, the wave frequency curves of FG nanoplates against wave number are drawn in Figure 11.5 for various amounts of magnetic field intensity to show the effect of external magnetic field on the mechanical response of the nanoplate. It can be concluded that, in using the nonlocal theory of Eringen, the differences can be better seen between the situations where zero or nonzero values are assigned to the

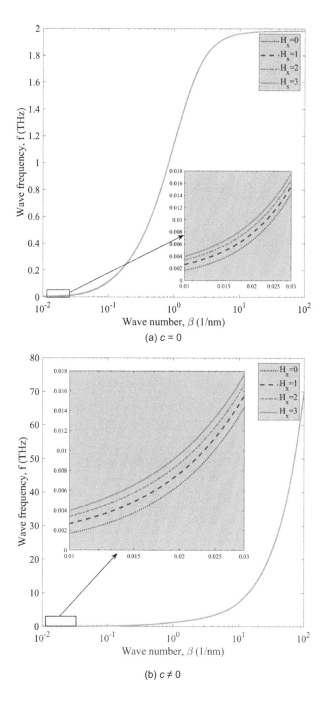

FIGURE 11.5
Variation of wave frequency of FG nanoplates versus wave number for various amounts of magnetic field intensity and scale factors ($p = 1$).

magnetic field intensity compared with the case of implementing the nonlocal stress-strain gradient elasticity theory. Indeed, the influence of induced Lorentz force on the frequency of the propagated waves can show only in the case of choosing small wave numbers, in particular, wave numbers lesser than $\beta = 0.05$ (1/nm). In the mentioned range of wave numbers, adding the intensity of the magnetic field results in an increase in the frequency of the waves scattered in FG nanoplates. This trend cannot be easily observed whenever the length-scale parameter is included as well as the nonlocal one. So, the amplifying role of the magnetic field intensity on the mechanical response of the continuous system can be observed once again as well as that of scale effects in this figure.

Next, mixed influences of small size and the existence of magnetic field on the mechanical behaviors of FG nanoplates attacked by elastic waves can be seen in Figure 11.6 plotting

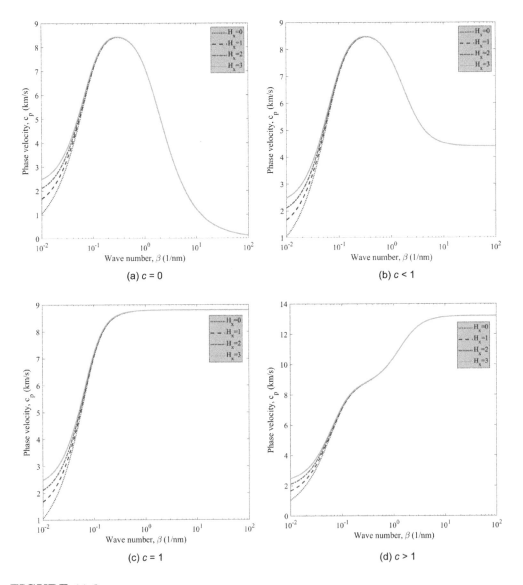

FIGURE 11.6

Variation of phase velocity of waves propagated in FG nanoplates versus wave number for various amounts of magnetic field intensity and scale factors ($p = 1$).

the variation of phase speed of dispersed waves versus wave number. According to the figure, waves can be propagated with a higher speed as the scale factor rises when wave numbers greater than $\beta = 0.2$ (1/nm) are selected. This increasing trend is the stiffness-hardening effect of the length-scale parameter of the nonlocal strain gradient elasticity on the mechanical response of the continua. On the other hand, it is clear that phase velocity shifts upward once the intensity of the magnetic field is assumed to be enlarged. Similar to the former illustration for the wave frequency of the FG nanoplates, in this figure, magnetically induced Lorentz force can enhance the dynamic response of the nanosize plate while small wave numbers are employed.

Furthermore, the effect of the applied Lorentz force on the phase velocity of the scattered waves is again studied in Figure 11.7 for various wave numbers and gradient indices. In fact,

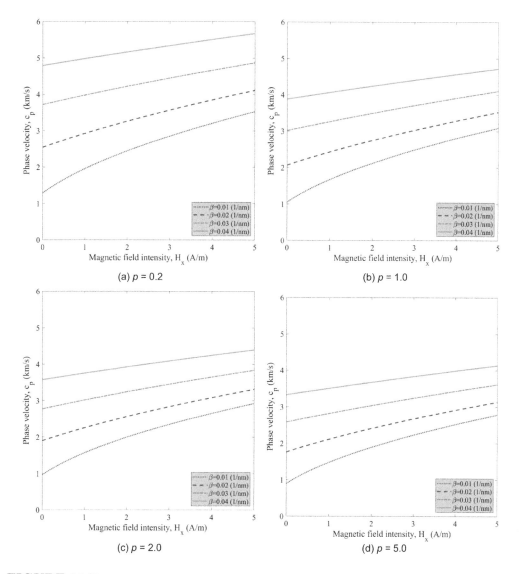

FIGURE 11.7
Variation of phase velocity of waves propagated in FG nanoplates versus magnetic field intensity for various wave numbers ($c < 1$).

due to the decreasing impact of gradient index on the equivalent stiffness of the implemented FGM (i.e., the total stiffness will be closer to that of the ceramic-rich phase with increase the gradient index), the phase velocity becomes smaller once a greater value is assigned to the gradient index. In addition, increasing the wave number results in achieving greater phase speeds. However, the slope of enhancement of phase speed with increment of magnetic field intensity becomes smaller as the wave number is added. It is worth mentioning that the enhancement of the dynamic response of the continuous system occurs with a nonlinear profile. Hence, former estimations can also be observed in this diagram (Figure11.8a).

Furthermore, the final numerical illustration shows the variation of phase velocity against gradient index for various magnitudes of the magnetic field intensity as well as different wave numbers. The wave numbers were selected from a range where the influence of magnetic

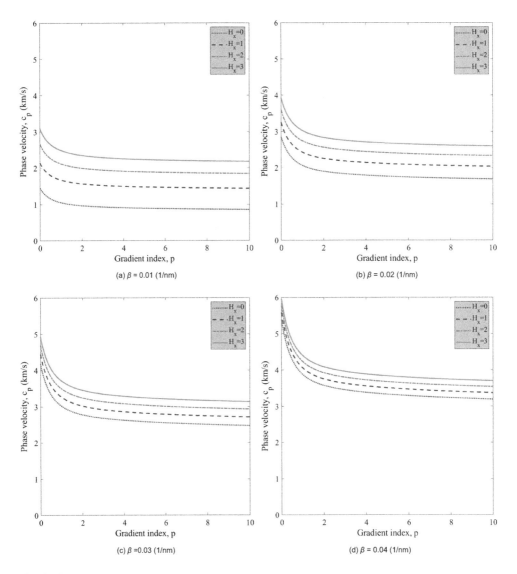

FIGURE 11.8

Variation of phase velocity of waves propagated in FG nanoplates versus gradient index for various magnetic field intensities ($c < 1$).

force can be seen. Obviously, the velocity of the scattered waves inside the composite nanoplate will be reduced when the gradient index increases. The physical reason for this was discussed in previous interpretations and we will not repeat it here. It can also be noted that the dispersion curves of nanoplate will be shifted to an upper bound as the intensity of the external magnetic field grows, which is in accordance with our previous illustrations. Moreover, it is perceived that the phase velocity of the waves traveling in the media becomes greater when the wave number is changed from $\beta = 0.01$ (1/nm) to $\beta = 0.04$ (1/nm). This trend can be verified by taking a brief look at Figure 11.6.

In conclusion, we can point to the increasing impact of the induced magnetic force on the dynamic dispersion response of the nanostructures. In fact, higher dynamic reactions can be estimated once the nanostructure is subjected to an external magnetic field. This trend can be observed in tiny wave numbers and cannot be seen once wave number is assumed to be great. Other results are consistent with those reported in previous chapters.

12

Conclusions

In this chapter, we will present the concluding remarks of this book. The present book was majorly arranged to cover the lack of a general reference for the wave propagation problem in small-scale continuous beam- and plate-type elements attacked by elastic waves. The most crucial issue presented in the present book was the formulation of the wave propagation problem within nanobeams and nanoplates in the framework of the nonlocal strain gradient elasticity hypothesis in association with both the classical and refined shear deformable beam and plate theories.

In Chapter 1, the preliminaries of the wave theory were presented to introduce to the readers the nature and the physics of the problem. In addition, the waves were categorized on the basis of a predefined criterion. Many examples were explained to clarify the crucial role elastic waves play in problems that either cannot be answered or whose solutions through other ways is hard to determine. At the end of the chapter, the well-known vibrating eigenvalue equation for the fully elastic wave problem was depicted, along with the well-known analytical solution which can be implemented to solve a wave propagation problem in beams and plates. The present solution was based on the well-known concept of the separation of the variables available in advanced engineering mathematics literature.

Chapter 2 was allocated to survey the size-dependent problems in the framework of continuum mechanics to bridge nanotechnology and mechanical engineering from a solid mechanics point of view. Nanosize elements were found to be of unbelievable significance in modern engineering designs. Also, it was demonstrated that, because of the growing implementation rate of tiny nanoscale devices in the modern industries, it is very important to collect sufficient data on the mechanical behaviors of nanostructures. Furthermore, the conventional continuum mechanics approaches were shown to have several shortcomings, which make them usable for investigating the mechanical behaviors of continuous systems in the nanoscale. Afterward, the major concept of scale-dependent continuum theories was discussed to provide more data about the reasons behind the development of the nonlocal elasticity theory of Eringen. Because Eringen's nonlocal elasticity could not underestimate the mechanical behaviors of nanostructures correctly, another size-dependent hypothesis was proposed. In fact, the second hypothesis, called the nonlocal strain gradient elasticity theory, could justify the stiffness-hardening behavior of the nanostructures, whereas Eringen's theory was only able to show the stiffness-softening behavior of nanoscale elements. Next, the mathematical framework for both of the abovementioned theories was presented for various elastic and smart materials to modify the constitutive equations of nanostructures present in these materials. It is worth mentioning that in this book the differential type of both Eringen's nonlocal elasticity and the nonlocal strain gradient theory is presented. However, the experiments have shown that the differential form of these theories cannot satisfy the boundary conditions of the static and dynamic problems. This statement cannot be assumed to be true for the wave propagation problem. In wave propagation problems, the length of the structure is assumed to be many times larger than the length of the dispersed waves, and therefore, the differential form has responses similar to those achieved from the integral form of the size-dependent theories.

Chapter 3 presents incredible features of the dispersion of elastic waves inside nanostructures. In this chapter, the differences in the analysis of nanosize elements with that of macroscale ones have been discussed to highlight the importance of tiny vdW forces, which are generally dismissed in the analyses of macroscale structures in the nanorealm. Next, the chapter discusses the crucial role of wave propagation analysis in nanostructures and its superiority to the vibration phenomenon in nanostructures. Dispersion of elastic waves inside nanostructures results in natural frequencies of the THz order, whereas only GHz-range natural frequencies can be obtained by generating a vibration in such nanostructures. The aforementioned issue is the most important aspect of the wave dispersion which makes it superior to the vibration phenomenon in the nanoscale structures. Thereafter, some examples of application of the wave propagation technique in various problems were presented to certify the efficiency of this technique in solving problems dealing with smart multi-tasking nanoscale devices. Finally, the variants that can affect the wave propagation responses of nanosize structures have been discussed. It was mentioned that the wave frequency and the speed of the scattered waves can be critically influenced, thus changing the wave number.

In Chapter 4, the wave propagation problems of nanosize beams and plates were investigated in an extended manner. The constituent material was assumed to be a functionally graded material (FGM) consisting of metal and ceramic. The homogenization procedure was explained in the framework of both power-law and Mori-Tanaka methods. Afterward, the kinematic relations of classical and refined sinusoidal beams and plates were developed to derive the motion equations of FG nanosize beams and plates in association with the dynamic form of the principle of virtual work. Next, the constitutive equations were modified for nanosize beams and plates. Thereafter, the motion equations were enriched and solved according to the well-known analytical method explained in Chapter 1. The results indicate the stiffness-softening role of nonlocality on the dynamic responses of the nanostructures, as well as the stiffness-hardening impact of the length-scale term on the dynamic behaviors of the same nanostructures. In other words, it was shown that the wave frequency or phase speed of the waves propagated in FG nanobeams and nanoplates can be amplified by adding the length-scale term, whereas the aforementioned parameters can be reduced if a greater value is assigned to the nonlocal parameter. It was also illustrated that the influence of the scale-dependent terms can be better observed when greater wave numbers are employed. This finding is in accordance with the final section of Chapter 3 dealing with the role of various terms on wave frequency and phase speed of the waves scattered in nanostructures. Moreover, it was shown that the difference between the implementation of the classical beam and plate hypotheses in reaching the dispersion curves of FG nanostructures is small enough to be ignored. On the other hand, it was shown that the phase velocity can be reduced when the gradient index of the FGM is added. The physical reason for this trend is the decreasing influence of the gradient index on the total stiffness of the material, which has a direct relation with the dynamic response of the system.

In Chapter 5, the influence of porosities in FGMs on the wave dispersion curves of FG nanostructures was analyzed. The major difference of this chapter with the previous one is in the modifications applied on the homogenization procedure used to obtain the equivalent material properties of the FGMs. The modifications were performed via two different approaches: first, by considering a volume fraction of the porosities and considering this phenomenon using the conventional form of the power-law FGM model; and second, is by capturing the dependency of the stiffness of the composite to its mass density in both porous and nonporous situations. In fact, the first model was a simple model giving a general sense to the readers of the book about how porosities can affect the stiffness of the FGM in a negative manner. On the other hand, the second model considers the coupled relation between the equivalent stiffness and the mass density while the employed

FGM is either porous or nonporous. Due to this fact, the second approach seems to be a more realistic homogenization method for achieving the effective stiffness and density of the composite material once there are porosities in the media. In this chapter, it was proven that, because of the stiffness-lessening in FGMs due to the presence of porosities, the dynamic response of the continuous nanostructures will experience a decrease compared with nonporous materials to fabricate the studied nanostructures. Again, the effect of wave number on the phase speed of the scattered waves was perceived as the volume fraction of the porosities increases. Indeed, the impact of the presence of porosities in the media could be better observed in wave numbers greater than $\beta = 0.05$ (1/nm).

After analyzing the mechanical characteristics of waves dispersed in porous FG nanostructures, we turned to probing smart piezoelectrically affected dispersion behaviors of waves traveling in FG piezoelectric nanostructures in Chapter 6. In this chapter, the dynamic form of the principle of virtual work was again extended for reaching the governing equations of smart piezoelectric nanoscale beams and plates with respect to the influences of electromechanical coupling on the features of waves propagated in smart nanostructures. For this, the nonlocal strain gradient piezoelectricity theory was presented to enrich the modified form of the constitutive equations of piezo nanostructures to apply the small-scale effects on the entire material properties of the smart material. Results of the studies performed in this chapter indicated the crucial impact of applied electric voltage on the wave propagation behaviors of smart piezoelectric nanostructures. It was illustrated that the dispersion velocity decreased as the magnitude of the applied electric voltage was intensified. In other words, the increase in the electric voltage acted in a way that the total stiffness of the nanostructure decreased, and because of this reduction in the stiffness, the phase speed of the propagated waves reduced than when no voltage was applied on the nanostructure. The effect of wave number on the efficiency range of the applied voltage was also shown in the results of this chapter. In fact, the reduction in the phase speed could be seen in wave numbers lower than $\beta = 0.05$ (1/nm), and as the wave number increased, no remarkable change could be reported in dispersion curves. In addition, the influences of gradient index and porosities in the smart piezoelectric FGMs were also discussed in this chapter.

In Chapter 7, we turned to the investigation of the effect of material's magnetostriction behavior on the wave propagation characteristics of smart magnetostrictive nanostructures. In this chapter, a simulation of magnetostrictive materials was presented at the beginning in the presence of a velocity feedback control system to control the dynamic behaviors of smart magnetostrictive nanostructures by changing a control gain. Again, the constitutive equations were modified on the basis of the initial smart constitutive equations of macroscale magnetostrictive materials for considering the influences of small-scale on the mechanical behaviors of the studied nanostructures. It was shown that magnetostrictive materials can damp the dynamic behaviors of waves propagated inside nanosize continuous systems by changing the control gain. Indeed, the phase speed and wave frequency of the scattered waves were reduced as the control gain of the feedback control system was added. This damping behavior could be better seen in the mid-range wave numbers for probing smart magnetostrictive nanobeams, whereas, for analyzing smart magnetostrictive nanoplates, this trend could be better observed whenever the wave number was large enough.

In Chapter 8, the smart behaviors of waves propagated in the FG magneto-electro-elastic nanostructures were investigated. Similar to Chapters 6 and 7 dealing with the smart aspects of waves dispersed in smart nanostructures, in this chapter, the most significant activity was to account for the consideration of the small-scale effects on the constitutive equations of the functionally graded magneto-electro-elastic (FG-MEE) nanomaterials to provide suitable relations for reaching the governing equations of smart nanostructures. This chapter also attempted to extend the Hamilton's principle for an MEE material to gather the Euler–Lagrange equations of the smart MEE continuous systems. The results of this chapter are approximately the same as those of Chapter 6; however, in this chapter, the magnetic potential acts in a way which leads to strengthening the stiffness of the nanostructure. In other words, the wave propagation curves of smart MEE nanostructures revealed that the dynamic responses of the waves dispersed in the smart nanostructures were increased as the magnetic potential was added. Therefore, the magnetic potential possesses an influence completely different from that of the applied electric voltage. Again, it could be realized that the effect of the magnetic potential appears in small wave numbers and cannot be seen as the wave number becomes large. This is the major reason that we did not present the escape frequency curves for piezoelectric and MEE smart nanostructures.

In Chapter 9, we discussed the effect of resting media on the wave propagation behaviors of nanostructures. We showed how the structural stiffness of the nanostructures can be varied with changes in the seat of the nanostructures, and due to the direct relation between the stiffness and the dynamic responses of a continuous system, the wave propagation curves of nanostructures were affected by the aforementioned changes. It was clear that the wave dispersion curves of nanostructures were intensified as the nanostructure was embedded on either Winkler or Pasternak elastic foundations. The reason for this is that the total stiffness of the nanostructure could be increased by embedding the structure on either linear or shear springs. Therefore, it is natural to see that, in this chapter, the illustrations denote a similar stiffness enhancement in the nanostructures as these coefficients are increased. However, it was observed that the wave dispersion curves shift downward as the damping coefficient of the viscoelastic medium was added. The physical interpretation of this trend was explained in this chapter, which is the appearance of the damping matrix seen in the final eigenvalue equations of the problem. It is interesting to point out that the effect of the foundation parameters can be observed only in very small wave numbers, and that these effects cannot be seen once either a mid-range or a large value is assigned to the wave number.

Next, the influences of subjecting the nanostructure to various types of thermal loading was investigated in Chapter 10. In this chapter, the constitutive equations for nanosize beams and plates were modified for including the effects of thermal environment on the stress-strain relationships. In addition, four various types of thermal loading were included to show the differences between the effects of these types on the wave propagation responses of FG nanostructures. These types were uniform, linear, sinusoidal, and nonlinear temperature increases. It was shown that the speed of the propagated waves decreased as the temperature was assumed to be raised. This issue is not strange and is in complete compatibility with our former estimations. The reason for this reduction in the dynamic responses of the system is that the stiffness of the FG nanostructure decreases as the temperature increases. Indeed, the increase in the temperature helps in the enhancement of the systems' equivalent compliance, and due to the reverse relationship between the compliance and the stiffness in the elastic region for linear elastic solids, it is natural to see a decrease in the dynamic response of the system due to the reduction in its total stiffness. It was shown that the uniform-type temperature increase was the most powerful type of thermal loading in reducing the phase speed or the wave frequency of the dispersed waves. It must be recalled that the effect of the thermal environment could be seen in all wave numbers, and that its impact cannot be summarized in small wave numbers.

Finally, the influence of the externally induced magnetic Lorentz force on the wave propagation behaviors of FG nanostructures was analyzed in Chapter 11. In this chapter, the magnetic induction relations of Maxwell were extended for a continuous system to reach the equivalent Lorentz force which will be generated in the nanostructure as a result of magnetic induction. The governing equations of nanosize beams and plates were developed after this regarding the induced Lorentz force as in the work done by external loadings. The results of this chapter reveal that the wave propagation responses of the nanobeams and nanoplates can be easily affected by increasing the magnitude of the magnetic field applied on the nanostructures. Similar to the findings of Chapter 8, the dispersion responses of the nanostructures intensified as the magnitude of the magnetic field intensity increased. In other words, it was shown that the Lorentz force has a hardening impact on the dynamic behaviors of the system. Hence, induction of Lorentz force can be presumed to be the same as any other phenomenon capable of enhancing the stiffness of the nanostructure. Clearly, the effect of this loading on dispersion curves can be easily seen in wave numbers smaller than $\beta = 0.1$ (1/nm). Thus, the influence of this term depends on the range of the wave number. Therefore, the graphs of the escape frequency were not surveyed for this problem because no change could be found in the escape frequency of the scattered waves by assigning a higher value to the magnetic field intensity.

Bibliography

[1] Gra, K. F. 1975. Wave Motion in Elastic Solids. Ohio: Ohio State University.

[2] Ebrahimi F. and Dabbagh, A. 2017. Wave propagation analysis of smart rotating porous heterogeneous piezo-electric nanobeams. *The European Physical Journal Plus*, 132(4), p. 153.

[3] Ebrahimi, F. and Dabbagh, A. 2018. On modeling wave dispersion characteristics of protein lipid nanotubules. *Journal of Biomechanics*, 77, pp. 1–7

[4] Ebrahimi, F., Barati, M. R. and Haghi, P. 2016. Nonlocal thermo-elastic wave propagation in temperature-dependent embedded small-scaled nonhomogeneous beams. *The European Physical Journal Plus*, 131(11), p. 383.

[5] Ebrahimi, F. and Haghi, P. 2017. Wave propagation analysis of rotating thermoelastically-actuated nanobeams based on nonlocal strain gradient theory. *Acta Mechanica Solida Sinica*, 30(6), pp. 647–657.

[6] Ebrahimi, F. and Dabbagh, A. 2017. Wave propagation analysis of embedded nanoplates based on a nonlocal strain gradient-based surface piezoelectricity theory. *The European Physical Journal Plus*, 132(11), p. 449.

[7] Ebrahimi, F. and Dabbagh, A. 2018. Wave propagation analysis of magnetostrictive sandwich composite nanoplates via nonlocal strain gradient theory. *Proceedings of the Institution of Mechanical Engineers, Part C: Journal of Mechanical Engineering Science*. doi:10.1177/0954406217748687.

[8] Ebrahimi, F., Barati, M. R. and Dabbagh, A. 2016. A nonlocal strain gradient theory for wave propagation analysis in temperature-dependent inhomogeneous nanoplates. *International Journal of Engineering Science*, 107, pp. 169–182.

[9] Ebrahimi, F., Dabbagh, A. and Barati, M. R. 2016. Wave propagation analysis of a size-dependent magneto-electro-elastic heterogeneous nanoplate. *The European Physical Journal Plus*, 131(12), p. 433.

[10] Eringen, A.C., 1972. Linear theory of nonlocal elasticity and dispersion of plane waves. *International Journal of Engineering Science*, 10(5), pp. 425–435.

[11] Eringen, A. C., 1983. On differential equations of nonlocal elasticity and solutions of screw dislocation and surface waves. *Journal of Applied Physics*, 54(9), pp. 4703–4710.

[12] Pradhan, S.C. and Murmu, T., 2010. Small scale effect on the buckling analysis of single-layered graphene sheet embedded in an elastic medium based on nonlocal plate theory. *Physica E: Low-Dimensional Systems and Nanostructures*, 42(5), pp. 1293–1301.

[13] Ansari, R., Arash, B. and Rouhi, H., 2011. Vibration characteristics of embedded multi-layered graphene sheets with different boundary conditions via nonlocal elasticity. *Composite Structures*, 93(9), pp. 2419–2429.

[14] Mahmoud, F.F., Eltaher, M.A., Alshorbagy, A.E. and Meletis, E.I., 2012. Static analysis of nanobeams including surface effects by nonlocal finite element. *Journal of Mechanical Science and Technology*, 26(11), pp. 3555–3563.

[15] Alzahrani, E.O., Zenkour, A.M. and Sobhy, M., 2013. Small scale effect on hygro-thermo-mechanical bending of nanoplates embedded in an elastic medium. *Composite Structures*, 105, pp. 163–172.

[16] Ebrahimi, F. and Salari, E., 2015. Thermo-mechanical vibration analysis of a single-walled carbon nanotube embedded in an elastic medium based on higher-order shear deformation beam theory. *Journal of Mechanical Science and Technology*, 29(9), pp. 3797–3803.

[17] Zenkour, A.M., 2016. Nonlocal transient thermal analysis of a single-layered graphene sheet embedded in viscoelastic medium. *Physica E: Low-dimensional Systems and Nanostructures*, 79, pp. 87–97.

[18] Ebrahimi, F. and Shafiei, N., 2017. Influence of initial shear stress on the vibration behavior of single-layered graphene sheets embedded in an elastic medium based on Reddy's higher-order shear deformation plate theory. *Mechanics of Advanced Materials and Structures*, 24(9), pp. 761–772.

[19] Ebrahimi, F. and Barati, M. R., 2018. Nonlocal and surface effects on vibration behavior of axially loaded flexoelectric nanobeams subjected to in-plane magnetic field. *Arabian Journal for Science and Engineering*, 43(3), pp. 1423–1433.

[20] Ebrahimi, F. and Karimiasl, M., 2018. Nonlocal and surface effects on the buckling behavior of flexoelectric sandwich nanobeams. *Mechanics of Advanced Materials and Structures*, 25(11), pp. 943–952.

[21] Fleck, N. A. and Hutchinson, J. W., 1993. A phenomenological theory for strain gradient effects in plasticity. *Journal of the Mechanics and Physics of Solids*, 41(12), pp. 1825–1857.

[22] Stölken, J. S. and Evans, A. G., 1998. A microbend test method for measuring the plasticity length scale. *Acta Materialia*, 46(14), pp. 5109–5115.

[23] Lim, C. W., Zhang, G. and Reddy, J. N., 2015. A higher-order nonlocal elasticity and strain gradient theory and its applications in wave propagation. *Journal of the Mechanics and Physics of Solids*, 78, pp. 298–313.

[24] Ebrahimi, F. and Hosseini, S. H. S., 2016. Thermal effects on nonlinear vibration behavior of viscoelastic nanosize plates. *Journal of Thermal Stresses*, 39(5), pp. 606–625.

[25] Ebrahimi, F. and Dabbagh, A., 2018. Thermo-magnetic field effects on the wave propagation behavior of smart magnetostrictive sandwich nanoplates. *The European Physical Journal Plus*, 133(3), 97.

[26] Ebrahimi, F. and Barati, M. R., 2018. Vibration analysis of nonlocal strain gradient embedded single-layer graphene sheets under nonuniform in-plane loads. *Journal of Vibration and Control*, 24(20), pp. 4751–4763.

[27] Ebrahimi, F. and Barati, M. R., 2018. Damping vibration analysis of graphene sheets on viscoelastic medium incorporating hygro-thermal effects employing nonlocal strain gradient theory. *Composite Structures*, 185, pp. 241–253.

[28] Ebrahimi, F. and Barati, M. R., 2018. Vibration analysis of graphene sheets resting on the orthotropic elastic medium subjected to hygro-thermal and in-plane magnetic fields based on the nonlocal strain gradient theory. *Proceedings of the Institution of Mechanical Engineers, Part C: Journal of Mechanical Engineering Science*, 232(13), pp. 2469–2481.

[29] Ebrahimi, F. and Dabbagh, A., 2018. Thermo-mechanical wave dispersion analysis of nonlocal strain gradient single-layered graphene sheet rested on elastic medium. *Microsystem Technologies*, 25, pp. 1–11.

[30] Ebrahimi, F. and Dabbagh, A., 2018. Effect of humid-thermal environment on wave dispersion characteristics of single-layered graphene sheets. *Applied Physics A*, 124(4), p. 301.

[31] Ebrahimi, F. and Barati, M. R., 2018. Vibration analysis of biaxially compressed double-layered graphene sheets based on nonlocal strain gradient theory. *Mechanics of Advanced Materials and Structures*, 26, pp. 1–12.

[32] Ebrahimi, F. and Barati, M. R., 2018. Hygro-thermal vibration analysis of bilayer graphene sheet system via nonlocal strain gradient plate theory. *Journal of the Brazilian Society of Mechanical Sciences and Engineering*, 40(9), p. 428.

[33] Ebrahimi, F. and Dabbagh, A., 2018. Wave dispersion characteristics of nonlocal strain gradient double-layered graphene sheets in hygro-thermal environments. *Structural Engineering and Mechanics*, 65(6), pp. 645–656.

[34] Ebrahimi, F. and Dabbagh, A., 2018. Viscoelastic wave propagation analysis of axially motivated double-layered graphene sheets via nonlocal strain gradient theory. *Waves in Random and Complex Media*, pp. 1–20.

[35] Ebrahimi, F. and Dabbagh, A., 2018. Wave dispersion characteristics of orthotropic double-nanoplate-system subjected to a longitudinal magnetic field. *Microsystem Technologies*, 24, pp. 1–11.

[36] Hernandez, C. M., Murray, T. W. and Krishnaswamy, S., 2002. Photoacoustic characterization of the mechanical properties of thin films. *Applied Physics Letters*, 80(4), pp.691–693.

[37] Philip, J., Hess, P., Feygelson, T., Butler, J. E., Chattopadhyay, S., Chen, K. H. and Chen, L. C., 2003. Elastic, mechanical, and thermal properties of nanocrystalline diamond films. *Journal of Applied Physics*, 93(4), pp. 2164–2171.

[38] Ramprasad, R. and Shi, N., 2005. Scalability of phononic crystal heterostructures. *Applied Physics Letters*, 87(11), p.111101.

[39] Sampathkumar, A., Murray, T. W. and Ekinci, K. L., 2006. Photothermal operation of high frequency nanoelectromechanical systems. *Applied Physics Letters*, 88(22), pp. 223104.

[40] Huang, G. L. and Song, F., 2008. High-frequency antiplane wave propagation in ultra-thin films with nanostructures. *International Journal of Solids and Structures*, 45(20), pp. 5368–5380.

[41] Hepplestone, S. P. and Srivastava, G. P., 2008. Hypersonic modes in nanophononic semiconductors. *Physical Review Letters*, 101(10), p. 105502.

[42] Parsons, L. C. and Andrews, G. T., 2009. Observation of hypersonic phononic crystal effects in porous silicon superlattices. *Applied Physics Letters*, 95(24), p. 241909.

[43] Chen, A. L. and Wang, Y. S., 2011. Size-effect on band structures of nanoscale phononic crystals. *Physica E: Low-Dimensional Systems and Nanostructures*, 44(1), pp. 317–321.

[44] Zhen, N., Wang, Y. S. and Zhang, C., 2012. Surface/interface effect on band structures of nanosized phononic crystals. *Mechanics Research Communications*, 46, pp. 81–89.

[45] Chong, K. P., 2008. Nano science and engineering in solid mechanics. *Acta Mechanica Solida Sinica*, 21(2), pp. 95–103.

[46] Wang, X., Gao, Y., Wei, Y. and Wang, Z. L., 2009. Output of an ultrasonic wave-driven nanogenerator in a confined tube. *Nano Research*, 2(3), pp. 177–182.

[47] Cha, S. N., Seo, J. S., Kim, S. M., Kim, H. J., Park, Y. J., Kim, S. W. and Kim, J. M., 2010. Sound-driven piezoelectric nanowire-based nanogenerators. *Advanced Materials*, 22(42), pp. 4726–4730.

[48] Akyildiz, I. F. and Jornet, J. M., 2010. Electromagnetic wireless nanosensor networks. *Nano Communication Networks*, 1(1), pp. 3–19.

[49] He, J., Qi, X., Miao, Y., Wu, H. L., He, N. and Zhu, J. J., 2010. Application of smart nanostructures in medicine. *Nanomedicine*, 5(7), pp. 1129–1138.

[50] Al-Hossain, A. Y., Farhoud, F. A. and Ibrahim, M., 2011. The mathematical model of reflection and refraction of plane quasi-vertical transverse waves at interface nano-composite smart material. *Journal of Computational and Theoretical Nanoscience*, 8(7), pp. 1193–1202.

[51] Abd-Alla, A. E. N. N., Eshaq, H. A. and Elhaes, H., 2011. The phenomena of reflection and transmission waves in smart nano materials. *Journal of Computational and Theoretical Nanoscience*, 8(9), pp. 1670–1678.

[52] Danlée, Y., Huynen, I. and Bailly, C., 2012. Thin smart multilayer microwave absorber based on hybrid structure of polymer and carbon nanotubes. *Applied Physics Letters*, 100(21), p. 213105.

[53] Voiculescu, I. and Nordin, A. N., 2012. Acoustic wave based MEMS devices for biosensing applications. *Biosensors and Bioelectronics*, 33(1), pp. 1–9.

[54] Shin, D., Urzhumov, Y., Jung, Y., Kang, G., Baek, S., Choi, M. et al., 2012. Broadband electromagnetic cloaking with smart metamaterials. *Nature Communications*, 3, p. 1213.

[55] Zhang, S., Gu, B., Zhang, H., Feng, X. Q., Pan, R. and Hu, N., 2016. Propagation of Love waves with surface effects in an electrically-shorted piezoelectric nanofilm on a half-space elastic substrate. *Ultrasonics*, 66, pp. 65–71.

[56] Lang, C., Fang, J., Shao, H., Ding, X. and Lin, T., 2016. High-sensitivity acoustic sensors from nanofibre webs. *Nature Communications*, 7, p. 11108.

[57] Elayan, H., Shubair, R. M., Jornet, J. M. and Mittra, R., 2017. Multi-layer intrabody terahertz wave propagation model for nanobiosensing applications. *Nano Communication Networks*, 14, pp. 9–15.

[58] Ebrahimi, F., Barati, M.R. and Dabbagh, A., 2018. Wave propagation in embedded inhomogeneous nanoscale plates incorporating thermal effects. *Waves in Random and Complex Media*, 28(2), pp. 215–235.

[59] Ebrahimi, F. and Rastgoo, A., 2008. An analytical study on the free vibration of smart circular thin FGM plate based on classical plate theory. *Thin-Walled Structures*, 46(12), pp. 1402–1408.

[60] Shen, H.S., 2009. A comparison of buckling and postbuckling behavior of FGM plates with piezoelectric fiber reinforced composite actuators. *Composite Structures*, 91(3), pp. 375–384.

[61] Huang, Y. and Li, X.F., 2010. A new approach for free vibration of axially functionally graded beams with non-uniform cross-section. *Journal of Sound and Vibration*, 329(11), pp. 2291–2303.

[62] Alshorbagy, A.E., Eltaher, M.A. and Mahmoud, F.F., 2011. Free vibration characteristics of a functionally graded beam by finite element method. *Applied Mathematical Modelling*, 35(1), pp. 412–425.

[63] Şimşek, M., Kocatürk, T. and Akbaş, Ş.D., 2012. Dynamic behavior of an axially functionally graded beam under action of a moving harmonic load. *Composite Structures*, 94(8), pp. 2358–2364.

[64] Thai, H.T. and Choi, D.H., 2012. A refined shear deformation theory for free vibration of functionally graded plates on elastic foundation. *Composites Part B: Engineering*, 43(5), pp. 2335–2347.

[65] Ebrahimi, F., 2013. Analytical investigation on vibrations and dynamic response of functionally graded plate integrated with piezoelectric layers in thermal environment. *Mechanics of Advanced Materials and Structures*, 20(10), pp. 854–870.

[66] Ghiasian, S.E., Kiani, Y., Sadighi, M. and Eslami, M.R., 2014. Thermal buckling of shear deformable temperature dependent circular/annular FGM plates. *International Journal of Mechanical Sciences*, 81, pp. 137–148.

[67] Şimşek, M., 2015. Bi-directional functionally graded materials (BDFGMs) for free and forced vibration of Timoshenko beams with various boundary conditions. *Composite Structures*, 133, pp. 968–978.

[68] Ghiasian, S.E., Kiani, Y. and Eslami, M.R., 2015. Nonlinear thermal dynamic buckling of FGM beams. *European Journal of Mechanics-A/Solids*, 54, pp. 232–242.

[69] Jafarinezhad, M.R. and Eslami, M.R., 2017. Coupled thermoelasticity of FGM annular plate under lateral thermal shock. *Composite Structures*, 168, pp. 758–771.

[70] Gharibi, M., Zamani Nejad, M. and Hadi, A., 2017. Elastic analysis of functionally graded rotating thick cylindrical pressure vessels with exponentially-varying properties using power series method of Frobenius. *Journal of Computational Applied Mechanics*, 48(1), pp. 89–98.

[71] Tang, Y. and Yang, T., 2018. Post-buckling behavior and nonlinear vibration analysis of a fluid-conveying pipe composed of functionally graded material. *Composite Structures*, 185, pp. 393–400.

[72] Eltaher, M.A., Emam, S.A. and Mahmoud, F.F., 2012. Free vibration analysis of functionally graded size-dependent nanobeams. *Applied Mathematics and Computation*, 218(14), pp. 7406–7420.

[73] Rahmani, O. and Pedram, O., 2014. Analysis and modeling the size effect on vibration of functionally graded nanobeams based on nonlocal Timoshenko beam theory. *International Journal of Engineering Science*, 77, pp. 55–70.

[74] Nazemnezhad, R. and Hosseini-Hashemi, S., 2014. Nonlocal nonlinear free vibration of functionally graded nanobeams. *Composite Structures*, 110, pp. 192–199.

[75] Ebrahimi, F. and Salari, E., 2015. Thermal buckling and free vibration analysis of size dependent Timoshenko FG nanobeams in thermal environments. *Composite Structures*, 128, pp. 363–380.

[76] Li, L., Hu, Y. and Ling, L., 2015. Flexural wave propagation in small-scaled functionally graded beams via a nonlocal strain gradient theory. *Composite Structures*, 133, pp. 1079–1092.

[77] Ebrahimi, F. and Barati, M.R., 2016. Hygrothermal buckling analysis of magnetically actuated embedded higher order functionally graded nanoscale beams considering the neutral surface position. *Journal of Thermal Stresses*, 39(10), pp. 1210–1229.

[78] Zamani Nejad, M., Hadi, A. and Rastgoo, A., 2016. Buckling analysis of arbitrary two-directional functionally graded Euler–Bernoulli nano-beams based on nonlocal elasticity theory. *International Journal of Engineering Science*, 103, pp. 1–10.

[79] Ebrahimi, F., Barati, M.R. and Dabbagh, A., 2016. Wave dispersion characteristics of axially loaded magneto-electro-elastic nanobeams. *Applied Physics A*, 122(11), p. 949.

[80] Ebrahimi, F., Dabbagh, A. and Barati, M.R., 2016. Wave propagation analysis of a size-dependent magneto-electro-elastic heterogeneous nanoplate. *The European Physical Journal Plus*, 131(12), p. 433.

[81] Ebrahimi, F. and Barati, M. R., 2016. Vibration analysis of nonlocal beams made of functionally graded material in thermal environment. *The European Physical Journal Plus*, 131(8), p. 279.

[82] Ebrahimi, F. and Barati, M. R., 2016. Wave propagation analysis of quasi-3D FG nanobeams in thermal environment based on nonlocal strain gradient theory. *Applied Physics A*, 122(9), p. 843.

[83] Barati, M.R. and Zenkour, A., 2017. A general bi-Helmholtz nonlocal strain-gradient elasticity for wave propagation in nanoporous graded double-nanobeam systems on elastic substrate. *Composite Structures*, 168, pp. 885–892.

[84] Ebrahimi, F. and Barati, M. R., 2017. Small-scale effects on hygro-thermo-mechanical vibration of temperature-dependent nonhomogeneous nanoscale beams. *Mechanics of Advanced Materials and Structures*, 24(11), pp. 924–936.

[85] Ebrahimi, F. and Barati, M. R., 2017. Hygrothermal effects on vibration characteristics of viscoelastic FG nanobeams based on nonlocal strain gradient theory. *Composite Structures*, 159, pp. 433–444.

[86] Ebrahimi, F. and Barati, M. R., 2017. A nonlocal strain gradient refined beam model for buckling analysis of size-dependent shear-deformable curved FG nanobeams. *Composite Structures*, 159, pp.174–182.

[87] Srividhya, S., Raghu, P., Rajagopal, A. and Reddy, J.N., 2018. Nonlocal nonlinear analysis of functionally graded plates using third-order shear deformation theory. *International Journal of Engineering Science*, 125, pp. 1–22.

[88] Natarajan, S., Chakraborty, S., Thangavel, M., Bordas, S. and Rabczuk, T., 2012. Size-dependent free flexural vibration behavior of functionally graded nanoplates. *Computational Materials Science*, 65, pp. 74–80.

[89] Wattanasakulpong, N. and Ungbhakorn, V., 2014. Linear and nonlinear vibration analysis of elastically restrained ends FGM beams withporosities. *Aerospace Science and Technology*, 32(1), pp. 111–120.

[90] Ebrahimi, F. and Mokhtari, M., 2015. Transverse vibration analysis of rotating porous beam with functionally graded microstructure using the differential transform method. *Journal of the Brazilian Society of Mechanical Sciences and Engineering*, 37(4), pp. 1435–1444.

[91] Ebrahimi, F. and Zia, M., 2015. Large amplitude nonlinear vibration analysis of functionally graded Timoshenko beams with porosities. *Acta Astronautica*, 116, pp. 117–125.

[92] Atmane, H. A., Tounsi, A. and Bernard, F., 2017. Effect of thickness stretching and porosity on mechanical response of a functionally graded beams resting on elastic foundations. *International Journal of Mechanics and Materials in Design*, 13(1), pp. 71–84.

[93] Ebrahimi, F., Jafari, A. and Barati, M. R., 2017. Vibration analysis of magneto-electro-elastic heterogeneous porous material plates resting on elastic foundations. *Thin-Walled Structures*, 119, pp. 33–46.

[94] Zenkour, A. M., 2018. A quasi-3D refined theory for functionally graded single-layered and sandwich plates with porosities. *Composite Structures*, 201, pp. 38–48.

[95] Gupta, A. and Talha, M., 2018. Static and stability characteristics of geometrically imperfect FGM plates resting on pasternak elastic foundation with microstructural defect. *Arabian Journal for Science and Engineering*, 43, pp. 1–17.

[96] Ebrahimi, F., Daman, M. and Jafari, A., 2017. Nonlocal strain gradient-based vibration analysis of embedded curved porous piezoelectric nano-beams in thermal environment. *Smart Structures and Systems*, 20(6), pp. 709–728.

[97] Ebrahimi, F., Jafari, A. and Barati, M. R., 2017. Dynamic modeling of porous heterogeneous micro/nanobeams. *The European Physical Journal Plus*, 132(12), p. 521.

[98] Ebrahimi, F. and Barati, M. R., 2017. Porosity-dependent vibration analysis of piezo-magnetically actuated heterogeneous nanobeams. *Mechanical Systems and Signal Processing*, 93, pp. 445–459.

[99] Ebrahimi, F. and Dabbagh, A., 2017. Wave propagation analysis of smart rotating porous heterogeneous piezo-electric nanobeams. *The European Physical Journal Plus*, 132(4), p. 153.

[100] Shahverdi, H. and Barati, M. R., 2017. Vibration analysis of porous functionally graded nanoplates. *International Journal of Engineering Science*, 120, pp. 82–99.

[101] Barati, M. R., 2018. A general nonlocal stress-strain gradient theory for forced vibration analysis of heterogeneous porous nanoplates. *European Journal of Mechanics-A/Solids*, 67, pp. 215–230.

[102] Eltaher, M. A., Fouda, N., El-midany, T. and Sadoun, A. M., 2018. Modified porosity model in analysis of functionally graded porous nanobeams. *Journal of the Brazilian Society of Mechanical Sciences and Engineering*, 40(3), p. 141.

[103] Saravanos, D. A., Heyliger, P. R. and Hopkins, D. A., 1997. Layerwise mechanics and finite element for the dynamic analysis of piezoelectric composite plates. *International Journal of Solids and Structures*, 34(3), pp. 359–378.

[104] Han, J. H. and Lee, I., 1998. Analysis of composite plates with piezoelectric actuators for vibration control using layerwise displacement theory. *Composites Part B: Engineering*, 29(5), pp. 621–632.

[105] Wang, Q. and Quek, S. T., 2000. Flexural vibration analysis of sandwich beam coupled with piezoelectric actuator. *Smart Materials and Structures*, 9(1), p. 103.

[106] Oh, I. K., Han, J. H. and Lee, I., 2000. Postbuckling and vibration characteristics of piezolaminated composite plate subject to thermo-piezoelectric loads. *Journal of Sound and Vibration*, 233(1), pp. 19–40.

[107] Wang, Q., Quek, S. T., Sun, C. T. and Liu, X., 2001. Analysis of piezoelectric coupled circular plate. *Smart Materials and Structures*, 10(2), p. 229.

[108] Wang, Q. and Wang, C. M., 2001. A controllability index for optimal design of piezoelectric actuators in vibration control of beam structures. *Journal of Sound and Vibration*, 242(3), pp. 507–518.

[109] He, X. Q., Ng, T. Y., Sivashanker, S. and Liew, K. M., 2001. Active control of FGM plates with integrated piezoelectric sensors and actuators. *International Journal of Solids and Structures*, 38(9), pp. 1641–1655.

[110] Gao, J. X. and Shen, Y. P., 2003. Active control of geometrically nonlinear transient vibration of composite plates with piezoelectric actuators. *Journal of Sound and Vibration*, 264(4), pp. 911–928.

[111] Liew, K. M., He, X. Q., Tan, M. J. and Lim, H. K., 2004. Dynamic analysis of laminated composite plates with piezoelectric sensor/actuator patches using the FSDT mesh-free method. *International Journal of Mechanical Sciences*, 46(3), pp. 411–431.

[112] Liu, G. R., Dai, K. Y. and Lim, K. M., 2004. Static and vibration control of composite laminates integrated with piezoelectric sensors and actuators using the radial point interpolation method. *Smart Materials and Structures*, 13(6), p. 1438.

[113] Lin, J. C. and Nien, M. H., 2005. Adaptive control of a composite cantilever beam with piezoelectric damping-modal actuators/sensors. *Composite Structures*, 70(2), pp. 170–176.

[114] Varelis, D. and Saravanos, D. A., 2006. Small-amplitude free-vibration analysis of piezoelectric composite plates subject to large deflections and initial stresses. *Journal of Vibration and Acoustics*, 128(1), pp. 41–49.

[115] Bian, Z. G., Lim, C. W. and Chen, W. Q., 2006. On functionally graded beams with integrated surface piezoelectric layers. *Composite Structures*, 72(3), pp. 339–351.

[116] Maurini, C., Porfiri, M. and Pouget, J., 2006. Numerical methods for modal analysis of stepped piezoelectric beams. *Journal of Sound and Vibration*, 298(4–5), pp. 918–933.

[117] Ebrahimi, F. and Rastgoo, A., 2008. Free vibration analysis of smart annular FGM plates integrated with piezoelectric layers. *Smart Materials and Structures*, 17(1), p. 015044.

[118] Azrar, L., Belouettar, S. and Wauer, J., 2008. Nonlinear vibration analysis of actively loaded sandwich piezoelectric beams with geometric imperfections. *Computers and Structures*, 86(23–24), pp. 2182–2191.

[119] Belouettar, S., Azrar, L., Daya, E. M., Laptev, V. and Potier-Ferry, M., 2008. Active control of nonlinear vibration of sandwich piezoelectric beams: A simplified approach. *Computers and Structures*, 86(3–5), pp. 386–397.

[120] Li, Y. and Shi, Z., 2009. Free vibration of a functionally graded piezoelectric beam via state-space based differential quadrature. *Composite Structures*, 87(3), pp. 257–264.

[121] Wu, N., Wang, Q. and Quek, S. T., 2010. Free vibration analysis of piezoelectric coupled circular plate with open circuit. *Journal of Sound and Vibration*, 329(8), pp. 1126–1136.

[122] Fu, Y., Wang, J. and Mao, Y., 2012. Nonlinear analysis of buckling, free vibration and dynamic stability for the piezoelectric functionally graded beams in thermal environment. *Applied Mathematical Modelling*, 36(9), pp. 4324–4340.

[123] Elshafei, M. A. and Alraiess, F., 2013. Modeling and analysis of smart piezoelectric beams using simple higher order shear deformation theory. *Smart Materials and Structures*, 22(3), p. 035006.

[124] Komijani, M., Reddy, J. N. and Eslami, M. R., 2014. Nonlinear analysis of microstructure-dependent functionally graded piezoelectric material actuators. *Journal of the Mechanics and Physics of Solids*, 63, pp. 214–227.

[125] Phung-Van, P., De Lorenzis, L., Thai, C. H., Abdel-Wahab, M. and Nguyen-Xuan, H., 2015. Analysis of laminated composite plates integrated with piezoelectric sensors and actuators using higher-order shear deformation theory and isogeometric finite elements. *Computational Materials Science*, 96, pp. 495–505.

[126] Barati, M. R., Sadr, M. H. and Zenkour, A. M., 2016. Buckling analysis of higher order graded smart piezoelectric plates with porosities resting on elastic foundation. *International Journal of Mechanical Sciences*, 117, pp. 309–320.

[127] Phung-Van, P., Tran, L. V., Ferreira, A. J. M., Nguyen-Xuan, H. and Abdel-Wahab, M., 2017. Nonlinear transient isogeometric analysis of smart piezoelectric functionally graded material plates based on generalized shear deformation theory under thermo-electro-mechanical loads. *Nonlinear Dynamics*, 87(2), pp. 879–894.

[128] Liu, C., Ke, L. L., Wang, Y. S., Yang, J. and Kitipornchai, S., 2013. Thermo-electro-mechanical vibration of piezoelectric nanoplates based on the nonlocal theory. *Composite Structures*, 106, pp. 167–174.

[129] Hosseini-Hashemi, S., Nahas, I., Fakher, M. and Nazemnezhad, R., 2014. Surface effects on free vibration of piezoelectric functionally graded nanobeams using nonlocal elasticity. *Acta Mechanica*, 225(6), pp. 1555–1564.

[130] Ebrahimi, F. and Salari, E., 2015. Size-dependent thermo-electrical buckling analysis of functionally graded piezoelectric nanobeams. *Smart Materials and Structures*, 24(12), p. 125007.

[131] Jandaghian, A. A. and Rahmani, O., 2016. Vibration analysis of functionally graded piezoelectric nanoscale plates by nonlocal elasticity theory: An analytical solution. *Superlattices and Microstructures*, 100, pp. 57–75.

[132] Ebrahimi, F. and Salari, E., 2016. Analytical modeling of dynamic behavior of piezo-thermo-electrically affected sigmoid and power-law graded nanoscale beams. *Applied Physics A*, 122(9), p. 793.

[133] Ebrahimi, F. and Barati, M. R., 2017. Buckling analysis of nonlocal third-order shear deformable functionally graded piezoelectric nanobeams embedded in elastic medium. *Journal of the Brazilian Society of Mechanical Sciences and Engineering*, 39(3), pp. 937–952.

[134] Ebrahimi, F. and Dabbagh, A., 2017. Wave propagation analysis of smart rotating porous heterogeneous piezo-electric nanobeams. *The European Physical Journal Plus*, 132(4), p. 153.

[135] Ebrahimi, F. and Dabbagh, A., 2017. Wave propagation analysis of embedded nanoplates based on a nonlocal strain gradient-based surface piezoelectricity theory. *The European Physical Journal Plus*, 132(11), p. 449.

[136] Ebrahimi, F. and Barati, M. R., 2017. Damping vibration analysis of smart piezoelectric polymeric nanoplates on viscoelastic substrate based on nonlocal strain gradient theory. *Smart Materials and Structures*, 26(6), p. 065018.

[137] Ebrahimi, F. and Shaghaghi, G. R., 2018. Nonlinear vibration analysis of electro-hygro-thermally actuated embedded nanobeams with various boundary conditions. *Microsystem Technologies*, 24(12), pp. 5037–5054.

[138] Ebrahimi, F. and Barati, M. R., 2018. Vibration analysis of piezoelectrically actuated curved nanosize FG beams via a nonlocal strain-electric field gradient theory. *Mechanics of Advanced Materials and Structures*, 25(4), pp. 350–359.

[139] Ebrahimi, F., Barati, M.R. and Dabbagh, A., 2016. Wave dispersion characteristics of axially loaded magneto-electro-elastic nanobeams. *Applied Physics A*, 122(11), p. 949.

[140] Claeyssen, F., Lhermet, N., Le Letty, R. and Bouchilloux, P., 1997. Actuators, transducers and motors based on giant magnetostrictive materials. *Journal of Alloys and Compounds*, 258(1–2), pp. 61–73.

[141] Olabi, A. G. and Grunwald, A., 2008. Design and application of magnetostrictive materials. *Materials and Design*, 29(2), pp. 469–483.

[142] Reddy, J. N. and Barbosa, J. I., 2000. On vibration suppression of magnetostrictive beams. *Smart Materials and Structures*, 9(1), p. 49.

[143] Hong, C. C., 2009. Transient responses of magnetostrictive plates without shear effects. *International Journal of Engineering Science*, 47(3), pp. 355–362.

[144] Yifeng, Z., Lei, C., Yu, W. and Xiaoping, Z., 2013. Variational asymptotic micromechanics modeling of heterogeneous magnetostrictive composite materials. *Composite Structures*, 106, pp. 502–509.

[145] Razavi, S. and Shooshtari, A., 2014. Free vibration analysis of a magneto-electro-elastic doubly-curved shell resting on a Pasternak-type elastic foundation. *Smart Materials and Structures*, 23(10), p. 105003.

[146] Shooshtari, A. and Razavi, S., 2015. Linear and nonlinear free vibration of a multilayered magneto-electro-elastic doubly-curved shell on elastic foundation. *Composites Part B: Engineering*, 78, pp. 95–108.

[147] Razavi, S. and Shooshtari, A., 2015. Nonlinear free vibration of magneto-electro-elastic rectangular plates. *Composite Structures*, 119, pp. 377–384.

[148] Xin, L. and Hu, Z., 2015. Free vibration of layered magneto-electro-elastic beams by SS-DSC approach. *Composite Structures*, 125, pp. 96–103.

[149] Shooshtari, A. and Razavi, S., 2016. Vibration analysis of a magnetoelectroelastic rectangular plate based on a higher-order shear deformation theory. *Latin American Journal of Solids and Structures*, 13(3), pp. 554–572.

[150] Vinyas, M. and Kattimani, S. C., 2017. Hygrothermal analysis of magneto-electro-elastic plate using 3D finite element analysis. *Composite Structures*, 180, pp. 617–637.

[151] Vinyas, M. and Kattimani, S. C., 2018. Finite element evaluation of free vibration characteristics of magneto-electro-elastic rectangular plates in hygrothermal environment using higher-order shear deformation theory. *Composite Structures*, 202, pp. 1339–1352.

[152] Kiran, M. C. and Kattimani, S., 2018. Buckling analysis of skew magneto-electro-elastic plates under in-plane loading. *Journal of Intelligent Material Systems and Structures*, 29(10), pp. 2206–2222.

[153] Vinyas, M., Nischith, G., Loja, M. A. R., Ebrahimi, F. and Duc, N. D., 2019. Numerical analysis of the vibration response of skew magneto-electro-elastic plates based on the higher-order shear deformation theory. *Composite Structures*, 214, pp. 132–142.

[154] Li, Y. S., Ma, P. and Wang, W., 2016. Bending, buckling, and free vibration of magnetoelectroelastic nanobeam based on nonlocal theory. *Journal of Intelligent Material Systems and Structures*, 27(9), pp. 1139–1149.

[155] Xu, X. J., Deng, Z. C., Zhang, K. and Meng, J. M., 2016. Surface effects on the bending, buckling and free vibration analysis of magneto-electro-elastic beams. *Acta Mechanica*, 227(6), pp. 1557–1573.

[156] Ghadiri, M. and Safarpour, H., 2016. Free vibration analysis of embedded magneto-electro-thermo-elastic cylindrical nanoshell based on the modified couple stress theory. *Applied Physics A*, 122(9), p. 833.

[157] Ebrahimi, F. and Barati, M. R., 2016. Dynamic modeling of a thermo-piezo-electrically actuated nanosize beam subjected to a magnetic field. *Applied Physics A*, 122(4), p. 451.

[158] Ebrahimi, F. and Barati, M. R., 2016. Magneto-electro-elastic buckling analysis of nonlocal curved nanobeams. *The European Physical Journal Plus*, 131(9), p. 346.

[159] Farajpour, A., Yazdi, M. H., Rastgoo, A., Loghmani, M. and Mohammadi, M., 2016. Nonlocal nonlinear plate model for large amplitude vibration of magneto-electro-elastic nanoplates. *Composite Structures*, 140, pp. 323–336.

[160] Ebrahimi, F., Dabbagh, A. and Barati, M. R., 2016. Wave propagation analysis of a size-dependent magneto-electro-elastic heterogeneous nanoplate. *The European Physical Journal Plus*, 131(12), p. 433.

[161] Ebrahimi, F. and Barati, M. R., 2016. Static stability analysis of smart magneto-electro-elastic heterogeneous nanoplates embedded in an elastic medium based on a four-variable refined plate theory. *Smart Materials and Structures*, 25(10), p. 105014.

[162] Ebrahimi, F. and Barati, M. R., 2016. A nonlocal higher-order refined magneto-electro-viscoelastic beam model for dynamic analysis of smart nanostructures. *International Journal of Engineering Science*, 107, pp. 183–196.

[163] Ebrahimi, F. and Barati, M. R., 2016. Magnetic field effects on buckling behavior of smart size-dependent graded nanoscale beams. *The European Physical Journal Plus*, 131(7), p. 238.

[164] Ebrahimi, F., Barati, M.R. and Dabbagh, A., 2016. Wave dispersion characteristics of axially loaded magneto-electro-elastic nanobeams. *Applied Physics A*, 122(11), p. 949.

[165] Ebrahimi, F. and Barati, M. R., 2017. Buckling analysis of piezoelectrically actuated smart nanoscale plates subjected to magnetic field. *Journal of Intelligent Material Systems and Structures*, 28(11), pp. 1472–1490.

[166] Ebrahimi, F. and Barati, M. R., 2017. Buckling analysis of smart size-dependent higher order magneto-electro-thermo-elastic functionally graded nanosize beams. *Journal of Mechanics*, 33(1), pp. 23–33.

[167] Ebrahimi, F. and Barati, M. R., 2017. Porosity-dependent vibration analysis of piezo-magnetically actuated heterogeneous nanobeams. *Mechanical Systems and Signal Processing*, 93, pp. 445–459.

[168] Ebrahimi, F. and Dabbagh, A., 2017. On flexural wave propagation responses of smart FG magneto-electro-elastic nanoplates via nonlocal strain gradient theory. *Composite Structures*, 162, pp. 281–293.

[169] Ebrahimi, F. and Dabbagh, A., 2017. Nonlocal strain gradient based wave dispersion behavior of smart rotating magneto-electro-elastic nanoplates. *Materials Research Express*, 4(2), p. 025003.

[170] Ebrahimi, F. and Barati, M. R., 2018. Vibration analysis of smart piezoelectrically actuated nanobeams subjected to magneto-electrical field in thermal environment. *Journal of Vibration and Control*, 24(3), pp. 549–564.

[171] Sahmani, S. and Aghdam, M. M., 2018. Nonlocal strain gradient shell model for axial buckling and postbuckling analysis of magneto-electro-elastic composite nanoshells. *Composites Part B: Engineering*, 132, pp. 258–274.

[172] Librescu, L., Hasanyan, D., Qin, Z. and Ambur, D. R., 2003. Nonlinear magnetothermoelasticity of anisotropic plates immersed in a magnetic field. *Journal of Thermal Stresses*, 26(11–12), pp. 1277–1304.

[173] Zheng, X., Zhang, J. and Zhou, Y., 2005. Dynamic stability of a cantilever conductive plate in transverse impulsive magnetic field. *International Journal of Solids and Structures*, 42(8), pp. 2417–2430.

[174] Lee, J. S., Prevost, J. H. and Lee, P. C., 1990. Finite element analysis of magnetically induced vibrations of conductive plates. *Fusion Engineering and Design*, 13(2), pp. 125–141.

[175] Kiani, K., 2012. Magneto-thermo-elastic fields caused by an unsteady longitudinal magnetic field in a conducting nanowire accounting for eddy-current loss. *Materials Chemistry and Physics*, 136(2–3), pp. 589–598.

[176] Murmu, T., McCarthy, M. A. and Adhikari, S., 2012. Vibration response of double-walled carbon nanotubes subjected to an externally applied longitudinal magnetic field: a nonlocal elasticity approach. *Journal of Sound and Vibration*, 331(23), pp. 5069–5086.

[177] Murmu, T., McCarthy, M. A. and Adhikari, S., 2013. In-plane magnetic field affected transverse vibration of embedded single-layer graphene sheets using equivalent nonlocal elasticity approach. *Composite Structures*, 96, pp. 57–63.

[178] Kiani, K., 2014. Vibration and instability of a single-walled carbon nanotube in a three-dimensional magnetic field. *Journal of Physics and Chemistry of Solids*, 75(1), pp. 15–22.

[179] Kiani, K., 2014. Free vibration of conducting nanoplates exposed to unidirectional in-plane magnetic fields using nonlocal shear deformable plate theories. *Physica E: Low-Dimensional Systems and Nanostructures*, 57, pp. 179–192.

[180] Karličić, D., Kozić, P., Adhikari, S., Cajić, M., Murmu, T. and Lazarević, M., 2015. Nonlocal mass-nanosensor model based on the damped vibration of single-layer graphene sheet influenced by in-plane magnetic field. *International Journal of Mechanical Sciences*, 96, pp. 132–142.

[181] Karličić, D., Cajić, M., Murmu, T., Kozić, P. and Adhikari, S., 2015. Nonlocal effects on the longitudinal vibration of a complex multi-nanorod system subjected to the transverse magnetic field. *Meccanica*, 50(6), pp. 1605–1621.

[182] Ghorbanpour Arani, A., Haghparast, E. and Zarei, H. B., 2016. Nonlocal vibration of axially moving graphene sheet resting on orthotropic visco-Pasternak foundation under longitudinal magnetic field. *Physica B: Condensed Matter*, 495, pp. 35–49.

[183] Ebrahimi, F. and Barati, M. R., 2017. Flexural wave propagation analysis of embedded S-FGM nanobeams under longitudinal magnetic field based on nonlocal strain gradient theory. *Arabian Journal for Science and Engineering*, 42(5), pp. 1715–1726.

[184] Ebrahimi, F. and Barati, M. R., 2018. Scale-dependent effects on wave propagation in magnetically affected single/double-layered compositionally graded nanosize beams. *Waves in Random and Complex Media*, 28(2), pp. 326–342.

[185] Ebrahimi, F. and Barati, M. R., 2018. Free vibration analysis of couple stress rotating nanobeams with surface effect under in-plane axial magnetic field. *Journal of Vibration and Control*, 24(21), pp. 5097–5107.

[186] Karami, B., Shahsavari, D. and Li, L., 2018. Temperature-dependent flexural wave propagation in nanoplate-type porous heterogenous material subjected to in-plane magnetic field. *Journal of Thermal Stresses*, 41(4), pp. 483–499.

[187] Dai, H. L., Ceballes, S., Abdelkefi, A., Hong, Y. Z. and Wang, L., 2018. Exact modes for post-buckling characteristics of nonlocal nanobeams in a longitudinal magnetic field. *Applied Mathematical Modelling*, 55, pp. 758–775.

[188] Ebrahimi, F. and Dabbagh, A., 2018. Wave dispersion characteristics of orthotropic double-nanoplate-system subjected to a longitudinal magnetic field. *Microsystem Technologies*, 24(7), pp. 2929–2939.

[189] Jalaei, M. H. and Ghorbanpour Arani, A., 2018. Analytical solution for static and dynamic analysis of magnetically affected viscoelastic orthotropic double-layered graphene sheets resting on viscoelastic foundation. *Physica B: Condensed Matter*, 530, pp. 222–235.

Index

Printed and bound by CPI Group (UK) Ltd, Croydon, CR0 4YY

18/10/2024

01776250-0015